Urban Geography

This extensively revised and updated fifth edition not only examines the new geographical patterns forming within and between cities, but also investigates the way geographers have sought to make sense of this urban transformation. *Urban Geography* is structured into three sections: 'contexts', 'themes' and 'issues' that move students from a foundation in urban geography through its major themes to contemporary and pressing issues. The text critically synthesizes key literatures in the following areas:

- an urban world
- changing approaches to urban geography
- urban form and structure
- economy and the city
- urban politics
- planning, regeneration and urban policy
- cities and culture
- architecture and urban landscapes
- images of the city
- experiencing the city
- housing and residential segregation
- transport, mobility and the city
- urban futures.

The fifth edition combines the topicality and accessibility of previous editions with extensive new material, including two new chapters on mobility and cities and urban futures, as well as a wealth of international case studies, extending its range of coverage across the field. This book features enhanced pedagogy including a range of new illustrations and tables, an abstract for each chapter, end of chapter essay questions and project activities, and annotated further reading from books, journals and websites. Written in an engaging, student-friendly style, this is an essential read for students and scholars of urban geography.

Tim Hall is Professor of Interdisciplinary Social Studies and Head of the Department of Applied Social Sciences at the University of Winchester, UK.

Heather Barrett is Deputy Head of the Institute of Science and the Environment and Principal Lecturer at the University of Worcester, UK.

Routledge Contemporary Human Geography Series

Series Editors:

David Bell, Manchester Metropolitan University

Stephen Wynn Williams, Staffordshire University

This series of texts offers stimulating introductions to the core subdisciplines of human geography. Building between 'traditional' approaches to subdisciplinary studies and contemporary treatments of these same issues, these concise introductions respond particularly to the new demands of modular courses. Uniformly designed, with a focus on student-friendly features, these books will form a coherent series which is up-to-date and reliable.

Existing Titles:

Cultural Geography
Mike Crang

Development Geography
Rupert Hodder

Political Geography
Mark Blacksell

Tourism Geography, 2nd edition
Stephen Williams

Geographies of Globalization, 2nd edition
Warwick Murray and John Overton

Urban Geography, 5th edition
Tim Hall and Heather Barrett

Urban Geography

Fifth edition

Tim Hall and Heather Barrett

Routledge
Taylor & Francis Group

LONDON AND NEW YORK

Fifth edition published 2018
by Routledge
2 Park Square, Milton Park, Abingdon, Oxon, OX14 4RN

and by Routledge
711 Third Avenue, New York, NY 10017

Routledge is an imprint of the Taylor & Francis Group, an informa business

First edition published by Routledge 1998
Fourth edition published by Routledge 2012

British Library Cataloguing-in-Publication Data
A catalogue record for this book is available from the British Library

Library of Congress Cataloging-in-Publication Data
A catalog record for this book has been requested

ISBN: 978-1-138-10182-1 (hbk)
ISBN: 978-1-138-10183-8 (pbk)
ISBN: 978-1-315-65259-7 (ebk)

Typeset in Times New Roman and Franklin Gothic
by Keystroke, Neville Lodge, Tettenhall, Wolverhampton

Contents

Figures

Tables

Acknowledgements

The authors would like to thank the students on the modules 'Understanding Urban and Rural Societies' (Winchester) and 'Urban Geography' (Worcester) for providing lively settings in which many of the ideas in this book were first raised and discussed. We would also like to thank Cath (Tim) and Martyn (Heather) for helping to keep us sane during the writing of this book. A great deal of thanks is also due to Andrew Mould and Egle Zigaite for their assistance in the production of this book and their patience as deadlines slipped by. Their efforts have been vital to the production of this volume.

Please can we pass on our thanks to the following for granting permission for material that has been reproduced here: Geographical Association (figure 3.3); Routledge (figures 3.6, 8.4 and 10.6; table 12.2); Alan Dixon (figure 4.7); Topical Press Agency (figure 6.3); FLC/ADAGP, Paris and DACS (figure 6.4); VEGAP, Madrid and DACS London (figure 7.3b); Tate Images, ADAGP, Paris and DACS, London (figures 9.1), Cástan Broto & Bulkeley/UN-HABITAT (table 13.1).

Section 1
Contexts

1 An urban world

Introduction

We live in an urban world, or, more accurately, many different urban worlds. In July 2007, for the first time in human history, the majority of the world's population resided in cities. This event, hailed as monumental in much media coverage, was, in itself, of little more than symbolic importance. The trends, most notably massive urban growth in the Global South, had been apparent for some time and show no sign of slowing down, let alone reversing. It is against this background, a growing, dynamic urban world characterized by increasing interconnection and inequality that faces challenges in the near future, including peak oil and probable climate chaos, that this book invites you into the world of urban geography. Urban geography texts always argue that their publication coincides with exciting and challenging times for the city. They are always correct. Whatever cities might be they are never boring.

The dynamic and diverse nature of the urban world presents a significant challenge for those attempting to write a textbook to guide students through its complexities. For a general textbook, the aim should be to provide the student with as comprehensive an overview as possible. However, this is always only ever partially fulfilled. Textbooks, such as this one, are written by authors who approach the study of the city in particular ways, drawing on their own set of knowledge and experiences. Who writes the book and where they are based matters. This has been a key issue raised about urban geographical writing on the city, where it has been pointed out that in reality universal ideas and theories about the city are only ever partial (see, for example, Robinson 2005a).

It is therefore an important starting point for researchers and writers to acknowledge and understand their own perspective and position in any piece of work. We are two urban geographers who were born and brought up in the United Kingdom (UK) and who have worked mainly in UK universities. Our professional discussions and experience have mainly been with others in Europe and North America. This has inevitably shaped our approaches to studying cities. While we have tried to move beyond the specifics of our urban experiences in this book, by focusing on exploring the broader processes shaping cities, where we make these abstract ideas concrete we will often draw on examples from our own experiences. Therefore, the coverage of examples used and issues raised will, like other textbooks, not reflect the urban world in all its diversity. This is where we invite you to build on what we have written here and to add your own perspectives and experiences. Throughout the book we have tried to offer you exercises and opportunities to reflect on your own knowledge of the urban and to consider the ways in which the urban realities that you inhabit and experience are shaped by these broader processes. So let us begin with your urban geographies …

Your urban geographies

This is a book written first and foremost for students. Its objective, therefore, is to equip you, the student with enough knowledge of cities and the ways that they have been thought about and researched, primarily but not exclusively from within urban geography, to allow you to understand key aspects of cities and to become an urban geographer in your own right.

As a student of urban geography, or one of its many cognate disciplines, you are likely to encounter cities and to address urban questions in many different ways. These may include abstract discussions of urban theory; essays and reports that ask you to pull together, synthesize and analyse a range of examples, typically in the light of theory or policy; assignments that require you to analyse secondary data and draw conclusions on the basis of this; or projects that ask you to go out and conduct some original research and collect your own data in one or more urban settings. Of the latter, the fieldtrip and the independent study or dissertation are among the most common, and typically, most rewarding, academic encounters with the city. Cities are such fascinating environments that it would be a great shame if this book did not encourage you to brave the weather and to get out and study the city, to perhaps look at the taken-for-granted urban environment that you pass through every day with fresh eyes. Alternatively, you might encounter the city through its many representations – films, novels, advertisements, media reports or computer games for example – and be asked to critically analyse the nature of these images and perhaps their significance. As you read this book think about what motivates you and about what you want your urban geographies to be, where they might take you and what they might contribute to the city. There is more to urban geography than just writing essays.

So, where do you begin? Well, for a start, it is unlikely that those of you reading this book have not encountered a city in some way or another, either as a resident of one or through reference to cities and urban life through a range of media, such as a book, television programme or film. It is worthwhile, therefore, asking you to reflect on what you already know about cities.

Exercise

A range of definitions, concepts and ideas associated with the terms 'city' and 'urban' exist. It is important that you are aware of the variety of ways in which urban areas can be defined and thought about. As a student developing your understanding of cities, it is useful to reflect on the ideas about urban areas that you already hold and how these link to broader ideas and beliefs. Either individually or in conversation with family, friends or classmates think about the following question (and do not read on before you have generated your own thoughts and reflections!):

What do the terms 'city' and 'urban' mean to you? Make a list of things that you think define 'the city' or 'the urban'.

Hopefully, the list you have generated is quite diverse, and this should give you an indication of the breadth of material that can be covered when examining cities. Your list may include things that define urban areas (population size, geographical boundaries, legal definitions), things urban areas possess (landscapes, buildings, infrastructure, activities) or attributes associated with the city (noisy, crowded, dangerous, creative, exciting, vibrant, polluted). It might also identify urban concerns at different levels, or geographical scales, from personal issues (conditions in your local neighbourhood) to things of global concern (the sustainability of urban growth). This indicates that there is not one city but many 'cities' and also many topics for urban geographers to study.

Developing your urban geographies

Your personal experiences of, and knowledge about, cities are an important starting point for developing your understanding of 'the urban'. However, as theories of learning suggest, personal experience in itself is not sufficient to develop thorough knowledge of an issue, and this experience needs to be built upon in order to develop a deeper understanding through a 'cycle of learning' (see Kolb 1984). So in order to develop your critical understanding of cities you need to reflect on your experiences and make sense of these by contextualizing your experience and knowledge in relation to other information about cities. Here you need to use your research skills to gather appropriate data/evidence on urban trends and issues – in the section below we outline some broad trends in contemporary urban development which will provide a starting point for thinking about these wider issues and setting your experiences in context, which will then be further developed throughout the book. The next stage of the 'cycle' in developing your critical understanding is to think about your experiences and this broader evidence and make sense of these through abstract conceptualization.

Here you will draw upon wider theories and concepts about urban development, change and experience in order to draw together these various strands of evidence and place them in the broader context of writing about cities. In the next chapter we will consider the development of urban geographical theory in order to provide a foundation for your own theorizing. Through this you will develop your critical knowledge and understanding about cities and urban life which will provide the foundation for your further experiences of and research into cities, so completing one round of the learning cycle.

In beginning to build on our more personal experiences of urban life and set these into a wider context we want to consider three important ideas underpinning the multiple geographies of the urban world which highlight some key trends in urban development and ways of thinking about cities. The first important idea is to place ourselves within the broader trends of urban development and change, or rather to consider the macro geographies of the urban world. Here it is useful to examine broad patterns in urban development at the global scale which emphasize the diversity in trends around the world. A second important idea to consider is the increasing connectedness of the world, where people and places are increasingly linked together in complex economic, political and cultural networks. Finally, it is important to consider how these broader processes are mediated by local contexts, thinking about the internal geographies of cities and the complexity of our urban lives and experiences. These key ideas about the urban world are introduced in the next section and underpin subsequent discussions about the urban which follow in the book.

Macro geographies of the urban world

A core question for anyone interested in studying cities is how many urban people there are in the world and where they live. Until the second half of the twentieth century significant urban development, or urbanization, was limited and spatially concentrated into a number of key regions, principally Europe, North America and Latin America. More recently, within these more urbanized societies, urban growth has been slow and the increases in urban populations relatively modest (figure 1.1). The most significant growth in the last thirty years has taken place in those parts of the world with low percentages of urban populations, with this predicted to increase in the near future. In particular, urban growth has been rapid in Asia, with China and India having particularly large and increasing urban populations. Growth has also been significant within Africa (figure 1.1).

Within these broad regional figures significant variation exists, and a more detailed examination of the recent trends in urbanization reveals that the urban world is far from uniform. Urban development is certainly changing the spatial organization of the world's economy and society, but at different rates in

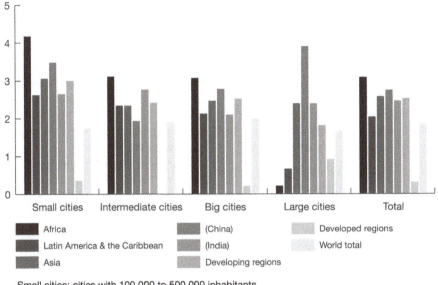

Small cities: cities with 100,000 to 500,000 inhabitants
Intermediate cities: cities with 500,000 to 1 million inhabitants
Big cities: cities with 1 million to 5 million inhabitants
Large cities: cities with 5 million or more inhabitants

Figure 1.1 *Annual growth rate of the world's cities by region and city size 1990–2000*

Source: Adapted from UN-HABITAT Global Urban Observatory (2008)

different places which leads to interesting questions for urban geographical research to examine. Globally there remains considerable variation in both the size and proportion of populations in urban places and the ways in which these populations are distributed, in terms of the number and size of cities (figure 1.1). For example, while much of China's growth has been in the form of large cities, urban growth within Africa has been predominantly of small and intermediate cities, with more significant urban growth confined to a small number of countries on this continent. Equally, the modest growth in cities in developing regions has been polarized, with gains in smaller towns and cities and something of a renaissance for some of the larger cities in these regions, which had been experiencing a decline in their populations.

The United Nations is a key source of data on the trends in urban growth, publishing its *World Urbanization Prospects* biannually and an annual *Demographic Year Book*. However, while being a useful source of information, the figures published should be viewed with caution as the number of people living in cities around the world is difficult to accurately define. In analysing current urban trends there is a fine line to be drawn between making useful general observations and a vague over simplification of the changes taking place (Clark 2003).

Exercise

Consider figure 1.1 again – what problems might there be in collecting and collating the data for a chart such as this? Think in particular about how data might be collected about the number of urban dwellers and how you might define what constitutes an urban area.

Once you have generated some ideas, visit the UN Global Urban Observatory website (https://unhabitat.org/urban-knowledge/global-urban-observatory-guo/) and compare your ideas with the information presented there about how their data sets have been compiled. This should highlight the variations around the world in the ways, frequencies, and so on in which census information is collected and how urban areas are defined.

From looking at the data an important issue emerges for those studying 'the urban', namely that most of the world's urban population live outside the developed world and also mainly in smaller or medium sized cities. Yet much of the urban geographical writing on cities that has been widely published has focused on larger cities, located within developed regions. It is therefore clearly a challenge to urban scholars to address these issues and to produce work that speaks to the urban world in its full diversity (Hubbard 2006; Robinson 2005a).

Connectivity, power and world cities

Another key urban trend has been the increase in the number and location of the world's megacities, defined as those with over 10 million inhabitants. It is predicted that the number of megacities in the world will increase to 27 by 2025, with the majority of these being located outside the developed world (see figure 1.2). The city of Mumbai is predicted to become the largest megacity after Tokyo, which will retain its top spot, while many megacities in developed regions, such as New York-Newark, will slip down the rankings. The increasing number and changing distribution of these large global cities has captured the imagination of commentators and researchers in recent years. The emergence of new megacities has prompted questions about the processes fuelling these changing patterns (urbanization), the varying role and status that cities in the world possess and the ways in which cities are connected to one another on a global scale. A fundamental question has been whether the rise of these new megacities heralds a shift in the distribution of the planet's most powerful and connected cities, or world cities.

Throughout the twentieth century, the colonial capitals and industrial cities of Northern Europe and North America were some of the world's largest cities and acted as key nodes through which goods, information and people flowed and as

2007

		Population (Thousands)
1	Tokyo	35,676
2	Mexico City	19,028
3	New York-Newark	19,040
4	São Paulo	18,845
5	Mumbai	18,978
6	Delhi	15,926
7	Shanghai	14,987
8	Kolkata	14,787
9	Buenos Aires	12,795
10	Dhaka	13,485
11	Los Angeles-Long Beach-Santa Ana	12,500
12	Karachi	12,130
13	Rio de Janeiro	11,748
14	Osaka-Kobe	11,294
15	Cairo	11,893
16	Beijing	11,106
17	Manila	11,100
18	Moscow	10,452
19	Istanbul	10,061

▨ : Cities located near a large water body (sea, river or delta)

2025

		Population (Thousands)
1	Tokyo	36,400
2	Mumbai	26,385
3	Delhi	22,498
4	Dhaka	22,015
5	São Paulo	21,428
6	Mexico City	21,009
7	New York-Newark	20,628
8	Kolkata	20,560
9	Shanghai	19,412
10	Karachi	19,095
11	Kinshasa	16,762
12	Lagos	15,796
13	Cairo	15,561
14	Manila	14,808
15	Beijing	14,545
16	Buenos Aires	13,768
17	Los Angeles-Long Beach-Santa Ana	13,672
18	Rio de Janeiro	13,413
19	Jakarta	12,363
20	Istanbul	12,102
21	Guangzhou, Guangdong	11,835
22	Osaka-Kobe	11,368
23	Moscow	10,526
24	Lahore	10,512
25	Shenzhen	10,196
26	Chennai	10,129

▨ : New megacities

Figure 1.2   *The world's megacities 2007 and 2025*

Source: Adapted from UN-HABITAT (2008)

centres where wealth was generated and power exercised. However, it is clear that the role and status of cities is shifting within our increasingly globalized world, where the speed, spread and depth of economic, political and cultural linkages is increasing and changing (Brenner and Keil 2006). The search for power and economic prosperity among a growing number of large global cities, acting within an increasingly unstable and unpredictable world, has generated intense competition between cities to gain status by encouraging growth through city marketing and planning activities. Here cities are seen to be acting increasingly entrepreneurially in order to attract the right activities and people with which to stimulate growth (Hall and Hubbard 1996, 1998).

Recent research has also sought to quantify the relative power and connectivity of cities on a global scale, most notably the work of the Global and World Cities (GaWC) Research Group based at Loughborough University in the United Kingdom (www.lboro.ac.uk/gawc/index.html). What this research suggests is that the most powerful and connected of the world's cities remain those of the Global North, despite their diminishing relative population size, and that the newly emerging megacities of much of the Global South lack significant global economic and political power despite their large size. It should be noted, however, that the criteria generally used to define status can be seen as 'capitalist' and 'western' in their conception, potentially excluding many cities defined as globally important by other criteria (Robinson 2005b). An alternative view would consider all cities as inherently global in some way but as differently positioned within a multitude of global networks (this is explored again in chapter two).

Internal geographies of cities

Finally, the trends in growth and change in urban populations around the world raise a number of significant issues for city dwellers and the managers of cities. At the heart of these concerns is the long-term sustainability of current urban trends and lifestyles, particularly the environmental impacts of city growth and problems of poverty and inequality within cities. For example, as noted above, Mumbai in India is predicted to become the world's second largest megacity. Its current growth and increasing global profile have led to the city becoming something of a key exemplar of the issues and concerns that could face many cities in the twenty-first century. In particular, its use in a number of recent books and films has led to wider global public awareness of the city and the issues it faces. For example, the widely acclaimed film *Slumdog Millionaire* released in late 2008 was set against the backdrop of Mumbai.

The film's story highlights both the magic and the horrors of life within this vast and rapidly changing city, and also some of the realities of everyday life in the city and the problems of poverty faced by many of its inhabitants. Consideration of films such as this that highlight life within cities should raise questions in our minds about why life portrayed in the city is the way it is and also the extent to which this is similar to or different from that with which we are familiar from our own urban experiences. Indeed, these are some of the key questions about urban life, or urbanism, that have been long-standing concerns of urban geographers, among others.

Yet, concerns about the impacts of rapid urban growth and change and the problems of life in cities are not merely a twenty-first century phenomenon. Anxieties about urban life have been evident since the rise of large industrial cities in Europe and North America in the nineteenth century. The rapid, largely unplanned, growth of these cities also generated fears about their social and environmental impact. Then, as now, these concerns were most eloquently expressed in some of the fictional writings of the period, such as the work of Charles Dickens on life in Britain's industrial towns in the nineteenth century. Much of his work graphically portrays the problems associated with poor living conditions within these cities:

It [Coketown] was a town of red brick, or of brick that would have been red if the smoke and ashes had allowed it; but as matters stood, it was a town of unnatural red and black like the painted face of a savage. It was a town of machinery and tall chimneys, out of which interminable serpents of smoke trailed themselves for ever and ever, and never got uncoiled. It had a black canal in it, and a river that ran purple with ill-smelling dye, and vast piles of building full of windows where there was a rattling and a trembling all day long, and where the piston of the steam-engine worked monotonously up and down, like the head of an elephant in a state of melancholy madness. It contained several large streets all very like one another, and many small streets still more like one another, inhabited by people equally like one another, who all went in and out at the same hours, with the same sound upon the same pavements, to do the same work, and to whom every day was the same as yesterday and tomorrow, and every year the counterpart of the last and the next.

Charles Dickens, *Hard Times* (1854)

This work, among other fictional and non-fictional writing, contributed to the formation of negative attitudes to urban life (anti-urbanism), where the city became associated with problems of poverty, disease, pollution, violence and alienation (Gold and Revill 2004). However, the impact of this urban growth, and the anti-urban reactions to it, did help generate both research into cities and moves to better manage them, which we consider later in the book. Yet despite these efforts, anti-urban representations of urban life remain prevalent. In particular, many visions of our planet's future within science fiction writing and film are set against backdrops of **dystopic** urban landscapes, which are either dark and menacing (such as in the film *Blade Runner*, 1982) or ruined and abandoned (such as in the film *Mad Max*, 1979) (Gold 2001). However, despite these numerous pessimistic visions, not everyone concurs with these nightmare scenarios and there is also optimism about the urban future. The idea of the city, and living within one, remains strong; one thing that we can be sure of is that cities are adaptable to new conditions and circumstances, both global and local in nature. Of immediate concern is how cities and urban populations can adapt to address important global concerns such as the use of the world's resources, climate change and world poverty and health. These are critical challenges facing the urban managers of today and tomorrow – we all have our part to play, perhaps yours will be a key one in the future!

Summary and structure of the book

Hopefully, this opening chapter has provided you with a stimulating start to, or continuation of, your urban geographical journey. There is a great diversity of cities in the world, which possess both similar and unique characteristics and concerns, and also many ways of looking at the city, both from above and below, from the official to the personal and from different cultural perspectives. We would hope that this has raised many questions for you about what is going on in the urban world, both locally to you and also further afield. The remainder of this book seeks to help you explore the issues identified in this introduction, and more, in greater depth, drawing on a wide range of information concerning cities, both from within urban geography and beyond.

The book is structured around three core sections; Contexts, Themes and Issues. The first section, of which this introductory chapter forms part, provides a series of Contexts, or foundations, for the study of urban geography. Also within this section, chapter two examines some of the significant theoretical and conceptual issues underpinning urban geographical study, while chapter three explores the diversity in the structure of the world's cities and how urban form varies over time and space. The section on Themes contains four chapters and examines some of the fundamental processes driving urban growth and change, which shape both cities and the lives of people within them. Chapter four examines economic processes, further examining the role of cities within the changing global economy and the impact that this has had on economic activities within cities. Chapter five considers politics and urban governance, examining the variety of ways in which cities are managed around the world and the balances of power that exist between various groups within cities. Chapter six develops the examination of city management by looking at the range of approaches and policies developed to plan the city and create better urban environments. Finally, in this section, chapter seven explores the social and cultural heterogeneity of cities and the ways in which this features within numerous dimensions of urban life. The final section of the book, Issues, considers some key questions about cities and aspects of urban life:

● Are cities built to be looked at or lived in? (Chapter eight considers the significance of architecture to the functional and symbolic form of cities and how this impacts on people's lives in the city.)
● Are images of cities important? (Chapter nine looks at a variety of representations of cities and the ways in which they are implemented in cultural politics and processes of urban development.)
● Are cities experienced as dreams or nightmares? (Chapter ten examines how we engage with the city in everyday life and whether this is a positive or negative experience for people.)
● Are cities able to provide a home for everyone? (Chapter eleven explores the fundamental role of the city as a residential space and considers why many cities cannot meet the housing needs of their populations.)
● Are cars killing cities? (Chapter twelve considers our mobility in cities and examines the transportation dilemmas facing many cities in the twenty-first century.)

- Are current trends in urbanization sustainable? (Chapter thirteen considers the impact of urban development on the environment and explores the key question running throughout the book of whether our increasing rates of urban growth, expanding cities and urban lifestyles are sustainable in the future.)

Why these questions? Well, they are issues that we consider important to discuss. It is not a definitive set, and we are sure that you could think of others. There is, therefore, a lot to explore! While this book will provide you with a comprehensive introduction to the study of urban geography, not everything there is to be known and explored is within these pages (even if we could achieve this, the resulting book would be too heavy and expensive to be of use to the majority of students!). The book, therefore, contains a number of features which will help to guide you to further sources of information and which are also designed to encourage you to engage actively with the book and its content. Each chapter includes:

- Highlighted terms – this identifies some of the specialist terms and concepts used in urban geography which are briefly defined and discussed in the glossary at the back of the book. The glossary can act as a useful reference point and basis for the further exploration of key themes and as an opportunity to test yourself on your knowledge and understanding of key ideas.
- Boxed case studies – these provide concise summaries and discussions of key, contemporary research on particular urban themes. Based on published papers, they are designed to highlight key authors and ideas and encourage you to follow-up and explore issues in greater depth, by using the case study article as the basis for further bibliographic searches.
- Follow-up activities – this section contains suggested questions, discussion topics, project ideas and research activities which you can use to develop both your knowledge of the key ideas and case studies introduced in the chapter and also to develop your study skills. The activities will make links with the further reading in the annotated bibliography at the end of the chapter.
- Annotated bibliography – this provides a list of some key books, journals and web resources which you can use as a starting point for your further investigation of a topic and to develop wider reading for projects or essays on a particular issue. Each source identified is accompanied by a brief note indicating what the source provides (note: summarizing sources that you read is a useful skill that will help you in organizing and producing your assignments).

So let us now continue our journey and begin to delve a little deeper into the world of urban geography ...

Follow-up activities

Essay title: 'What are some of the key issues facing cities around the world in the twenty-first century?'

Commentary on essay title

An effective answer would outline some of the key urban trends and issues introduced in this opening chapter and would provide some examples of these issues from cities around the world drawing on academic research into cities. It might also look to suggest which of these issues are the most challenging ones facing cities. An excellent answer would look to move beyond this extended list and critically explore why these issues face cities and why they present particular challenges. It would also look to set discussion more widely within academic writing on the city, evaluating different perspectives, or lenses, adopted to look at urban issues.

Project idea

Develop the idea of 'your urban geographies' introduced in this chapter. Develop a case study of the city you live in or a city that is familiar to you, gathering evidence to examine this city from the three perspectives outlined in this chapter: the macro geographies of your city, the connectivity of your city and the internal geographies of your city. What types of evidence can you gather to explore your city (for example, population statistics, economic data about companies operating in your city, field research, writings about your city, your own personal experiences)? Evaluate the evidence you gather and think about the benefits and problems of using different sources of data and in examining the city from a variety of perspectives.

Further reading

Books

- Bell, D. and Jayne, M. (2006) *Small Cities: Urban Experience Beyond the Metropolis*, Abingdon: Routledge
  Much writing on urban geography and the city more generally tends to focus on a relatively small number of 'big' cities. By contrast this is an interesting collection of essays focusing on those (smaller) cities and more mundane urban spaces not normally featured in urban geography's mainstream literatures.

- Jonas, A.E.G., McCann, E. and Thomas, M. (2015) *Urban Geography: A Critical Introduction*, Chichester, Wiley-Blackwell

A useful and comprehensive introduction to the field of contemporary urban geography.

- LeGates, R.T. and Stout, F. (2016) *The City Reader*, 6th edn, Abingdon: Routledge
  Brings together a wide range of essential writings on cities. Its scope, both in terms of time and perspectives on the city is broad. It has established itself, over the course of its six editions, as a landmark collection of writings on the city. An essential resource for the student of urban geography.

- Pacione, M. (2009) *Urban Geography: A Global Perspective*, 3rd edn, Abingdon: Routledge
  A key urban geography textbook; comprehensive coverage and written in a student-friendly way.

- UN-HABITAT (2012) *State of the World's Cities 2012/13: Prosperity of Cities*, Nairobi: UN-HABITAT
  A comprehensive overview of trends in urbanization around the world and the issues facing cities by this key global organization. Part of a regular series of reports on the world's cities by the UN-HABITAT programme.

Journal articles

- Derudder, B., Taylor, P., Ni, P., De Vos, A., Hoyler, M., Hanssens, H., Bassens, D. and Huang, J. (2010) 'Pathways of change: shifting connectivities in the world city network, 2000–08', *Urban Studies*, 47(9): 1861–1877
  Based on GaWC research ideas, the article considers the degree of connectivity between particular world cities.
- Hewitt, L. and Graham, S. (2015) 'Vertical cities: representations of urban verticality in twentieth century science-fiction literature', *Urban Studies*, 52(5): 923–937
  An interesting recent example of the analysis of the city through its representation in, in this case, literature. It is also of interest as it reflects the growing recent interest in the verticality of cities. This is a dimension of cities which has been somewhat ignored within traditional urban geography compared to their horizontal development.

- Nijman, J. (2007) 'Comparative urbanism', *Urban Geography*, 28(1): 1–6
  A key author in recent debates about the need to extend geographical research to include more comparative studies, especially concerning cities beyond the Global North.

- Robinson, J. (2011) 'Cities in a world of cities: the comparative gesture', *International Journal of Urban and Regional Research*, 35(1): 1–23
  A key article which challenges the 'western' focus of much urban geographical writing and theorizing. Robinson's other writings in this vein, including her 2005 book *Ordinary Cities: Between Modernity and Development* are also well worth checking out.

- Shearmur, R. (2010) 'Editorial – A world without data? The unintended consequences of fashion in geography', *Urban Geography*, 31(8): 1009–1017 An article highlighting the need for quantitative data to underpin urban analysis. Since this was published the interest in the potentials of 'big data' for understanding, managing and shaping cities has grown significantly. However, as chapter thirteen shows, this may not offer the panacea it first appears.

Websites

- Gapminder (www.gapminder.org/) – a very useful site containing a wealth of data on global socio-economic trends and variations which can be mapped to reveal global diversity and inequalities.

- Global and World Cities Research Centre (GaWC) (www.lboro.ac.uk/gawc/) – a wealth of research material into world cities and the connectivity of cities. Also displays some interesting maps showing world city connectivity.

- UN-HABITAT website (www.unhabitat.org/) – global organization concerned with the issues facing the world's cities. Contains a wealth of research and information.

- Urban Geography Research Group (UGRG) of the Royal Geographical Society/ Institute of British Geographers (www.urban-geography.org.uk/) – website of the research group of academic geographers in Britain. Contains some useful links to other sites and some good book reviews of urban geography texts.

② Changing approaches

Introduction

A discussion of the history of urban geography, how it has changed through time, is a standard aspect of all urban geography textbooks. Urban geography has been characterized by a number of radical shifts in its theoretical underpinnings, the aspects of the city that geographers are interested in and the methods they employ to study the city. It is important to be aware of this history and to realize that the ways that you will approach the study of cities are products of this history. These discussions tend to take one of two forms. Either they offer a broad historical review of the subject's evolution (this was the approach adopted in previous editions of this book) or a focus on current issues or debates with reference to the subject's evolution (see Robinson 2005a). These accounts all have a strongly chronological flavour. We do not aim to replicate these here. We want to take a more thematic approach. While the approaches that urban geographers have taken have been dynamic and changing, we can recognize a set of fundamental questions that represent enduring concerns of urban geographers.

> Historically then, the study of cities is identifiable with continuities and discontinuities – continuities in terms of the basic questions cities pose, discontinuities in terms of how they have been studied and theorised.
>
> (Paddison 2001: 4)

It is these 'basic questions' that we are primarily interested in considering in this chapter.

Exercise

What do you consider to be the most important 'basic questions cities pose' (Paddison 2001: 4)? Before reading on see if you can think of four or five examples. Can you find examples from the academic literature of geographers studying these questions? Try to think about the different approaches that geographers have taken to these questions. Think in terms of the methods they have employed and any evidence of their theoretical stances on these issues. Is this how you would approach these questions? Can you think of alternative approaches?

Fundamental urban questions

Brian Berry (1964: 147) concisely captured the conceptualization of cities, as reflected in the range of concerns of urban geographers, in his oft quoted phrase: 'cities as systems within systems of cities'. This captures much of what urban geography has been concerned with since its emergence as an academic discipline in the late nineteenth and early twentieth centuries (Hubbard 2006). Namely, an interest in exploring many facets of the internal lives of cities while recognizing that cities are part of wider contexts, be they systems or networks of cities or more general political or economic contexts.

We want to expand this conception of urban geography a little here. We would argue that there are at least two further areas of interest that are not so easily captured by Berry's quotation above. In our view, urban geographers have been interested in four fundamental, enduring themes. These are:

● the internal geographies of cities of various kinds;
● the relationships between cities and their wider contexts;
● exploring and accounting for global urban diversity;
● different ways of thinking about, defining, theorizing and researching the city.

We will use these four themes to guide our discussion of changing approaches to urban geography and its cognate disciplines within the remainder of the chapter.

The internal nature of cities

Urban geographers have continually sought to make sense of the city's internal structure, to discern order within the seemingly chaotic. The origins of this impulse can be found in the birth of urban studies at the Department of Sociology at the University of Chicago, which was founded in 1913. The theoretical and methodological foundations laid down by pioneering urban scholars such as Robert Park and Ernest Burgess, among others, have been fundamental to a wide range of disciplines, including urban geography, concerned with the city.

Rather than plough through the detail of the successive attempts that geographers and others have made to understand the structure of cities (for such discussions please see the suggested reading at the end of the chapter), we simply want to isolate one dimension that urban scholars have looked at and which would seem to underpin the internal geographies of cities. This is the operation of power within the city and its associated processes of competition and conflict. In doing so we will discuss the different ways in which successive 'schools' or paradigms of urban geography thought about the operation of power in the city and how this influenced their views of the processes that shaped the internal structure of cities. We would not argue that this is the only dimension that one could choose, but for us it is illustrative of key changes in thinking across the history of urban geography.

Drawing on ideas from the natural sciences, the Chicago School talked of human ecology and explored competition between groups of people with differing abilities to pay economic rent for land. From this they argued that the land use patterns and patterns of residential segregation reflected equilibria between the abilities of different groups to pay economic rent, their needs and inter-group competition. Despite their interest in economic power these researchers were not blind to culture as a force shaping cities, noting ethnicity and other facets of lifestyle as a factor producing communities within cities. From this view of the city as an arena of competition stemmed a number of urban models that represented the first attempts to systematically understand the structure of the city and to tie this to underlying causal processes (Pacione 2009). Widely known and applied models to date from this period included Burgess' concentric zone model and Hoyt's sector model (see Pacione 2001a and chapter three).

In the 1950s and 1960s the social sciences became increasingly concerned with producing rigorous, statistical explorations of society that pursued a **nomothetic** search to unearth regularities, general laws and patterns of human behaviour. This was facilitated by significant advances in computer technology and the desire of the social sciences to attain credibility and relevance (Hubbard 2006). This approach was particularly influential in geography that began to be described as a **positivist**, spatial science. In urban geography, this partly took the form of testing the urban models either from the Chicago School or subsequent ones that were much influenced by their work. Influential work from this period included explorations of the social areas of cities (Shevky and Bell 1955) and later factorial ecology that sought to identify the socio-economic and cultural factors that underpinned urban spatial patterns (see Knox and Pinch 2010: 67–73 and chapter eleven).

Despite the undoubted technical sophistication of this work it was subject to a number of criticisms by the late 1960s. These included criticisms of inherent flaws, such as its unrealistic views of human beings and their knowledge of the environment and consequently the poor predictability of urban models (Pacione 2009). However, ultimately more damning were criticisms of spatial science's lack of ability to say anything about a number of emerging urban crises linked to poverty, inequality and conflict in cities of the Global North at the time. Put simply, positivist urban geography lacked relevance and engagement with the topical urban issues of the time. Despite the efforts of an influential behavioural offshoot (see Goodey and Gold 1985; Walmsley and Lewis 1993; Kitchin 1994), which by drawing on environmental psychology aimed to more accurately model the processes of human perception and cognition, this particular brand of urban geography saw its influence wane into the 1970s.

Interestingly, while the origins of the Chicago School of urban studies lay in explorations of one rapidly industrializing city, the origins of the approach that came to replace spatial science in urban geography can be traced back in part to Friedrich Engels' (1844) revelations of the horrors of another, Manchester, during his time there in the 1840s. This, and Engels' subsequent work with **Karl Marx**, were key influences on the emergence of **structuralist** approaches across the social sciences in the early 1970s. Engels, and structuralists more generally, saw power in rather different terms to the Chicago School and their followers. Emphasizing inequality and the exploitation of the working class, structuralism within urban geography focused on class as the key dimension of urban life and saw social and spatial outcomes as the consequences of structural changes within the capitalist **regime of accumulation** (Pacione 2009).

Structuralist, or neo-Marxist, urban geographies then were very different to the positivist ones that had preceded them. These were urban geographies underpinned often by a strong normative impulse, a sense of what should be (Hubbard 2006; Pacione 2009). In some cases, these convictions prompted researchers to take on activist roles or positions overtly critical of planning and government policy, which were felt to be instrumental in the maintenance of unequal class relations and the propping-up of capitalism. Some of the key works from this period included *Social Justice and the City* (1973) by David Harvey (a 'reformed' positivist) and *The Urban Question* (translated into English in 1977) and *The City and the Grassroots* (1983) by Manuel Castells.

Many of the key works in this paradigm explored the dynamics of investment in urban property markets. They interpreted these patterns as attempts to resolve periodic crises within capitalism, charting the social and economic consequences of this dynamic through processes such as gentrification (Smith 1996), suburban development, deindustrialization and urban abandonment. Although subject to a number of criticisms, often referring to its failure to adequately incorporate human agency into its analysis (see Savage et al.'s (2003: 52–53) criticisms of David Harvey for example), it retained a prominence within urban geography and indeed its influence is still felt today.

An alternative perspective that arose primarily within urban sociology focused not on power in terms of class positions and conflict but rather in terms of the ability of key individuals, urban managers and gatekeepers, to control access to resources in the city (Pacione 2009).

> Such actors included housing managers, planners, estate agents, mortgage lenders, financiers, police, councillors and architects. Collectively and individually, it was argued these actors could deny certain social groups access to particular property markets (and hence particular parts of the city).
>
> (Hubbard 2006: 32–33)

Despite the focus on individual actors their actions tended still to be interpreted in terms of the maintenance of class relations.

The urban managerialist approach, in emphasizing conflict, racism and inequalities in wealth and power in less abstract terms than neo-Marxist forms of structuralism made some important contributions to urban studies (see Pahl 1970; Rex and Moore 1974; Rex and Tomlinson 1979). In highlighting the barriers to access to resources, such as housing, for many in the city, they destabilized positivist models of the city in which such constraints were largely absent (Hubbard 2006) (see also chapter eleven).

The desire to unpack the internal structure of the city remains a strong impulse within urban geography, with recent contributions including attempts to model the post-industrial, post-modern or global metropolis (see discussion below and chapter three). However, work in this vein has tended to shift in one of two directions, either looking at how specific processes, such as gentrification for example, play out within cities or, alternatively, adopting more ethnographic approaches or ones that seek to explore more the meanings of urban spaces (Pacione 2009). We will look at some of this work later in the chapter and throughout the book.

Cities and wider contexts

As Brian Berry's (1964) quotation (see p. 18) reminds us, one of the key interests of urban geographers has been, and remains, the question of the relationships between cities and wider contexts, however these may be construed. Although not a central concern of positivist approaches it was present in their interest in urban systems, for example, Christaller's **central place** theory (Hubbard 2006: 32). However, this dimension of the urban has been much more a concern of other approaches and is something we have seen a growing interest in recently. Again, the question of power, in this case the operation of power upon cities and the power of cities themselves, provides a useful lens through which to examine the ways in which urban geographers have approached this in a number of different ways.

A recurrent concern among structuralist urban geographers has been the impacts of structural changes on cities, the wider context in this case seen as the capitalist system or the global economy. The city, in being regarded as a crucial site of the resolution of periodic crises within capitalism, was often considered as 'victim'. Work in this vein emphasized the destructive impacts of uneven development in processes like deindustrialization and urban development. Later work, however, argued that cities were not as helpless as this view might suggest and explored the ways in which cities and space were also active in shaping the processes of ongoing development (Massey and Meegan 1982; Massey 1984).

More recently an interest in the geographies of globalization has focused attention on a number of global cities (London, Paris, New York, Los Angeles and Tokyo among others) as crucial lynchpins within the global economy, highlighting the connectivity between these cities and, concomitantly, the implications for those cities that are not part of these networks (Brenner and Keil 2006; Kim 2008). The notion of the links between even the most mundane urban spaces has gained a significant foothold theoretically and analytically within urban geography following Doreen Massey's arguments for a 'global sense of place' (1994). Here she argues that places should not be seen as closed, bounded, coherent entities, but rather as open, complex and interconnected to ranges of other spaces through links associated with travel, migration, trade, commerce and culture, as well as more personal biographies and memories. Here again though, Massey recognizes power at play, arguing that while some spaces are the originators of many connections, shaping the geographies of other spaces around the world, other, perhaps less 'powerful', spaces, tend to be, overwhelmingly, the receivers of connections or connected in ways that do not allow them the power to shape spaces elsewhere. In this case, they are more shaped than shapers. Since Massey first introduced the idea of places as networked in this way there has been a significant interest in the application of a variety of network approaches in human geography which we are seeing increasingly applied to analysis of cities (Amin 2002b; Amin and Thrift 2002). We are likely, therefore, to see a growing interest in research based around notions of connectivity within urban geography in the future (see discussions on this subject later in this book).

Global urban diversity

A number of commentators (Eade and Mele 2002; Hubbard 2006; Robinson 2002, 2005a; Roy 2009) have noted something of a crisis in urban theory recently. We have witnessed the emergence of theoretical perspectives (feminist, post-modern, post-colonial, for example) that are both radically different to, and explicitly critical of, theoretical perspectives on the city that stemmed from positivist and structuralist traditions and which, until recently, dominated urban thinking across a number of disciplines. The rise of these alternative perspectives can be interpreted as a failure of 'traditional' urban theory to reflect, or be able to engage with, the realities of global urban diversity. Jennifer Robinson (2002, 2005a) has gone as far as arguing that urban theory risks becoming irrelevant if it remains narrowly focused on certain aspects of life in cities of the Global North, at a time when current trends in urbanization are profoundly shifting the distribution of urban populations towards the Global South (see chapter one). Not only has urban theory failed, thus far, to respond to these trends but it has also failed to embrace the diversities of city life globally, remaining too fixated on a narrow set of social divisions (especially class based divisions) that were more applicable to cities in a particular place (the Global North) at a particular time (the twentieth century) than they are to the diversity of cities and urban lifestyles around the world now.

Robinson is particularly critical of the tendency of urban theory to transfer or 'universalise' 'located and parochial assumptions' (2005a: 6) from the small set of cities in which they were devised to other, very different, cities elsewhere. The effects of this have been either that the assumptions are irrelevant to the cities to which they have been applied, or, that in not fitting into these assumptions and models, certain cities are relegated and defined as 'other' or simply seen in terms of under-development or in terms only of what they lack in relation to other cities. The application of western urban theory then to cities elsewhere may be disempowering, a tendency regarded as colonial in its effects. There is a danger then, following this, that this application of urban theory can create hierarchies, categories and divisions to which cities are consigned. Much of the impulse behind the emergence of new perspectives and alternative urban theories is to resist this colonial, disempowering impulse and to produce urban theories that are able to recognize cities on their own terms and to accommodate global urban diversity. As Robinson argues: 'we need a form of theorising that can be as cosmopolitan as the cities we try to describe' (2005a: 3).

Hubbard (2006) is similarly critical of urban theory's attempts to account for recent trends in urbanization in cities of the west. These cities are being increasingly affected by a number of economic, political, social and cultural factors that were neither anticipated nor included in positivist and later structuralist models and theories of the city and which appear to raise questions about some key aspects and assumptions of these perspectives. These processes, often referred to under the banner of post-modern urbanization, have included deindustrialization, the rise of entrepreneurial forms of urban governance, increasing levels of social polarization and fragmentation and the reconfiguration of both individual and group identities in new, multiple and complex ways through practices such as consumption, migration and leisure that are increasingly central to the urban experience. They have exposed earlier urban theory as overly rigid, crude and inflexible. It is not that recent urban theory has failed to notice these changes. Indeed, there has been much written about them by many urban geographers and others from cognate disciplines (see Harvey 1989a; Soja 1989, 1996; Davis 1990; Watson and Gibson 1995; Dear and Flusty 2005), a significant proportion of which is based upon analysis of Los Angeles, a city that has become constructed as the archetypal post-modern city (see also chapter three). Rather, much of what has been produced has been done from within the theoretical straightjacket of twentieth century urban theory that has been unable to offer sufficiently cosmopolitan, to use Robinson's term, urban theory through which to speak of these changes (Hubbard 2006: 42–55).

Criticisms of these attempts to theorize the post-modern city have included questions about the representativeness of Los Angeles. As with earlier manifestations of urban theory we have seen here a tendency to universalize insights derived from the specific analysis of Los Angeles. Further, these accounts have been accused of producing top-down perspectives that have failed to include the grounded multiple realities of the cities under scrutiny (Ley and Mills 1993). This, it has been argued, is both disempowering and reductionist, erasing the subtle contours of social difference and experience within different cities.

Critics are far from uniform

Exercise

Look at some of the writing on Los Angeles from writers such as Davis, Soja or Dear and Flusty. How transferable, do you think, are their views of post-modern urbanization to urban settings with which you are familiar? How different, and in what ways, would our view of post-modern urbanization be had it been based more extensively around explorations of cities other than Los Angeles?

Despite an avowed engagement with the complexities of global urban diversity and the processes of post-modern urbanization, these accounts seem to have failed to escape the tendencies to abstraction and reductionism typical of twentieth century urban theory.

> While there is a widely noted dissatisfaction with the 'will to abstraction' which forced the city to conform to abstract models, categorizations and languages, urban scholars have often fallen back on these very forms of abstraction in their attempts to comprehend new forms of urbanity.
>
> (Hubbard 2006: 55)

Thinking about the city

Having read our brief gallop through the history of urban geography you may be asking, quite rightly, how this is relevant to your own practice as a budding urban geographer. We want to stress three points here. First, it is important to reiterate that the urban geography that you will undertake is the product of a long evolution of theory, methods and concerns. Your urban geographies then will be shaped by this history. Second, it is important to stress that debate, even disagreement, is very much alive within urban geography. The gauntlet thrown down by Jennifer Robinson and other post-colonial theorists, for example, is evidence of this. You have choices as an urban geographer then. Debates to weigh up and maybe participate in. Think critically about your urban geographies, where they come from and what alternatives exist. This will make you a more effective and incisive urban geographer and will bring the subject alive to you. This is something that all of the urban geographers that you will read about in this and other books have gone through, and continue to do so. Third, thinking in terms of theory is a vital part of urban geography. While empirical investigation of cities is important, the significance of case studies is only really apparent when they are connected to bigger questions. What then do the case studies that you look at and issues that you research yourselves tell you about how the city works? In what ways do the findings of these investigations connect to the more fundamental issues that we have outlined here and what do they reveal about them? Thinking theoretically about cities is an important skill for any urban geographer, it will produce richer, more critical urban geographies rather than ones that are naïve or descriptive.

Summary

This chapter has not attempted to offer a detailed, comprehensive account of changing approaches within urban geography. It has said nothing, for example, about humanistic approaches (see, for example, Relph 1976, 1987 and chapter ten). It has tried to offer a flavour of the way in which the subject has evolved around a number of key questions, some reasons why and the contributions it has drawn from cognate disciplines. It should be seen as an introduction to this important field that can be explored in greater depth through the readings outlined below.

Follow-up activities

Essay title: 'We need a form of theorising that can be as cosmopolitan as the cities we try to describe' (Robinson 2005a: 3). Discuss Robinson's challenge to urban theory and the ways that we might go about constructing an alternative, 'cosmopolitan' urban theory.

Commentary on essay title

An effective answer would reiterate the key points of Robinson's critique of urban theory, perhaps making reference to other writers who have raised similar points. This would cover both her criticisms of the colonial tendencies of this theory and the basis that she proposes for an alternative. An excellent answer would then move beyond Robinson's work to consider the work of others who have proposed alternatives to prevailing forms of urban theory. The papers by Rao and Wolch in the further reading below would be relevant here as well as some of the critiques outlined by Hubbard.

Project idea

'Current trends in urbanization are making the cities of the Global North a peripheral part of the urban world.' To what extent do you agree with this statement? Can you gather evidence to support your position? What types of evidence have you used (for example, population statistics, economic, political or cultural power)? Do all of these forms of evidence tell the same story? What alternatives could you choose and how would this affect your argument?

Further reading

Books

- Amin, A. and Thrift, N. (2002) *Cities: Reimagining the Urban*, Cambridge: Polity Press
  A radical reimagining of the boundaries and concerns of urban geography. Fuel to the fire of recent debates.

- Chen, X., Orum, A.M., Paulsen, K.E. (2013) *An Introduction to Cities: How Place and Space Shape Human Experience*, Chichester: Wiley-Blackwell (Part 1: 'The Foundations')
  A very sound and accessible review of many of the key perspectives on the study of the city. The focus is very much around key theorists and their contributions.

- Edensor, T. and Jayne, M. (eds) (2012) *Urban Theory Beyond the West: A World of Cities*, Abingdon: Routledge
  An important edited collection that reflects much of the recent shift in writing about urban theory which takes seriously perspectives on urban theory emerging from non-western cities. See also the recommended works by Ferenčuhová, Robinson, Rao and Roy.

- Hubbard, P. (2006) *City*, Abingdon: Routledge (Chapter 1: 'Urban theory, modern and post-modern')
  Phil Hubbard offers one of the most readable and comprehensive accounts of the history of urban theory currently available.

- Pacione, M. (2009) *Urban Geography: A Global Perspective*, London: Routledge

Pacione's book is about the most comprehensive overview of urban geography that is currently available. Chapter 2 'Concepts and theory in urban geography' is particularly relevant to the discussion in this chapter.

- Robinson, J. (2005) *Ordinary Cities: Between Modernity and Development*, Abingdon: Routledge
  Jennifer Robinson offers some fundamental criticisms of the urban theory that has tended to hold sway across a range of disciplines. Her arguments are most fully developed in this book. Her arguments are also advanced within a number of journal articles, one of which is recommended below and one at the end of the previous chapter.

Journal articles

- Ferenčuhová, S. (2016) 'Accounts from behind the Curtain: history and geography in the critical analysis of urban theory', *International Journal of Urban and Regional Research*, 40(1): 113–131
  Another paper, like many recommended here, which recognizes the limitations of urban theory and which attempts to 'decentre' it. Here the perspective is from scholars writing about post-Soviet cities. Even if your interest is not specifically in these cities, it is worth reading for the more general critiques of urban theory that this perspective highlights.

- Van Meeteren, M., Derudder, B. and Bassens, D. (2016) 'Can the straw man speak? An engagement with post-colonial critiques of "global cities research"', *Dialogues in Human Geography*, 6(3): 247–267
  A critical review of global cities research from a post-colonial perspective.

- Rao, V. (2006) 'Slum as theory: the South/Asian city and globalization', *International Journal of Urban and Regional Research*, 30(1): 225–232
  An excellent review of recent writing on reimagining urban theory. This, like the paper by Roy below, is an example of a very highly cited and influential contribution to recent debates on urban theory. As with much of this recent work, following the influence of Jennifer Robinson's work, it asks that we think of theory beyond purely Western perspectives.

- Robinson, J. (2002) 'Global and world cities: a view from off the map', *International Journal of Urban and Regional Research*, 26(3): 531–554
  An early articulation of some of Robinson's key challenges to urban theory and now something of a key paper in enlivening and broadening contemporary debates around urban theory.

- Roy, A. (2009) 'The 21st-century metropolis: new geographies of theory', *Regional Studies*, 43(6): 819–830
  Calls for a re-examination of urban theory relevant to the twenty-first century drawing on critical perspectives from the Global South.

3 Urban form and structure

Introduction

Of all the attributes associated with the city, its physical presence exerts a significant influence on us. The form or shape of the city, known as its morphology, can be seen as the tangible outcome of a complex mix of socio-economic forces and the ideas and intentions of groups and individuals acting both from within and outside a city. In his textbook *The Study of Urban Geography*, first published in 1972, Harold Carter located his chapter considering urban form towards the end of the book, arguing that the reader needs to have engaged with the rest of the subject matter of urban geographical study in order to understand the complexity of urban form. Yet, its consideration at the end, rather than the beginning, of Carter's book was perhaps also a reflection of the lack of centrality of studies of urban form to mainstream urban geography at the time he was writing. Since then, the examination of urban form has moved more centre stage in urban geographical research, and urban theory has begun to pay more attention to the 'materiality' of the city (Lees 2001; Hubbard 2006). We would therefore contend that consideration of urban form provides an important introductory context for urban geographical study.

In this chapter we provide a brief introduction to some of the ways in which geographers have sought to analyse urban form and understand the dynamic and changing structure of cities over time and space. We begin by examining the development of ideas associated with urban morphological study. We then adopt a broadly chronological approach by examining key stages in the historical development of cities and their form (table 3.1). This broad three-fold division is a traditional and appealing way of considering the changing nature of urban form within geography. While this remains a useful starting framework, it must be remembered that this simple typology is based principally within western research traditions (see chapter two) and consequently masks a considerable variety in trajectories of urban development around the world.

Studying urban form and structure

The study of urban form is an important root of urban geography, stretching back to the nineteenth century. Many early studies of cities were essentially

Table 3.1 Key phases of urbanization

	Pre-industrial Mercantile	Industrial Modern	Post-industrial Post-modern
Key theorists	Sjoberg (1960) Vance (1977)	Marx/Engels (1840s) Chicago School: Burgess (1920s) Hoyt (1930s)	California School: Scott (1988) Soja (1989, 1995) Davis (1990)
Mode of production	Mercantile capitalism Craft/workshop Trade in goods	Industrial capitalism Large factory/Fordist Manufacturing	Late capitalism Small factory/ post-Fordist services
Mode of regulation	Private-elite Laissez-faire development Limited state control	Public sector State planning Welfare/state control	Public-private partnership Entrepreneurial Flexible intervention
Form	1 Densely packed structure often walled 2 Prestige buildings/ homes in core surrounded by occupational/ethnic quarters 3 Plots in multiple use	1 Commercial CBD surrounded by industrial zone 2 Zones of suburbs extending out from the centre 3 Increase in social status from centre to periphery	1 Fragmented or 'patchwork' form 2 Post-suburban 'edge city' developments 3 Increased socio-economic polarization and 'fortress' landscapes

descriptive 'site and situation' studies (Carter 1995: 3), concerned primarily with examining physical characteristics as the determining factor in the location and development of settlements. Later research not only undertook more detailed studies of urban forms but also considered the forces creating them. This urban **morphogenetic** study developed particularly strongly in German universities in the early twentieth century, drawing on work by geographers, historians and architects. The traditions developed here have had a continuing influence on the development of urban morphological study in many European countries and in North America through both the spread of ideas and the movement of key researchers (Whitehand 1987).

A key legacy of this German tradition was the development of approaches to the study of urban form within geography, principally through the work of M.R.G. Conzen, who was a student in Germany in the early twentieth century. Conzen moved to Britain in the 1930s and undertook a number of detailed urban studies of British towns in the 1950s and 1960s, with his studies of Whitby, Alnwick and Newcastle being viewed as most significant (Whitehand 1987). These studies established some important foundations for the study of urban form which have become widely accepted and which have underpinned much subsequent work in this area:

Basic principles for urban morphological study

- The three-fold division of the urban landscape into plan, building form and land use, as a way of unpacking its complexity.
- The sub-division of a city's plan into streets, plots/blocks/open spaces and building/block plans as a way of understanding the key elements of the city's two-dimensional form.
- The recognition of the individual plot, or land parcel, as the fundamental unit of analysis.

Conceptualization of developments in the urban landscape

- Recognition that different elements of the urban landscape change at different speeds over time, with the plan more resistant to change than building forms and land uses. Existing forms then provide constraints on subsequent development.
- Conceptualization of cycles of development at the micro scale within the plot (Conzen identified the **burgage cycle** as a particular variant expressing a more general phenomenon of gradual plot infilling – see pre-industrial city section).
- Conceptualization of the phases of growth of the city at the macro scale through the idea of **fringe belts** – see industrial city section.

Despite this work being well received at the time of publication, Conzen's ideas have not been widely utilized until relatively recently. Many studies of urban form remained essentially descriptive and came in for criticism in the 1960s as geography moved to embrace scientific and structural approaches. Studies of urban development and the 'shape' of the city became dominated by studies of function and land use, drawing on the work of the Chicago School (see chapter two). Here, form and use became conflated and buildings and spaces were viewed primarily as containers of activities, if they were examined at all. Consideration of the city as a complex physical entity almost disappeared from mainstream urban geographical research and urban geography textbooks.

Since the 1980s, there has been a resurgence in studies of urban form (Whitehand 1992a). The impetus for this resurgence has come from scholars in a variety of different disciplines around the world, including geography, planning, architecture, urban design and urban history. Within geography three principal strands of current interest are evident. First, work based on Conzen's ideas has continued in Britain (Whitehand 2001) and in the United States (Conzen 2001). Key areas have been development of the fringe belt concept (Whitehand and Morton 2006), the plan analysis of medieval towns (Lilley 2000) and examination of agents of urban change (Whitehand 1992b). Second, renewed interest in culture in geography has focused attention on the symbolic qualities of the urban landscape (Hubbard 2006). Finally, researchers have become interested in the emergence of the new urban forms linked to post-modernism and changes in the dominant forms of architecture in cities (Knox 1993) (see also chapter eight). This recent research has begun to explore the common ways in which cities are built and transformed, and has widened research beyond the traditional areas of study, showing the applicability of many earlier concepts to modern urban environments and cities beyond Europe (Conzen 2009).

Early urban forms

Urban origins

The development of the first cities is viewed as an important watershed in human history, although they made a relatively late appearance in terms of human occupancy of the planet. The first cities appeared in Mesopotamia between 5,000 and 6,000 years BP, with further early urban growth occurring in other parts of the world, specifically the Indus Valley (4,000–5,000 years BP), Egypt (5,000 years BP), China (4,000 years BP), Central Andes (2,500 years BP), Mediterranean (2,500 years BP), Mesoamerica (1,000 years BP) and West Africa (1,000 years BP). However, the origins of this early urban growth, and the reasons for this, continue to generate much debate (see for example, Scarre 2005).

Generally, early urban development was linked to changes in human societies beginning in the Neolithic Period (6,000 to 10,000 years BP) and associated with the development of human control of the environment through the domestication of animals, crop production and irrigation. Early theories proposed by archaeologists suggested that the production of agricultural surpluses based on irrigated farming, and the management of this by a social elite, provided the preconditions for urban development. However, this idea of agricultural primacy has been questioned, with the development of urban areas and populations seen to have provided the stimulus for agricultural development. The reality was probably a complex inter-relationship between these two elements and technological change, in conjunction with other factors such as defensive and religious needs, rather than a simple causal link (Scarre 2005; Chant 2008). What is important from considering these debates is recognition of the important link between socio-economic transformations and processes of urbanization, with the associated emergence of particular urban forms.

While these transformations in human societies and associated urban developments emerged in different parts of the world, there is little evidence to suggest the diffusion of urban ideas between them, and it is thought that most of these early regional empires developed independently of each other (Scarre 2005). Only with the expansion of Greek urban influences and the subsequent rise of the Roman Empire around 2,100 years BP is there evidence for the diffusion of city-building ideas, in particular the use of a grid layout for planned new settlements. While legacies from some of these early cities have survived as the basis for modern cities, many of these early sites were abandoned as environments changed and empires waxed and waned, becoming principally the preserve of archaeological discovery and discussion.

The form of the pre-industrial city

The diversity in the origins of early cities in time and space produced a great variety in their form. The forms of earliest cities were often intimately linked to the religious and cultural beliefs of particular societies, providing specific symbolic links between the population and their wider world view (figure 3.1 and case study on p. 34).

Despite this variety in early urban forms, similarities in the principal functions and socio-political structure of early cities led Gideon Sjoberg (1960) to propose a model of the structure of the pre-industrial city (figure 3.2). Within Sjoberg's

Figure 3.1 *Tulum, Mexico, reflecting the religious and cultural beliefs of the Mayan civilization*

Source: Author's photograph

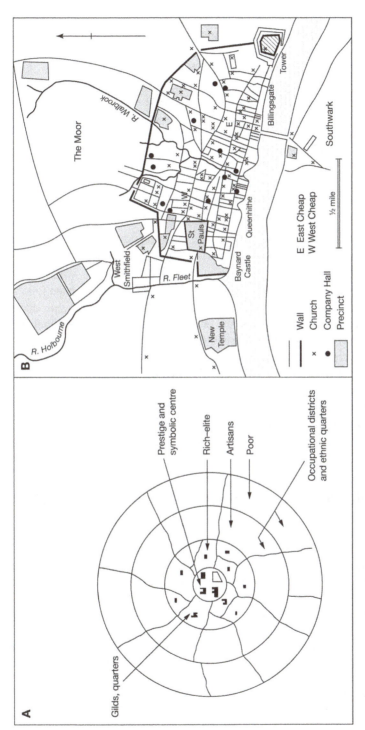

Figure 3.2 *(a) A model of the pre-industrial city, (b) Pre-industrial London*

Source: Adapted from Herbert and Thomas (1982: 63)

model, prestige buildings, religious complexes and the residences of the social elite were located in the centre, indicating the pre-eminence of the core and the greater prominence of these activities over economic concerns. Concentric rings of decreasing social status span out from the centre. These zones of artisans and unskilled workers were further differentiated by virtue of occupation or ethnicity. His model was developed as part of the critical discussions of the work of the Chicago School and sought to demonstrate that the social structure proposed for modern cities (see below) was the reverse for pre-industrial cities. Variations to Sjoberg's model, such as Vance's consideration of the mercantile city (1977), emphasized the importance of occupational sub-districts in pre-industrial cities based on economic rather than purely religious concerns and questioned the extent of a zone of lower class labourers, seeing these scattered throughout the sub-districts.

In common with other well-known urban models, Sjoberg's is based on socio-economic structure and does little to illuminate understanding of the physical structure of pre-industrial cities. As noted above, while the physical form or 'look' of pre-industrial cities was quite specific, related to localized building traditions, similarities in the functions of early cities produced some commonalities in the layout and building types contained within them. Early cities were 'pedestrian' cities and therefore relatively compact in size with fairly narrow streets, reflecting available transport technologies. The plan of pre-industrial cities is often referred to as 'organic' in structure, implying that it is irregular and 'natural' rather than formal and planned. However, the description of the form of pre-industrial cities as 'unplanned' is somewhat false as there is clear evidence of planning in many early cities, for example in many medieval European cities (see case study).

Case study: urban form and symbolism in the medieval city

In his article 'Cities of God? Medieval urban forms and their Christian symbolism', geographer Keith Lilley combines morphological and iconographic approaches to examining urban landscapes to explore how urban forms in medieval Europe conveyed Christian symbolism. He draws together the morphological traditions of mapping medieval townscapes using detailed cartographical analysis and the examination of documentary and archaeological evidence with a reflective and critical reading of these maps and sources to highlight the ideas and values of those who produced them, utilizing iconographic approaches to studying urban forms adopted by cultural geographers. Lilley contends that the built form of the medieval city can be read in the same way as medieval religious art and architectural forms, such as cathedrals, which were seen to embody both literal and mystical meanings of Christian Scripture.

continued

From examining textual and visual representations of the medieval city he argues that the city was viewed as a scaled-down version of the world, or microcosm, standing as a 'map' of Christian belief and meaning. The city literally and mystically stood between 'man' (sic) and God, connecting the earthly and heavenly worlds in writings of the time. Evidence of the increasing use of geometry in the planning of cities highlights how cosmic symbolism was materially expressed in the layout and physical form of medieval urban landscapes, particularly through the use of urban forms based on the circle and the square, quartered to make the sign of the cross, reflecting the idealized symbolic form of Jerusalem. Equally, the straightening up of urban layouts in the twelfth and thirteenth centuries can be linked to ideas about beauty and design being associated with divine design of the world and straight lines offering a path to God. Indeed God himself was widely depicted as a geometer, designing the cosmos, and many surveyors of the time were seen to be engaged in inscribing Christian values into urban landscapes. This Christian symbolism was further inscribed into the urban life of the medieval city and communicated to urban populations more widely through religious processions which often went from the periphery of the city to the main religious building at the centre, so symbolizing the route to God and the centrality of God within the city and also the wider Christian world.

Source: Lilley (2004)

In line with their defensive and control functions cities were often surrounded by fortifications, particularly walls. These walls acted as important barriers to outward expansion (Conzen 1960), creating a densely packed plan form as building was contained within this boundary. This densely packed structure is also evident in the multiple use of plots and buildings in the pre-industrial city for both commercial and residential purposes. Conzen's (1960) examination of **burgage plots** in Alnwick reveals a typical structure where homes, workshops and storage areas were combined in a single plot, which over time was increasingly infilled with buildings due to development pressures. Cities also acted as trading centres, containing spaces for the exchange, storage and processing of goods. In particular, cities contained places for the exchange of goods in the form of market places or shop or craft streets, usually at the core of the city. Sometimes these were zones where all trading took place, while in larger centres a number of distinctive market places or craft zones were evident, dealing with trade in particular goods, such as livestock or crops (see the example of London, figure 3.2).

While these generalized forms can be identified in many cities with pre-industrial origins, it should be noted that differences in religious codes and the organization of social relations produce some distinctive form variants. For example, in traditional Islamic cities, the site of the main mosque has an important influence on urban form, while extended family living arrangements and ideas about the distinction of public and private domestic space produce distinctive layouts in residential districts (Bianca 2000). Equally, the alignment of buildings in traditional Chinese cities often followed the principles of *feng shui* where possible (Whitehand and Gu 2007).

Exercise

Consider the origins and early development of the city you live in, or a city close to where you live. Try to discover what information exists on the early development of your chosen city, such as old maps, historical documents and texts or archaeological evidence. Good places to begin your search would be your university or college library, local libraries or archives or on-line information sources such as local government or history group websites. The more sources you find the more comprehensive an account of your chosen city's development you can produce (you might try this as a group activity with other people in your class).

Review the information you collect and consider whether your city has pre-industrial origins, the nature of its pre-industrial form and if so whether any urban forms from that period survive into the present layout of the city. If they do not, or the forms are different from those discussed above, then think about why this is (what are the origins of your city and the processes underpinning its development)? Think also about the problems and limitations involved in gathering information about the early development of cities.

The modern city

In parallel to the development of the first cities, the rise of the modern cities that we know today was precipitated by two further significant phases in the development of human history, namely industrialization and colonialism.

The complex series of innovations and period of economic and technological change that has become commonly referred to as the Industrial Revolution had a profound effect on urban development particularly in Europe and North America from the eighteenth century onwards. Many pre-industrial cities increased significantly in size and had their form altered by redevelopment and the addition of new urban forms associated with new economic and socio-cultural impulses. In addition, many new urban centres emerged based on the exploitation of new resources, such as coal, new transportation links, such as railways, and new industries.

In parallel to industrialization a number of European countries embarked on a period of global exploration and colonization. This was led by countries such as Spain, Portugal, the Netherlands and Britain from the fifteenth century onwards. Initial quests for trade routes developed into the claiming and settlement of land on the continents of North and South America, Africa and Asia. This resulted in the export of European ideas of city building to the 'New World', although traffic was not all one way and some design ideas from colonized countries also filtered back into European cities (King 1990).

Many new cities were founded to act as political centres for colonial control and as centres for colonial trade. Where existing cities were utilized, separate colonial extensions were often added to existing indigenous urban forms.

The industrial city

The rise of the industrial city stimulated much research into these new and changing cities. As noted in chapter two, the work of urban sociologists at the University of Chicago was particularly influential. Chicago was a fairly new city at the time; it had grown rapidly in the late nineteenth century, owing much of this growth to industrialization, its emerging role as a railway hub and large scale in-migration. Classical models of modern urban structure, namely Burgess' concentric zone model (1925) and Hoyt's sector model (1939) were based on this research (figure 3.3).

For Burgess, the city was a 'social organism' and he used the ecological analogy of invasion and succession to conceptualize the process of suburbanization. In his model, recent migrants to the city would lack money and would seek out the cheapest accommodation in the city, namely older housing in the inner city. As these migrants became wealthier and able to make wider housing choices, they would move out to the next zone of more expensive housing. Thus social status and wealth increased towards the edge of the settlement, with the best housing and wealthiest groups on the edge of the city. The model also identified a central business district, where commercial functions dominated, and a transitional zone, where industry mixed with poor housing. The model therefore identifies the increasing functional segregation of the industrial city, with an increasing separation of home and work. Hoyt's model was based on urban land economics. He disputed the notion of a concentric zonal structure for the industrial city and argued that residential areas could be more properly understood as a pattern of sectors. He identified a sectoral pattern of high rent areas and suggested that high status sectors could be found along routes radiating out from the centre and away from industrial zones. However, his model also includes an element of filtering, with wealthier residents moving out from the centre.

Subsequent 'testing' of these models has revealed the complexity of urban socio-economic structure and the range of factors influencing this. An important early variation to Burgess and Hoyt's models was the proposal of a multiple nuclei model by Harris and Ullman (1945). Here they argued that cities rarely develop around a single nucleus and that a range of local conditions were important to the location and clustering of various land uses. Other important

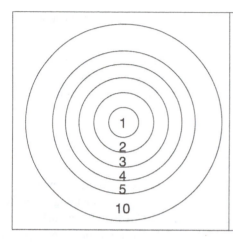

District

1 Central business district
2 Wholesale light manufacturing
3 Low-class residential
4 Medium-class residential
5 High-class residential
6 Heavy manufacturing
7 Outlying business district
8 Residential suburb
9 Industrial suburb
10 Commuters' zone

District

1 Central business district
2 Wholesale light manufacturing
3 Low-class residential
4 Medium-class residential
5 High-class residential
6 Heavy manufacturing
7 Outlying business district
8 Residential suburb
9 Industrial suburb
10 Commuters' zone

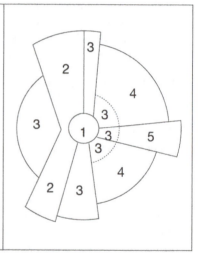

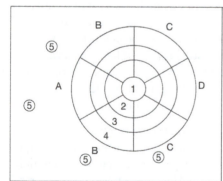

A Middle-class sector
B Lower middle-class sector
C Working-class sector (and main sector of council estates)
D Industry and lowest working-class sector

1 CBD
2 Transitional zone
3 Zone of small terraced houses in Sectors C and D; larger by-law housing in Sector B; large old houses in Sector A
4 Post-1918 residential areas, with post-1945 housing on the periphery
5 Commuting-distance 'dormitory' towns

Figure 3.3 *Models of the industrial city*

Source: Adapted from Park and Burgess (1925); Hoyt (1939) and Mann (1965)

variations were identified in industrial cities where capitalist and free market conditions had not been given an unfettered rein. In the UK, analysis of British cities revealed a more complex pattern of zones and sectors, exemplified by Mann's (1965) model (figure 3.3), with lower status local authority housing on the edge of cities, a result of state intervention in the housing market from the early twentieth century onwards. Intervention by the state also gave rise to very different forms in cities developed under communism in Eastern Europe and China. Some Soviet cities displayed a distinctive 'camel back' structure with a high density core, surrounded by a low density industrial zone, with high rise housing projects on the periphery (Pacione 2009: 184). However, many cities in communist countries were only partial examples of this 'idealized' form and with the collapse of centrally controlled land markets many have become more like European modern cities (Rudolph and Brade 2005).

Again, while useful, these classical urban models only provide a partial insight into the form of the industrial city, based as they are on considerations of socio-economic structure alone. They provide no indication of the physical forms associated with the dynamic growth and change of the industrial city. Work by urban morphologists has sought to provide more insight into the macro scale form of the industrial city resulting from its outward expansion, using the **fringe belt** concept which links the long-term development of urban areas to economic fluctuations, the role of innovation and cycles of building in particular (Whitehand and Morton 2006). Morphological research has demonstrated that while the physical expansion of the industrial city has been successively outward, the phases of development have not been uniformly smooth. Urban expansion is rarely continuous, but rather cyclical with periods of rapid outward growth alternating with periods of standstill. These periods of stability, or limited movement, have produced fixation lines either in the form of a physical barrier, such as the town wall in the pre-industrial city, or a metaphorical barrier resulting from a lack of economic dynamism. On the edge of the city, during these slumps, land is cheaper and more available and it becomes feasible to build at lower densities, including houses on large plots, large institutional buildings, such as hospitals or universities, large commercial complexes, such as industrial and retail parks, and to include extensive land uses such as sports pitches. This leads to a belt of land, or fringe belt, which has a different more open form to those areas of more intensive development either side of the belt. Larger cities can contain a number of fringe belts, with the plan and functions of the inner and middle belts becoming fixed as the next building boom moves the city beyond them (see figure 3.4 and case study below). Recent comparative research has demonstrated the general applicability of the fringe belt concept, using research in a number of cities to demonstrate their existence in urban areas around the world (Conzen 2009).

Much of the outward expansion of the industrial city consisted of residential development, and a key feature of the industrial city is the series of distinct

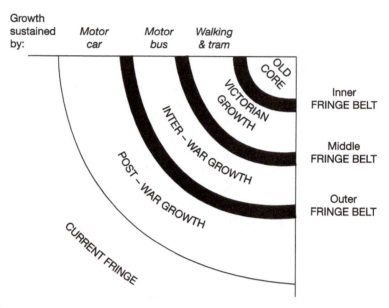

Figure 3.4 *Model of fringe belt development*

Source: Adapted from Whitehand (1994: 11)

Case study: fringe belt change in Birmingham, UK

Drawing on research into the early twentieth century fringe belt in Birmingham, UK, Whitehand and Morton's (2006) article examines the pressures on these distinctive areas, particularly their green open space, resulting from recent socio-economic change. The fringe belt examined in Birmingham marks the edge of the city as it existed at the time of a major hiatus in housebuilding just before, during and just after the First World War (1914–1918). This fringe belt was heterogeneous in plan, building form and land use, containing a range of extensive land users including institutions, public utilities, allotment gardens, parks, recreational areas and some industry. Like other fringe belts its ground plan characteristics differentiate it from the housing areas on either side of it.

Previously, research had suggested that fringe belts were most vulnerable to redevelopment as they became embedded in the built-up area, thereafter becoming less susceptible to such redevelopment as institutions occupying fringe belts invested in their estate and became established, deflecting other land use pressures. However, recent socio-economic and political changes in the UK have led to renewed pressures for the redevelopment of these areas. Examination of development control records for a selection of case study sites within the fringe belt suggested that since the 1960s there had been substantial pressure for redevelopment for both housing and other purposes within the area. The research also suggests that negotiation of this pressure became increasingly

continued

contentious and protracted within the planning process.

Key pressures have been the increasing encouragement from central government for the building of more dwellings within urban areas on previously developed, or 'brownfield', land and the growing commercialization of public services, such as higher education, health and leisure, many of which are strongly represented in fringe belts. Despite growing concern for the protection of green spaces in urban areas, ambiguity in the definition of what counts as brownfield land has led to increasing pressures to redevelop the open land in fringe belt areas and intensify its use, particularly for housing. The detailed case study research reveals a changing attitude among landowners, both commercial estates and charitable organizations, to promote more commercial redevelopment of their land holdings. While the research also reveals increasing consideration of environmental matters in planning discussions, these only occasionally outweighed commercial development considerations. However, despite these considerable pressures the fringe belt that came into existence over a century ago has remained substantially intact and continues to be an enduring urban landscape feature. Overall, the research highlights the often conflicting goals of higher density development and environmental conservation within planning as they are played out in these distinctive fringe belt zones.

Source: Whitehand and Morton (2006)

landscapes created by successive phases of suburban expansion. Despite being burdened with an image that suggests monotonous regularity, the suburban areas of cities display great variety. This variety derives in a large part from the different periods during which suburbs have developed (Harris and Larkham 1999; Vaughan et al. 2009). In particular, the pace and style of suburban development is related to booms and slumps in house building cycles, the adoption of various innovations in transport technology, and the architects, developers and planners involved.

The variety of suburban landscapes makes generalization on form difficult. However, many early suburban developments in the nineteenth century in Europe and North America displayed a fairly regular grid or geometric form. This resulted from a mix of early ideas on city design and planning and property development concerns based on the need to easily parcel up land for sale and maximize the number of buildings on available land. In the twentieth century, suburbs began to adopt more open and organic forms, as planning and design was influenced by ideas from Garden City movements in Europe and North America which advocated lower building densities with extensive open space (see also chapter six). The type of dwelling within suburbs has been relatively uniform until recently, with single family dwellings dominating. Key changes have been the incorporation of garages into these buildings as the ownership of motor cars has increased, and more recently a greater variety of dwelling types, including smaller homes and apartments, as the demographic structure of many

Exercise

A useful web resource for students interested in examining the form of cities is Google Earth (www.google.co.uk/intl/en_uk/earth/index.html). Using images from satellite photographs it allows you to zoom in on particular cities or areas of cities (although global coverage is variable) and look at the configuration of streets, buildings and open spaces in the city. Additionally, its street view function allows you to look at some streetscapes from ground level.

Using Google Earth look at the city that you live in or a city close to where you live. Focus in on particular parts of the city and compare the layouts of the suburban residential districts. Can you recognize the different forms, linked to different periods of suburban growth discussed in this chapter? Can you see any fringe belt areas (areas with more open land/green space)? Compare your findings with the points discussed in this chapter – do your city's suburban areas appear similar to or different from the form of the modern city outlined in this chapter? If any differences do exist think why this might be (for example, different timings of suburban processes or cultural differences).

western societies has changed to include more single-person households. Recent research has also begun to highlight processes and forms of adaptation to suburbs at the micro scale, including intensification of plot use through infill and alterations to single family dwellings, including extensions and conversions to apartments (Whitehand and Carr 1999).

Industrialization also produced significant changes in the internal plan and form of city centres. Specifically, the scale and pace of urban change increased, with cities moving from mainly incremental change through the accretion and addition of forms, exemplified by the burgage cycle, to more comprehensive large scale redevelopment and planning. New urban forms also emerged, particularly new commercial forms such as factories, office buildings and retail premises (department stores), civic buildings, resulting from the developing role of urban governments, and transportation and storage infrastructure such as railway stations and warehouses. In his work on British industrial cities, Conzen (1978) identified some of the changes to the plan of cities prompted by industrialization. He highlighted changes to the street plan caused by street widening, straightening and new **breakthrough streets** to accommodate new transport forms, such as trams, and increasing volumes of traffic. He also recognized changes to the plot pattern within central areas, where older smaller plots were either amalgamated into larger units for new commercial functions or reconfigured by comprehensive redevelopment schemes, such as slum clearance or transport developments. The increasing size of plots was also accompanied by an increasing scale in building forms. Perhaps the most significant addition to the silhouette of the city was the 'skyscraper'. Domosh

(1989) outlines the combination of factors underpinning the emergence of skyscrapers in New York in the late nineteenth century, including technological innovations, competition for land in a buoyant business environment and the personal ambitions of the owners of the companies commissioning the buildings.

The colonial and post-colonial city

As noted above, the urban systems of many places beyond Europe were influenced by successive phases of colonial expansion by a number of European countries. The specific impact on the form of settlements in those countries colonized by European nations depended on the timing of colonization and the country doing the colonizing. In many cases colonizers founded new settlements to act as control points for trade and rule, exporting certain building traditions from their own countries. For these settlements a grid plan layout was the norm, for ease of land allocation and development, occasionally modified by local topography. These new settlements and systems of control led to the abandonment of indigenous cities, as in Latin America (see figure 3.1 above), and only occasionally were existing settlement sites utilized, as in Mexico City where the Spanish colonial city supplanted the indigenous city (Massey et al. 1999). In other instances, colonizers built new accommodation alongside indigenous settlements, seeking to retain a distance between populations, which led to the juxtaposition of different urban forms, sometimes referred to as a 'dual city'. For example in North Africa, French colonial settlements, containing commercial functions, government buildings and residences, were built alongside the existing Islamic medinas (figures 3.5a and 3.5b). Similar dual city forms are also evident in countries influenced by British colonization, such as India, where cities such as Delhi and Kolkata contain an indigenous city alongside a British built 'new town' inspired by the Garden City idea (Drakakis-Smith 2000). However, as Drakakis-Smith notes, the idea of a distinct dualism is too simplistic and in reality hybrid zones of indigenous and colonial forms exist. Allied to this, it should be noted that often the architecture of buildings involved a fusion of indigenous and colonial traditions (figures 3.5a and 3.5b).

As former colonies have moved into post-colonial independence, their cities have been shaped by two key processes linked to their colonial pasts. First, for many cities, the immediate post-colonial period has been driven by a desire to 'modernize' unfettered by colonial restrictions. Many places sought to develop their economies to incorporate new industrial and commercial functions, adding many of the forms of the industrial cities of Europe and North America to their landscapes, such as high-rise CBD office districts and factory areas. This has led some commentators to suggest that we are witnessing a period of urban convergence, where urban environments around

Figure 3.5 *The dual city, juxtaposing the traditional Islamic medina (left) and colonial extension (above), Sfax Tunisia*

Source: Author's photographs

the world are becoming increasingly similar and 'placeless' (Brenner and Keil 2006), although others argue that much variety still remains (Grant and Nijman 2002). Second, many post-colonial cities have developed socio-economic structures that are different from western industrial cities. Key features of the 'classic' socio-economic models of the so-called 'Third World city' are the presence of large areas of low status squatter settlement housing on the edges of the city, with higher status enclaves remaining in the core of the city, often now gated residential developments (Pacione 2009). This structure mirrors the structure of the pre-industrial city rather than the suburban structures of the industrial city. This division stems from the stark socio-economic inequalities in these urban populations, high rates of immigration into these primate urban centres from poorer rural hinterlands and the lack of housing provision in economies that are still relatively poor in global terms, frequently still in dependent relationships with the major post-industrial economies.

The post-modern city

A great deal has been written about the apparent transformation of the form of cities in Europe and North America at the turn of the twenty-first century. This debate has focused on the emergence of post-modern, **post-Fordist** or post-industrial urban forms. These discussions about the changing geographies of the city are linked to wider debates about the apparent transition from modernity to post-modernity in advanced capitalist societies which have taken place in many disciplines from architecture and urban studies to film, literary criticism and fashion. However, while visual evidence of fundamental transformations in urban landscapes around the world might seem compelling, the extent to which changing economic and governance structures and architectural styles represent fundamentally different processes underpinning urbanization is a matter of debate (Harvey 1989a; Dear 2000).

Much of the debate about the changing form of cities has been based on a small number of cities, mainly in North America, which have become constructed as archetypes of post-modern urbanization (see also chapter two). In particular Los Angeles has assumed a position comparable to that of Chicago in the early twentieth century. Los Angeles has been home to an influential school of urban studies since the 1970s, known as the 'California School'. Based on their observations of changes taking place in Los Angeles, commentators have argued that the idea of the structure of the modern city outlined by the Chicago School is outdated and that we have been witnessing the emergence of fundamentally different city forms (see Sassen 2008 for a critical consideration of the ideas of the Chicago and California Schools).

A major theme of this work has been the idea of the fragmentation of urban form and its associated social and economic geographies, namely that the city is ceasing to exist as the recognizable single coherent entity that it was as the modern city. The post-modern city is seen to be more chaotic in structure, fragmenting into a series of independent settlements, economies, societies and cultures. The city then consists of a number of large spectacular residential and commercial developments with economically and environmentally degraded space in between. This is expressed in the idea of the 'galactic metropolis' (Lewis 1983), which describes urban form as resembling a pattern of stars floating in space rather than a unitary metropolitan development growing steadily outward from a single centre. This idea of fragmentation is represented in Dear and Flusty's (1998) 'patchwork' model of the post-modern city which is very much at odds with the centred and ordered models of urban ecology (Hubbard 2006) (figure 3.6).

Underpinning these ideas about the structure of the post-modern city is the recognition of a number of emergent trends in the processes shaping urban landscapes. Ed Soja summarized these trends as 'six geographies of restructuring' (Soja 1995: 129–137):

- the restructuring of the economic base of urbanization (with emerging sectors based on producer services, high-technology, cultural, entertainment and knowledge based industries);

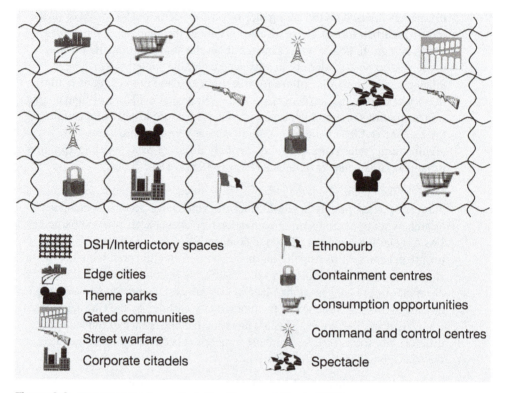

Figure 3.6 *The post-modern city (after Dear and Flusty 1998)*
Source: Hubbard (2006: 55)

- the formation of a global system of world cities;
- the emergence of 'edge cities' or 'exopolis';
- the changing social structure of urbanization (with heightened economic, social and cultural inequalities expressed in new patterns of segregation and polarization in cities);
- the rise of 'paranoid' or **carceral** architecture based on protection, surveillance and exclusion;
- a radical change in urban imagery (the increasing presence of simulation in urban landscapes, imaginations of alternatives to the 'dreadful reality' of actual cities, e.g. themed shopping malls).

Within these new geographies the existence of spectacular, themed flagship developments, fortress landscapes, socially polarized residential communities and edge cities are most often highlighted as evidence of emerging post-modern urban forms (see case study below). Located both in downtown redevelopment zones and on the edge of cities, spectacular, or flagship developments encompass a variety of commercial development types (such as office and retail complexes, convention centres, sports stadia, museums and cultural centres). Commonly, however, they are large scale and have an emphasis on eye-catching, decorative, spectacular or innovative, typically post-modern, architecture. They are often designed to have an iconic quality and act as a symbolic focus for redevelopment initiatives (Knox 1993, see also chapter eight).

The term 'fortress landscapes' is something of a catch-all term for landscapes which are designed around security, protection, surveillance and exclusion. While security was a concern in the modern city, an increasing obsession with control and protection is seen as characteristic of the post-modern metropolis, eloquently described in Mike Davis' seminal book *City of Quartz* (1990). In many cities, impenetrable façades and the gating and electronic control of access in luxury residential schemes are increasingly common development forms. Finally, a key feature of post-modern urban form has been the 'decentring' of the city, with the development of multiple urban 'centres' and the rise of 'edge cities' which rival, or replace, downtown dominance. Within Dubai, for example, new mega project developments have been planned as 'cities within the city', creating a fragmented urban layout (Pacione 2005: 262) (see also case study below). Within the post-modern city, 'edge cities' are seen to represent post-suburban landscapes which include extensive private masterplanned developments that are radically different from the traditional residential suburb. The design of individual dwellings and the landscaping of developments are often closely based on anglophile ideas of tradition and rurality (Knox 1993). These developments come with a wide range of amenities normally found in urban centres planned into them, including town centres, public squares, theatres and schools. These masterplanned residential developments often surround, or are found in close proximity to, concentrations of offices, shops and other formerly downtown commercial functions which have decentralized (Soja 1995).

Despite the growing influence of cities like Los Angeles internationally, these so-called archetypal cities of post-modernity are not necessarily representative of the experiences of contemporary urbanization in other cities around the world (see also chapter two). The problems of applying these models of the post-modern city are very similar to the problems of trying to apply earlier models of the industrial city. In most cases the overall form of our cities still reflects modern urbanization processes of industrial capital and planning. Yet, within these largely modern structures, new urban forms have emerged and there is growing evidence that the internal spaces of many cities have been resorted or reorganized. However, post-modern urbanization, like previous phases of urbanization, is emerging as a complex series of trajectories mediated locally rather than a simple, universal trajectory of development. As urban geographers, the question for us is not whether these new ideas and models should be applied to analysing cities, but rather how they should be applied and to consider the extent to which they can be regarded as blueprints for urban change in the twenty-first century.

Case study: splintered urbanism – Dubai, United Arab Emirates

Dubai can be seen as a splintered, patchwork metropolis with a fragmented urban morphology and unequal social demographic. Planned around the car, the development of the city from the 1980s onwards has spread out from the old urban core along planned growth corridors, notably the Sheikh Zayed Road, creating a diffuse business district rather than a distinct downtown. As a relatively new but ambitious city, Dubai has attempted to ascend in the world urban hierarchy and establish itself as the image of the twenty-first century metropolis through a range of spectacular urban developments, such as the Palm Jumeirah residential island development, the sail-like Burj Al Arab hotel and the Burj Khalifa, the tallest building in the world (at time of writing). However, as a result of a development slow-down following global financial crisis in 2008, these spectacular developments are often separated by desolate landscape patches awaiting development. Many of these spectacular developments represent fortress landscapes, with gated residential enclaves and access-controlled hotels. Officially sanctioned public spaces can be found in shopping malls, public beaches and parks and outdoor cafes, although as Elsheshtawy notes, to describe these as publically inclusive is debateable given that access is controlled via security guards and charging for entry, and by the fact that they are located at a distance from the work and accommodation districts for poorer city residents, with limited access by public transport.

Dubai is therefore composed of multiple, disconnected centres of spectacular enclaves, abandoned development land and poorer districts separated by multi-lane highways. In this respect, Dubai is also illustrative of 'splintered urbanism' (Graham and Marvin 2001), with urban transportation infrastructures, such as the Metro or bus routes, either bypassing or cutting through many poorer areas within the city, precluding meaningful movement and interaction and leading to the ghettoization and demonization of these marginalized groups and areas as 'lawless'. While the mobile elite transverse the city in air-conditioned cars or the new Metro system to access privatized retail and leisure spaces, poorer migrant worker groups endure expensive taxi rides or long waits at bus stops, queues and multiple exchanges on buses to move between work, informal socialisation areas in the old city centre and distant suburban housing zones. However, in informal spaces next to bus stops or within abandoned lots, Elsheshtawy notes that marginalized migrant workers, mainly from South East Asia, are able to establish alternative transnational spaces, where globalized connections are maintained through social interactions and a range of support settings such as ethnic restaurants, places of worship, money exchanges and internet/mobile phone shops. However, in contrasting the gleaming spaces of the new Metro and its stations to the empty, desolate landscapes around it, Elsheshtawy reflects that the distances between the city's spectacular elite districts and its marginalized social groups seem more visible than ever.

Source: Elsheshtawy (2012) see also Acuto (2010)

Summary

This chapter has sought to provide a brief overview of the varied forms that cities have taken over time, linked to key phases in the development of human society. Examining the physical form of the cities around you, either through the study of maps and documents or through direct field observations, provides an important foundation for further research. The key is to move beyond this observation and description to examine the reasons behind the forms identified. Many of the ideas about urban form introduced here provide a foundation for more detailed consideration of the key processes shaping the city, such as economic change and planning, and key elements of the city's structure, such as architecture and housing, later in the book. Why our cities look the way they do and what they might look like in the future remain important concerns for urban study.

Follow-up activities

Essay title: 'In what ways can the study of urban form help explain the dynamic and changing nature of urbanization?'

Commentary on essay title

An effective answer would outline what constitutes urban form and highlight the link between urban form and the socio-economic and cultural forces shaping the city. It would perhaps utilize the three-phase model of urbanization outlined at the beginning of the chapter and examine how particular models of the city reflect key phases in the development of cities. It would also highlight the dynamic nature of urban change, exemplified in plot change and redevelopment and the presence of fringe belts, pointing out that changes in urban forms are variable, with the plan being the most persistent legacy. An excellent answer would develop this discussion by pointing to the significance of local forces and the actions and intentions of a variety of actors in mediating these broader processes. Drawing on recent critical work in urban geography it would seek to question the general applicability of urban models and general theories considering urban form, pointing to the diversity of forms around the world.

Project idea

A traditional activity for urban geographers examining the form of the city is to go out and look at urban landscapes and to examine how these vary across the city by undertaking a transect. On a map of your city, or a city that is local to you, set out a

continued

transect line to follow, such as along a major route, out from the centre of your city to the suburban edge. In the field follow your transect, mapping the forms of the buildings around you in major zones (height, types, materials, approximate general age) and the predominant land uses. Following your fieldwork, come back and on a base map of the plan layout of your city produce two annotated maps of your transect showing changes in building form and land use. Examine the variations in form and land use from the city centre and compare your findings with the theoretical patterns described in this chapter. How would you account for any variations between these theoretical ideas and your observations?

Further reading

Books

- Graham, S. (2016) *Vertical: The City from Satellites to Bunkers*, London: Verso
  Not about urban form per se, this intriguing book asks us to look at the city in three dimensions and the increasing emphasis on verticality in terms of its organization.

- Kropf, K. (2016) *The Handbook of Urban Morphology*, London: John Wiley & Sons
  Intended as a practical manual of urban morphological analysis, including a guide to methods of analysis and key terms and concepts.

- Larkham, P.J. and Conzen, M.P. (eds) (2014) *Shapers of Urban Form: Explorations in Morphological Agency*, New York: Routledge
  A range of disciplinary and geographically focused contributions considering the dynamics of urban change and the processes and agents responsible for this. Dedicated to Whitehand, it provides an update on his classic text, *The Making of the Urban Landscape*.

- Lilley, K.D. (2009) *City and Cosmos: The Medieval World in Urban Form*, London: Reaktion Books
  An innovative and interdisciplinary look at the form of early cities and how Medieval Christians infused their urban surroundings with meaning.

- Oliveira, V. (2016) *Urban Morphology: An Introduction to the Study of the Physical Form of Cities*, Cham, Switzerland: Springer
  Recent comprehensive text considering the development of urban morphology as a discipline and outlining key concepts, theories and methods in urban morphology with detailed case studies of New York, Marrakesh and Porto.

- Vaughan, L. (ed.) (2015) *Suburban Urbanities: Suburbs and the Life of the High Street*, London: UCL Press
  A collection of chapters looking at suburbs, anchored in the space syntax morphological approach but spanning a range of disciplinary perspectives and geographical horizons.

Journals

- Barau, A.S., Maconachie, R., Ludin, A.N.M. and Abdulhamid, A. (2015) 'Urban morphology dynamics and environmental change in Kano, Nigeria', *Land Use Policy* 42, 307–317
  Linking together analysis of human and natural systems, the paper considers the impact of the changing morphology of the city in terms of its resilience and sustainability.

- Conzen, M.P., Whitehand, J.W.R. and Gu, K. (2012) 'Comparing traditional urban form in China and Europe: a fringe-belt approach', *Urban Geography*, 33(1): 22–45
  Useful paper demonstrating the cross-cultural application of the fringe-belt concept as a frame of reference for examining the physical structure and historical development of cities.

- Meyer, W.B. and Esposito, C.R. (2015) 'Burgess and Hoyt in Los Angeles: testing the Chicago models on an automobile age American City', *Urban Geography*, 36(2): 314–325
  Interesting paper that tests the applicability of using the 'Chicago model' to study contemporary US cities such as Los Angeles, and argues that this model remains applicable to studying urban change in the newer, automobile-age city.

- Sassen, S. (2008) 'Reassembling the urban', *Urban Geography*, 29(2): 113–126
  A critical discussion of the debates surrounding the differences between urban models based on Chicago and Los Angeles, part of a special edition of the journal on this theme.

- Xie, P.F. and Gu, K. (2015) 'The changing urban morphology: waterfront redevelopment and event tourism in New Zealand', *Tourism Management Perspectives*, 15: 105–114
  The paper considers the changing morphology of Auckland's waterfront landscape and the role of regeneration strategies and event tourism as drivers of change of these areas into 'post-modern' urban landscapes.

Websites

- Google Earth (www.google.co.uk/intl/en_uk/earth/index.html) – a downloadable tool that allows you to view a range of satellite and other image data to look at cities.

- International Seminar on Urban Form (www.urbanform.org/) – international umbrella organization for academics interested in studying urban form. Contains useful reference and bibliographic information and further links.

Section 2
Themes

4 City economies

Introduction

In chapter three, we highlighted the fundamental link between processes of economic change and processes of urban development. Indeed, many regard economic forces as the dominant influence on urban change (Pacione 2009: 4). Equally, cities have played an important role in the history of economic development and can be viewed as 'crucibles of economic change' (Short 1996: 13). Economic activities are therefore important in both defining cities and in shaping the lives of people within them. Geographical research into the economies of cities has encompassed both of the core concerns of urban study identified by Berry (1964: 147) – 'cities as systems within systems of cities' – which were outlined in chapter two. Geographers have been concerned both with the distribution of economic activities within cities and the external economic relations of cities. However, while these twin foci have endured in studies of the economies of cities, the approaches employed by geographers have shifted considerably. We begin by providing a brief overview of key approaches, particularly considering the important insights provided by **structuralist** approaches. We then examine key trends resulting from recent global economic changes and the impact this has had on relations between cities within a global network of cities. This is followed by exploration of some of the key changes to economic activities within cities around the world which have happened as a result of these broader economic shifts.

Changing approaches to studying the economies of cities

Neo-classical economics

Economic explorations of the city assumed a prominence during the quantitative revolution in geography in the 1960s (see chapter two). Geographers began to utilize **neo-classical economic** locational theories, adapting and reworking classical land use models, such as those developed by Von Thunen, Christaller, Weber and Losch. Two key strands of research emerged from this: work on urban hierarchies and spheres of influence and work concerned with the distribution of urban land uses (see Carter 1995 for a good overview of this research).

Work on urban hierarchies was based primarily on Walter Christaller's **central place theory**. Christaller argued that within a region the number, size and distribution of towns was determined by the range and threshold of goods and services. The range of a good is the distance people will travel to purchase it, which is greater for higher value goods. The threshold of a good is the minimum population required to support its continued supply, with higher value goods needing a larger population to support them. Consequently the provision of these goods and services is found within the larger settlements of the hierarchy. Subsequent empirical work, based on these ideas, sought to test whether national, regional and local urban systems existed in clear hierarchies and to examine the spheres of influence of settlements, based on examination of a range of the different functions and services they offered. For example, Berry (1967) looked at the evolution of the central place system in Iowa based on access to a range of retail and service functions, such as clothing and furniture purchases.

Work on urban land use built on the development of early urban models, such as Hoyt's, linking the distribution of land uses to urban land rents, particularly through the concept of the bid-rent curve, where high value commercial land uses were seen to occupy prime accessible locations in the urban centre through their ability to outbid other uses with less need for central locations (Alonso 1964). Developing these ideas, Whitehand (1972) applied the concept of bid-rent to the formation of **fringe belts** (see chapter three), arguing that the willingness and ability of users to bid varied over time, specifically between housing and institutional users, creating a dynamic rather than a static land rental market.

However, as these approaches established themselves at the core of urban geography textbooks, two key developments changed the trajectory of economic research in urban geography. First, there was criticism of the assumptions underpinning neo-classical economic models, linked to wider critiques of positivist approaches in urban geography (see chapter two). Additionally, the work of spatial scientists was criticized by radical, structuralist geographers who felt that this work lacked the ability to say anything about the economic problems faced by cities. This critique was allied to the second key development which overtook urban economic research, namely significant transformations to the world economy. Processes of economic globalization and associated processes of deindustrialization undermined many of the assumptions of previous neo-classical work and presented new problems for cities and issues for analysis (Hubbard 2006).

Structuralist approaches

As we noted in chapter two, structuralist approaches have had a profound impact on urban geography. In seeking to understand the fundamental global economic transformations and problems affecting many cities, geographers put analysis of structural changes in capitalism at the heart of explanations of urban change, drawing on the work of Karl Marx for their analytical framework. For geographers adopting a structuralist perspective, Marx's conceptualization of the capitalist **mode of production** provided a key foundation for analysis. For Marx, history was characterized by the rise and fall of different modes of production, particularly the move from feudalism to capitalism. Central to Marx's analysis of the development of the capitalist mode of production was the idea of the development of labour power as a commodity. Marx argued that those buying labour took advantage of the difference between the price of labour required to produce goods and the price of those goods that could be obtained in the market place. Any surplus value accrued acted as a source of a capitalist's profit. If these profits were reinvested in new technologies and the production of new commodities, further growth and profit could be produced. However, crucially, Marx believed that this capitalist cycle of profit and investment was prone to periodic crises, where rising costs (particularly wages), technological obsolescence, over-production or the falling demand for commodities would result in falling profits. Significantly for geographers, Marx noted the importance of space in overcoming these periodic crises, through the exploitation of new markets and populations. Cities can be seen to play a key role in this respect, as 'instruments of production' bringing together pools of exploitable (cheaper) labour and providing markets for commodities (Hubbard 2006: 37).

Urban researchers have continued to explore this idea of capital as both a creative and destructive force in the search for new investment opportunities and profit, considering the impact of these cycles of change on landscapes, divisions of labour and social relations. The work of David Harvey has provided important insights in this respect. Harvey's work represents an attempt to read historical cycles of urban development as a reflection of the resolution of these crises of over-accumulation within various 'circuits of capital' (figure 4.1).

In his book *The Limits to Capital* (1982) Harvey argued that investment in, and hence production of, the built environment (the secondary circuit – labelled (2) in figure 4.1) occurred when an over-accumulation of capital in manufacturing and commodity production (the primary circuit – labelled (1) in figure 4.1) caused returns in this sector to fall. This made land and property an attractive alternative investment. Providing that a framework existed to facilitate it, these conditions caused capital to switch from the primary to the secondary circuit. However, this investment in the secondary circuit eventually leads to an over-accumulation in this sector, causing returns on investment to fall. The result of this is that capital will either switch back to the primary circuit or seek more profitable investment opportunities in the secondary circuit (i.e. newer developments). Although not

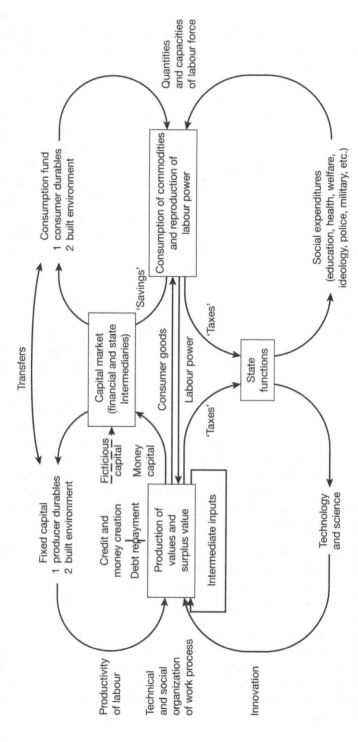

Figure 4.1 The circuits of capital

Source: Adapted from Harvey (1989b: 67)

without its caveats (see Castree and Gregory (2006) for a critical overview of Harvey's work), Harvey's conception of capital switching between investment opportunities in different economic sectors and geographical locations has provided an important insight into the changing economic fortunes of cities. Physically, the switching of capital is manifest in cyclical patterns of urban growth both within and between cities, with built environments becoming abandoned in the wake of capital movement to more profitable opportunities elsewhere. For Harvey, this idea of capital switching could therefore be utilized both to explore intra-urban changes to the built environment and to understand regional and international patterns of uneven development as the product of the global capitalist economy.

While overt reference to Marxist analysis has all but disappeared from many core economic geography texts (see for example Dicken 2007 and Knox et al. 2008), many of the core ideas and key terms adopted by this work have continued to be employed in analyses of economic change. One reason for declining interest in a Marxist take on urban economic change was a feeling that, as a theory based on analysis of industrial capitalism, it lacked relevance in explaining urban changes associated with post-modernism and post-industrialization. However, Harvey (1989a) in his book *The Condition of Postmodernity* responded to this challenge by arguing that post-modernism represented a reordering of, rather than a break from, previous manifestations of capitalism. This argument was important in reinvigorating critical urban analysis which has sought to examine the impacts of current shifts to advanced capitalism and the consequences of increasing economic globalization for cities (Brenner and Keil 2006).

Drawing on a range of post-modern approaches, more recent research in urban geography has adopted a pluralistic analytical perspective to examining restructuring, questioning the economic determinism of earlier structuralist perspectives (Fyfe and Kenny 2005: 107). For many, while structuralist analyses provided broad insights into economic change within and between cities, the operation of economic structures alone could not be simply linked to particular spatial outcomes. Those critical of prioritizing economic interpretations of the city have pointed to the importance of recognizing the **embeddedness** of capitalist economic processes within specific societies with particular traditions and histories, which are central to the functioning of the market and which vary over time and space. Consequently, while there is some agreement among urban geographers over the broad nature of restructuring and the reconfigurations of urban space taking place in the twenty-first century city, there is disagreement over the relative importance of economic, socio-cultural and political processes in explaining this restructuring, and increasing recognition of the ways in which global economic forces are mediated by local factors, such as the nature of local urban governments, economies and cultures, creating particular expressions of economic advantage and disadvantage both within and between cities (Fyfe and Kenny 2005: 107).

The city and the global economy

Globally, the capitalist economy has undergone a significant period of instability, change and reorganization in recent years, which has stimulated a considerable volume of research into the impact of these trends on cities. Central to these trends has been a fundamental shift in the capitalist **regime of accumulation**, specifically a transition from industrial to post-industrial or advanced capitalism, involving the rise of new economic sectors and new spatial relations of production. This has been linked to the increasing globalization of the world economy through the formation of global production networks, global markets and, significantly, global financial systems. Urban researchers have been interested in the reordering of urban systems, and the relationships between cities resulting from these trends. Critically, research has highlighted the emergence of a 'new urban order' (Short 1996: 89) within a global system of cities, which has seen declining status for some established industrial cities, urban growth in newly industrializing locations and the rise of a small number of world or global cities which operate as the command and control centres for the global economy.

Global economic change and the new urban order

Global shift and the new international division of labour

In his seminal work on global economic change, *Global Shift*, first published in 1986 and now in its seventh edition, Peter Dicken notes that the world economic map has undergone significant shifts in the last fifty years. The end of the post-Second World War industrial boom in the early 1970s signalled a crisis for industrial capitalism, with falling profitability in a number of economic sectors, precipitated by a range of factors including rising oil prices, increasing inflation, international monetary instability, falling investment values and increasing labour and production costs for firms. Central to responses to this economic crisis was the restructuring of existing business operations to maintain profitability by reducing costs, involving both the reorganization of industrial processes and the relocation of business operations, and a switch to investment in newly emerging economic sectors. In particular, production processes based on **Fordist** mass production ideas gave way to **post-Fordist** modes of industrial organisation (Dicken 2007).

Table 4.1 A simple model of TNC business operations (after Short 1996: 83)

	Urban location	Locational requirement	Urban changes
Company headquarters	Metropolitan areas	Need for face-to-face contact	Telecommunication developments produce limited dispersal
		Close to business services	
		Close to government functions	Some suburbanization Increasing concentration in a few global cities
Research and development functions	Suburban areas and small cities	Good environment to attract workers	Growth in small to medium sized towns in amenity rich areas
		Close to centres of knowledge generation (e.g. university)	
Routine manufacturing plants	Small cities and rural areas	Cheap labour Low taxes	Growth in sun belt and newly emerging economies

What have emerged from these changes are intricate production networks controlled by large complex firms which operate in multiple locations, increasingly internationally, commonly referred to as trans-national corporations (TNCs). Significantly, innovations in transport and communications have enabled TNCs to become more spatially dispersed, locating different parts of their operations in the most economically advantageous locations. Typically, this has produced a division between the location of headquarters and research and development functions, which have remained within the world's wealthiest economies, and the location of routine manufacturing, which follows the availability of cheap labour around the globe (see table 4.1).

The operation and structure of these TNCs has been seen to have led to global shifts in manufacturing and also service employment, which have created a **new international division of labour (NIDL)** in which more routine business operations are carried out in the global economic periphery (Bryson 2007; Knox et al. 2008). These shifts are seen to account for a number of major, worldwide urban economic changes. These include the deindustrialization of cities in the Global North, the growth of cities in newly industrializing countries in the Global South and the growth of global cities as the control and command centres of an interconnected world economy (see below).

Long waves and economic growth

Another key factor in the emergence of a new urban order associated with the transition to advanced capitalism has been the growth of new economic sectors as key drivers of global economic development. As we have already noted, capitalist growth and development is cyclical in nature. Over long periods of time, the world economy has been seen to change in a series of cycles or waves of growth and stagnation. Of about fifty years' duration, these are known as Kondratiev waves. They are associated with phases of significant technological innovation, which are linked to other innovations in production, distribution and organization and which then spread through the economy (Dicken 2007; Knox et al. 2008). The first Kondratiev wave was associated with the original Industrial Revolution (linked to the mechanization of textile production and improved iron production). The current period of global economic change is associated with the end of the fourth Kondratiev wave – associated with Fordist mass production and based principally on the vehicle, consumer durable and petrochemical sectors – and the early stages of the fifth. The fifth Kondratiev cycle is associated with post-Fordist flexible production based principally on micro-electronics, digital technologies, robotics, biotechnologies and the knowledge, creative and service sectors (Dicken 2007).

Of key significance is that these cycles do not just involve technological changes but are also associated with particular forms of industrial organization and have a specific geography as technological leadership changes between places. It seems that each wave displays its own distinct urban geography, being largely associated with different types of city to the previous one. In effect, 'the locus of the leading edge innovative industries has switched from region to region, from city to city' (Hall and Preston 1988: 6, cited in Dicken 2007: 77). While the locus of innovation during the fifth Kondratiev wave has remained broadly within the advanced industrial economies of the world, the places within these economies that have benefited from the emergence of these new sectors are different to those associated with the previous wave. In particular, smaller cities, linked to educational and research establishments, or associated with cultural innovation or within amenity and environmentally rich areas (so-called 'sun belt' areas) have benefited from the emergence of these new industrial spaces (see below).

Economic globalization

The final process shaping the emergence of a new urban order is globalization. Globalization is a term that is all around us, commonly used in both academic and popular discussions of contemporary change. Short and Kim (1999: 3) argue that it is the new 'big story' for the twenty-first century, eclipsing discussions about Marxism and modernism that dominated the previous century. However, globalization is not a single entity, but is a term describing a complex of linked processes which are leading to increasing interconnectedness in the world and to an intensification in the speed and depth of economic, cultural and political relations (Murray 2006). While there are a number of forces underpinning globalization, the dominant force is generally regarded as economic (Dicken 2007). There has been an increasing globalization of production, consumption and exchange relations which has transformed a 'world economy', based on trade between firms in different nation states, into what can be viewed as a truly global economy which works as a unit in real time on a planetary scale (Pacione 2009). In particular, increasing economic globalization is evident in the hyper-transferability of capital across national boundaries. Recent financial crises, the repercussions of which have seemingly spread rapidly around the world, have alerted us to the increasing volume and speed of financial flows which are a very visible and important indicator of increasing globalization (Short and Kim 1999; Dicken 2007). The economic fortunes of individual cities are therefore becoming less autonomous and are increasingly bound up with the economic fortunes of other urban areas and with processes operating at wider geographical scales.

This increasing global economic interdependence is the result of a number of related processes, specifically the emergence of TNCs as key shapers of economic flows, as noted above; the deregulation of national financial markets; and telecommunications advances 'gluing' together spaces and creating the 'space of flows' between points in the global economy. While these three factors have, on the one hand, led to the incorporation of more and more places into the capitalist global economy, through the wider dispersal of production activities, they have also precipitated the counter tendency of increasing centralization of the command and control functions of this global economy. As the global replaces the national as the significant level for understanding general economic trends, the role of cities as the hubs through which this global economy functions has come under close scrutiny. Commentators have highlighted the emergence of a network of world cities which are critical to the control and reproduction of global capitalism (Brenner and Keil 2006).

At the heart of these contradictions of dispersal and concentration in the global economy have been technological advances in travel and communication. We live in an era of time-space compression characterized by a speeding up of communication and rapid circulation of data, knowledge and ideas (Murray 2006). In particular, advances in telecommunications – telephone, e-mail, computer networks and the internet – have reduced or eliminated time delays in communication between distant places and increased the sophistication of exchange available. Consequently, space should matter less and many economic activities are more able to geographically disperse. However, despite these advances, economic activities remain 'sticky' and continue to concentrate in clusters (Dicken 2007). This stickiness operates in relation to those businesses at the heart of the control of global capitalism which have become increasingly concentrated into particular cities. This is, in part, a result of the spatial, sectoral and social unevenness of access to the international electronic communications economy (see case study below). This electronic/informational economy is predominantly focused in a small number of major world cities which contain high concentrations of these technologies and which act as key nodes for the information flows central to the reproduction of advanced capitalism (Castells 1996). As one moves down the urban hierarchy, concentrations of advanced

---

Case study – the digital divide, Sydney, Australia

Research by Baum et al. (2004) examines the idea of a growing digital divide, whereby urban societies are being divided into information rich and information poor sectors. Their paper focuses on the existence of a digital divide across the Sydney metropolitan area, which here is used to signify the gap between those who have access to ICT technology and those who do not. Key issues determining use of computers and the Internet are socioeconomic and demographic factors, such as income, age, educational attainment and ethnicity, and the availability of infrastructure, particularly access to broadband services. The research is based on an analysis of computer and Internet use by households across Sydney, employing a factorial ecology analysis using census data.

While Sydney is generally seen to be a 'wired city' that is information rich, this is based on aggregate level data which misrepresents the true picture whereby some communities are information rich while others are forming an information underclass. Unsurprisingly, there is a general pattern whereby high socioeconomic status is associated with high home computer and Internet usage. However, patterns of usage are more variable when considering family and ethnic status, with traditional and non-traditional family status areas, and English speaking background and non-English speaking background areas, displaying both high and low usage, with levels of usage determined by the combination of these characteristics with income. Patterns of high and low home computer and Internet usage are therefore geographically dispersed, and are linked to the way in which the spatial social structure of Sydney has developed over time. This spatial clustering of information poor groups has implications for policy and access to infrastructure and services, where a deepening digital divide could compound the social disadvantage already associated with residential location. Given the spatial as well as socio-economic variables associated with computer and internet use, the authors conclude by raising the question of whether place based or people based programmes should be employed in tackling digital disadvantage within cities.

Source: Baum, Van Gellecum and Yigitcanlar (2004)

communications technologies decline, as does the degree of interconnection with the global economy. Cities with low concentrations of this infrastructure are consequently falling behind, such as a large number of older industrial cities and cities in the Global South – effectively moving them further from the core of the global economy. This is creating a geographical pattern of a small number of 'information rich' global cities forming a highly interconnected trans-national urban system, surrounded by large 'information poor' hinterlands with which they are becoming less well connected.

The global urban system and world cities

As we noted at the beginning of the chapter, until fairly recently research into urban systems assumed that cities were neatly enclosed within national hierarchies. However, the processes of contemporary global economic restructuring outlined above made 'the idea of a mosaic of separate urban systems appear anachronistic and frankly irrelevant' (Taylor 2003: 21). In response to this, urban research has become focused on mapping the global urban system and particularly on considering the formation of world, or global, cities (Brenner and Keil 2006). Short and Kim (1999) note two principal strands to this research: attributional and linkage. Attributional research has sought to rank cities globally according to the presence and extent of certain attributes, defined as significant in determining global status. Linkage based research has focused on the interactions between cities to identify key nodes in the global urban network.

The coining of the term 'world city' has been attributed to the Scottish urbanist Patrick Geddes (1915). Geddes used the term to describe those cities where a disproportionate amount of the world's trade was carried out. This has proved a useful base definition for subsequent research, highlighting the significance of economic power and control in defining a world city. These ideas were reworked by Peter Hall (1966) where he defined world cities as those that were key centres of political power, containing key national and international institutions, key port and airport hubs, banking centres and globally important cultural loci. This work was important in providing foundational criteria which could be used to define those cities of key global significance. These criteria were further developed by John Friedmann (1986) whose work provided a major stimulus to research in this area. Friedmann identified a world city hierarchy of thirty centres based on seven indicators: status as a major financial centre, status as a major manufacturing centre, numbers of headquarters of TNCs, the rate of growth in business service sector, presence of international institutions, status as a global transport node and population size (figure 4.2).

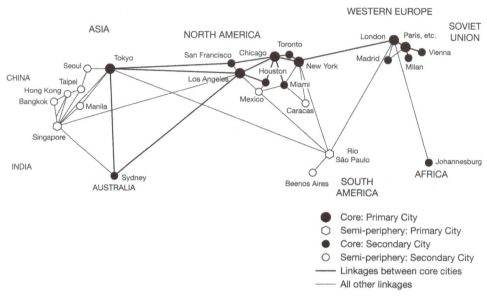

Figure 4.2 *The hierarchy of world cities*

Source: Adapted from Friedmann (1986: 71)

These criteria were further developed by Saskia Sassen (2002) who refined earlier definitions by arguing that it was the presence of the headquarters of financial institutions and advanced producer service companies which was central to the definition of globally powerful cities. One thing all these firms share is a dependence on specialist knowledge and personnel and therefore they are embedded in knowledge rich environments, which are provided in world cities. As Brenner and Keil (2006: 75) note:

> according to Sassen, this newly consolidated producer and financial services 'complex' represents the real economic foundation of global cities, because it provides transnational corporations with a host of essential services that enable them to implement, manage and regulate their production and investment networks on a global scale.

Identification of key criteria defining global cities has spawned a wide range of attributional research which has sought to identify global urban hierarchies. While this research is quite varied, most studies identify three cities at the top of the global hierarchy – London, New York and Tokyo – which are often referred to as the 'global triad' (Short and Kim 1999). They are the principal nodes of a global circuit of information, capital and investment flows and represent centres of global power. Crucially, these centres are key nodes in the global financial market and contain key banks and stock exchanges, are centres for major, advanced producer service companies and are global transport hubs, particularly with regard to air transport. There is, however, little consensus in the research literature about which cities should be included in the next tier (Short and Kim 1999). The second tier broadly consists of major regional centres in key global arenas, such as Los Angeles in North America, Paris and Frankfurt in Europe and Osaka and Singapore in Asia. Cities in this tier compete for dominance in economic terms and in some cases edge close to the global triad in influence and power. Below this a third and fourth tier are usually identified, which consist of major cities of lesser global economic importance, and nationally important cities with some strong international links, respectively (Pacione 2009). Knox (1995) adds a fifth tier where local leadership has sought to develop distinctive niches in the global market place. A desire to boost their global standing has led some city administrations to promote particular developments and initiatives which seek to enhance their world city attributes (see also chapters six and nine). These cities are described in the literature as 'wannabe' world cities. Taylor (2004) identifies two types of wannabe city; inner wannabes, those cities which seek to change existing global hierarchies and to come out of the shadow of a dominant world city in their region, and outer wannabes, those traditionally defined as Third World cities which seek to develop their economies and to enter the world economy.

Short and Kim (1999) provide a useful overview of research in this attributional tradition and also offer their own analysis of global hierarchies and world city status based on some of the principal indicators that have been used in previous analyses, specifically command functions (utilizing data on the location of the headquarters of the global top 500 industrial and service firms as an indication of economic decision making power), financial markets (using data on the location of the headquarters of the world's top 100 banks and the value and number of listed companies of major stock markets) and producer services (using data on the location of the headquarters of major firms in key sectors such as advertising, law and accountancy). While their consideration of financial status broadly identifies Tokyo, New York and London at the top of the hierarchy, their examination of change over time reveals a more complex and dynamic picture of global influence. While the stock markets of the global triad dominate, these three plus Paris, Frankfurt, Osaka and Beijing emerge as key global banking centres. Interestingly, over time the relative position of London and New York as key centres declines while the position of the other centres strengthens. Short and Kim's (1999) consideration of advanced producer services reveals the wider complexity of this type of analysis of world cities, highlighting differing patterns of influence depending on the firms selected for analysis and the sectors looked at. Their analysis of advertising firms highlights the dominance of New York and Tokyo at the top of the world rankings, but a more complex regional structuring in the second tier. Finally, their consideration of telecommunications reveals the limits to this type of analysis, specifically the limits to the availability of data to explore world city rankings. While telecommunications infrastructure and information flows are noted as important in world city formation, they have been little measured due to the limited data available.

Ultimately, attributional research has revealed that definition by using particular indicators does not guarantee a precise measurement of a city's position in the global economy. Many have argued that world cities have been predefined, with data selected to confirm their status rather than being used to develop criteria to define their status (Robinson 2005a). Short et al. (1996) termed this the 'dirty little secret' of world cities research, commenting critically on this pre-selection and the lack of comparable data on cities around the world with which to make more informed comparisons.

Exercise

Look at the Fortune data from the top 500 global firms: (http://money.cnn.com/magazines/fortune/global500) – this is published annually. Review this data by location and consider which cities emerge as the main locations for the top 500 companies. Look at the data for other years – has the pattern of company headquarters shifted over time? Also consider the types of company that make up the top 500 – are they concentrated into particular economic sectors? Think about the criteria used by Fortune in compiling their list of companies – can you think of any assumptions that underpin these criteria and can you think of any alternative criteria? What differences in patterns might emerge?

In attempting to develop world cities research, linkage based approaches have focused on the idea of cities as nodes in multi-layered networks of interactions, emphasizing the functional connectivity of the global urban system. Consideration of world cities in this way has refocused research from world cities as the container of things (case study approach) to their connection to other cities (relational approach) (Hubbard 2006: 180). Smith and Timberlake (1995) highlight a variety of ways in which intercity flows can be measured, specifically identifying a typology based on four functions of flows (economic, political, cultural and social reproduction) and three forms that the flows may take (human, material and information). However, while the typology indicates a rich variety of intercity linkages, analysis is limited by data availability.

The development of linkage based, or relational, research has drawn on Manuel Castells' (1996) work on informational cities which identifies the world city network as an important network in the global 'space of flows' (1996: 412) and which views global cities as places where these flow processes are embedded

(1996: 415). Castells' ideas have influenced the work of Peter Taylor and others who are part of the Global and World Cities (GaWC) research network based at Loughborough University in the United Kingdom (see www.lboro.ac.uk/gawc). Much of the initial research by members of this centre has focused on the measurement of economic linkages between world cities, particularly focusing on the operation of leading advanced producer service firms. Empirical research into these networks has utilized data on the global location strategies of firms, that is, where their offices are, what functions they perform and their importance (Taylor 2004). In their recent research (Taylor et al. 2010) an interlocking network model has been devised, based on the study of 75 of the top producer service and banking/finance firms in 526 cities. The offices of these firms are scored from 0 being no presence in a city to 5 being the headquarters. This produces the *service value of a city to the firm* as a measure of the connectivity of the city, based on the assumption of more significant flows occurring between the major offices. From these measures a hierarchy of cities ranked as alpha, beta or gamma and either plus or minus has been produced, with alpha++ being the most connected world cities (London and New York) (figure 4.3).

While research into global city networks has become more nuanced in recent years, a number of limitations to these analyses remain. As Brenner and Keil (2006) note in their edited reader on global cities, ultimately this

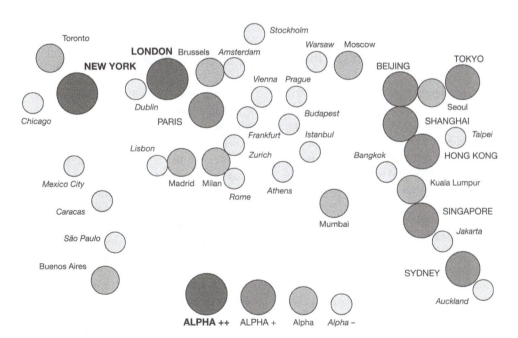

Figure 4.3 *Cartogram of alpha world cities in 2008*

Source: Adapted from GaWC (2010)

research represents an analysis of power, namely the centrality or exclusion of cities within the global economy. As figure 4.3 demonstrates, many major world cities simply do not feature in the alpha world map, with African cities particularly 'invisible'. Robinson (2005a), among others, has highlighted the limits of current world city research, based on continued narrow economic definitions of 'world cityness' and assumptions about the nature of relations between firms within cities (see also case study below). She argues that there is a need to rethink the indicators used (if NGO offices are used, Nairobi emerges as top 'world city'), a need to gather more qualitative information on the relations between cities and a need to rescale research to consider the way in which cities function as territories. Similarly, Hall (2010a) notes that in writing on globalization the focus is entirely on the legitimate/legal economy, ignoring the fact that around 15 per cent of global GDP is linked to illegal and criminal trades. Developing these critiques, Hubbard (2006: 165) highlights the important questions for urban researchers that have been raised about taken-for-granted hierarchical notions of scale within geography. In particular, he points to the need to consider the intransitivity of cities, whereby cities can be seen to exist on multiple scales simultaneously rather than simply occupying an allotted place in a system of space. Here the implication is that all cities are global, and while geographers may use 'global city' to identify particular cities we cannot talk of non-global cities.

Case study – globalization in cities of the Global South

In their paper, Grant and Nijman (2002) offer a comparative analysis of the spatial impacts of global economic restructuring in two cities in the Global South, Accra in Ghana and Mumbai in India, cities not normally considered within global cities research. Grant and Nijman's analysis focuses on the changing foreign corporate presence in the two cities as a key measure of economic globalization and a key component in the changing corporate geographies of the two cities. While Accra and Mumbai are different in size and cultural setting, they share similar global economic positions, a similar historical British colonial experience, act as gateway cities for both countries' economies and

both experienced significant socio-spatial transformation from the 1980s onwards as a consequence of economic liberalization policies and the arrival of significant foreign direct investment. The analysis is based on a survey undertaken by the authors of foreign companies in each city, mapping the location of company headquarters of foreign and domestic firms and assessing their global links (no easy task, as the authors explain in their methodology!).

During the colonial period, the landscapes of both cities exhibited high levels of segregation of 'foreign' and 'native' commercial enterprises and residential areas (the classic 'dual city' – see chapter

continued

three), although these boundaries were eroded and became blurred in the early twentieth century. In the wake of independence, the foreign corporate and residential presence declined in relative terms and former European commercial zones became nationalized by the development and spread of emerging domestic companies. Economic liberalization in the 1980s heralded a significant increase in foreign corporate activity in both cities, particularly finance and producer service companies. In both cities, finance and producer service firms display a high degree of spatial concentration. However, while domestic owned firms are concentrated into the old native and European commercial districts, new foreign owned firms are disproportionately located in new global commercial business districts, in suburban locations on roads leading to the airport in Accra and on reclaimed land close to the commercial and political heart of the city in Mumbai. As in other world cities these companies value the propinquity to other foreign companies. Additionally, the locational decisions of new foreign firms are conditioned by the operation of local land markets, and the availability of new building land through the free market, which tends to be away from the traditional business centres.

Therefore, within both cities three central business districts exist, each different in terms of their importance to foreign and domestic companies and in their links to the global economy: a district centred on the old native town with a mix of residential uses and mainly small domestic businesses; a district centred on the former colonial European town, mainly in commercial use with larger domestic companies and some foreign firms; and finally a newly developed business district, dominated by producer and financial services with a large number of global foreign firms. Domestic firms tend to be more connected to the global economy if they are located in business districts with a disproportionately large share of foreign companies. The authors conclude that these corporate geographies are quite different to those of globalizing cities of the West which warrant further investigation by urban researchers.

Source: Grant and Nijman (2002)

Changing economic geographies of the city

The **structuralist** slant on the city, and a focus on change at a global scale, can provide many important insights. However, this perspective can frequently lead to a lack of consideration of the changes on the ground within cities, specifically how economic changes impact on particular cities and urban populations. In this section we examine the ways in which urban economies are being reconstituted as a result of the interplay between the changing global economic forces discussed above and local economic, political and cultural factors. The potential scope of the topics which could be covered here is vast, so some selection is necessary and therefore we have focused discussion on a few of the significant economic changes that have impacted on cities and their populations recently. This section examines changes in manufacturing industry, the rise of service and creative economies in cities and growth in informal economic activities, highlighting the implications of these changes for the internal geographies of different cities, particularly their form and labour markets.

Industrial change and the city

Deindustrialization and the city

The rise of manufacturing industry represented a key phase in the history of many major cities. In particular, many of the major cities of Europe and North America were founded on, or became closely associated with, the expansion of industrial capitalism from the mid-nineteenth century onwards. However, by the early 1980s the majority of these older industrial cities were experiencing severe economic problems, particularly unemployment linked to the decline of the manufacturing sector. It would not be an overstatement to say that deindustrialization was the most significant economic process to affect large cities in Europe and North America in the late twentieth century (table 4.2).

The deindustrialization of the manufacturing cities of Europe and North America may be attributed to three factors: factory closure, the migration of jobs to other areas of the country or abroad and the replacement of jobs by technology. However, it is difficult to generalize about the impact of deindustrialization upon different industrial cities as not all were equally affected by this process. The process of deindustrialization displayed a number of important dimensions:

Table 4.2   The decline of manufacturing employment in the UK 1975–2009

Year	Number of employees (thousands)
1975	7,351
1980	6,801
1985	5,254
1990	4,994
1995	3,918
2000	3,951
2004	3,282
2009	2,571

Source: *Labour Market Trends/Office for National Statistics*

- Regional – cities in older industrial regions were affected more extensively;
- Sectoral – concentrated on older manufacturing sectors;
- Social – impacting on particular sections of the labour force;
- Temporal – resulting in long term unemployment;
- Urban – decline concentrated in the inner zones of cities.

These impacts varied between different cities and depended upon the size of the city, the composition of individual urban economies and national and local government actions. Larger cities with more diverse economies enjoyed better fortunes during the period of deindustrialization. At the top of the urban hierarchy, the largest cities, such as London and New York, lost their status as major manufacturing centres, as new employment growth was dominated by service employment. However, cities within manufacturing heartlands experienced more extensive job losses which were only partially offset by new service employment. For example, cities in the Midlands/North and Celtic fringe in the UK were particularly badly affected (cities such as Birmingham, Sheffield, Manchester, Liverpool, Newcastle, Glasgow and Belfast), as were those in the so-called 'rust belt' in the North East US (cities such as Baltimore, Pittsburgh, Cleveland, Detroit and Chicago) (Knox et al. 2008).

The social consequences of deindustrialization have also displayed distinct concentrations. Those who lost jobs within manufacturing were predominantly male, either young or late-middle-aged and from ethnic minority backgrounds. Many of those who lost jobs in this sector remained long-term unemployed as comparable employment opportunities failed to emerge. Some of the implications of these changes have included the polarization of income opportunities with the erosion of the intermediate income layer within labour markets. Here the concept of a dual city (Sassen 2002) has been invoked to describe the increasingly polarized nature of urban labour markets between high-paid managerial and low-paid menial employment as traditional manufacturing jobs have disappeared. Equally, the employment market in many rust belt areas has been qualitatively transformed as the new service sector jobs which have emerged have tended to differ significantly from those in manufacturing. The impacts of this on labour markets have included the rise of part-time working and flexible working practices, the feminization of the workforce and a growth of informal and criminal economies. These trends have led to significant changes in social and cultural relations within urban economies in these older manufacturing centres. This has been highlighted both by geographical research and in popular culture. Massey (1994) highlights some of the profound changes in gender relations precipitated by industrial change in manufacturing regions in the UK, where traditional male employment was lost and more women began to enter the workforce. Within popular culture, the award-winning British film *The Full Monty* (1997) is a bittersweet tale which chronicles the problems of male unemployment in the steel manufacturing city of Sheffield.

Finally, deindustrialization had particular spatial impacts on cities. One of the most dramatic demonstrations of the loss of economic dynamism in older industrial cities was the selective out-migration of urban populations and industries from the inner areas of these cities which has resulted in declining inner-city populations and something of an economic vacuum in these areas. This decline in inner city districts created an 'urban doughnut' effect, with the inner industrial areas of these cities characterized by abandoned landscapes (figure 4.4a) and a spatially polarized urban population, with those with poor income and lifestyle opportunities trapped in poor housing areas in the core. Increasingly, many downtown and inner city areas/populations have found itself/themselves disconnected from the formal economy. In both the UK and US, the existence of these declining inner city areas has promoted the development of policy initiatives by both central and local governments to attempt to regenerate these zones (figure 4.4b) (see also chapter six).

New industrial spaces

The flexible production systems of the post-Fordist economy and the technological innovation driving the fifth Kondratiev wave have combined in particular places, creating new industrial spaces. The nature, organization and markets of these activities mark them out as sufficiently different from those of previous eras for them to be considered new industries with new geographies of industrial location. Generally, discussion of these new production spaces in the academic literature refers to the emergence of new industrial zones within advanced capitalist economies, based on new technologies such as e-technology and biotechnology (Dicken 2007; Knox et al. 2008). However, in considering these emerging industrial zones, the links to new industrial spaces in many parts of the Global South, particularly in Export Processing Zones (EPZs), also need to be examined. Central to the emerging geographies of new industrial spaces are the location decisions of TNCs in relation to their research and development functions and routine manufacturing operations.

Castells and Hall (1994) identify four main types of new industrial space, or 'technologically innovative milieux': high-tech industrial complexes (where research and development and manufacturing are linked), science cities (research complexes with no direct link to manufacturing), technology parks (designated by governments or other organizations to attract new high-tech industrial growth) and technopoles (again designated, usually by governments, to aid regional growth by attracting new high-tech firms). These new industrial spaces are dominated by three locational requirements: need for access to a highly qualified and functionally flexible labour force; an environment that facilitates constant innovation and cross-company and industry co-operation;

Figure 4.4 *(above) Sheepcote Street canal area, Birmingham in the late-1980s prior to redevelopment (below) The same area today, part of the Convention Quarter redevelopment close to the National Indoor Arena.*

Source: Author's photographs

and good infrastructural and telecommunication linkages to corporate headquarters, universities, other research institutions, linked companies and international markets. In addition, these industries (or rather the people who work in them) tend also to be attracted by the vibrant cultural mix and pleasant environmental characteristics of places.

Generally, the growth of these new industrial spaces has been largely distinct from areas of industrial decline, focusing on so-called emergent 'sun belt' locations in environmentally attractive locations or older university towns. Some older urban areas have benefited from these developments, where cities have a perceived cultural vibrancy, often associated with major universities. The most often cited example of this new industrial space is Silicon Valley in California. From a largely agricultural area in the 1950s, it has developed to become a major centre for high-technology activities, based on a flow of innovations and highly qualified personnel from local universities. Based on the foundation of the Stanford Industrial Park, it has grown to a self-sustaining zone of high-technology innovation (Scott and Storper 2003). There have been many subsequent attempts to replicate the success of Silicon Valley in other locations through the development of technology parks or the designation of technopoles, such as Cambridge in the UK (Castells and Hall 1994) and Bangalore in India (Audirac 2003). These industrial parks are often decentred and located on the edge of cities, next to the neighbourhoods from which their knowledge rich workers are drawn.

While the majority of these new industrial spaces are opening up in very different locations to old industrial spaces, a key question for policy makers is whether their success can be translated into these older industrial areas in order to generate new employment opportunities. Within some cities, new industrial clusters based on innovative, knowledge based, technology intensive activities such as computer graphics, software design and multi-media businesses are emerging as part of a new economy of the inner city. Their growth is linked to parallel developments of cultural industries (see below) which have been seen to physically and culturally revitalize inner city neighbourhoods. Older manufacturing cities have attempted to foster and attract these new enterprises through marketing and regeneration activities in order to offset the consequences of decline in traditional manufacturing activities (see chapters six and nine).

New industrial spaces are also emerging within rapidly urbanizing parts of the world. Here cities, typically large port cities, perform 'gateway' functions, linking city regions to global economic forces. In areas such as South East Asia, local administrations have encouraged industrial and urban expansion by setting up export processing zones (EPZs). EPZs are centres for labour intensive manufacturing, based on the import of raw materials and the export of factory products. Within these EPZs, TNCs benefit from low wage costs, government tax concessions and freedom from many legislative ties. Over 90 countries had established EPZs by the end of the twentieth century as a way to stimulate industrial development and to generate foreign export revenue (Pacione 2009). Many locations are coastal, as in China, although one of the largest areas of EPZs is along the Mexico–US border. Within these areas, sharp distinctions can emerge between modern internationally orientated business sectors and traditional locally orientated sectors of the economy (see also Grant and Nijman 2002 – case study above). This pattern of foreign direct investment induced urbanization within economically developing countries is referred to as exo-urbanization, where smaller settlements expand as a result of inward investment and rural to urban migration, and merge to form urban regions. This process of urbanization is fuelled both by direct policy intervention from above drawing in large scale foreign direct investment (Shen et al. 2002) but also by growth from below in smaller factories and settlements (Ma 2002).

Cultural industries

Cultural or creative industries are an important part of the larger emerging twenty-first century knowledge economy (Jayne 2005; Peck 2005). This post-industrial economy has been described as 'dematerialized' or 'weightless', where knowledge and creativity are the raw materials of economic prosperity (Hubbard 2006). While creativity can be defined quite widely, in terms of considering creative industries a more narrow definition and focus is used, generally including industries such as fashion, media, film production, design, creative technology-based activities and music. Cultural quarters or spaces dedicated to the development of these industries have become a common feature of cities throughout Europe and North America (Bell and Jayne 2004). Clusters of these industries exist at a variety of scales from areas as small as a street to being a key sector within a number of world cities. Development of the cultural and creative industry sector has been seen as particularly significant in the regeneration of some deindustrialized inner areas (Landry 2000). However, in analysing the development of the cultural industries sector in cities it is difficult to generalize, as the sector displays enormous diversity. Some spaces are the result of organic development as artists or small media companies have moved into abandoned industrial districts to take advantage of extensive empty properties and low rents. In other cases the development of cultural quarters is a carefully managed aspect of urban regeneration programmes originated and run by local, regional or central governments.

A key issue is why cultural industries and creative individuals are drawn to particular cities and parts of cities and how these clusters of activity develop and grow, which is important in underscoring discussions about how best to foster creativity and harness it in cities. The clustering of cultural and creative industries in cities is seen to be underpinned by the idea of the city as a cultural crucible (Hall 1998). Hubbard (2006: 212) notes that the link between urbanity and creativity is an important myth about cities. While it is clearly a myth that creativity only exists in cities, it is an important myth that underpins many city marketing and economic development strategies (see chapter nine). The work of Richard Florida (2004) is frequently cited in discussions about how cities acquire competitive advantage in a global economy and has been influential in policy circles. Examining cities in the US, Florida highlights the importance of a concentration of a creative class (highly educated, knowledge rich workers) within a city which feeds the creation of new business ideas and commercial products. He also emphasizes the significance of the socio-cultural diversity of urban life that attracts these workers to particular cities. To be competitive cities must attract and retain these creative individuals, which feeds into city marketing campaigns which emphasize the cosmopolitan, vibrant and chic aspects of particular cities and urban areas (again see chapter nine).

While this work flags up why creative people may be drawn to particular cities, research has rarely elaborated 'how creative individuals are stimulated by the city as a living environment' (Hubbard 2006: 218). Hubbard notes that to date there has been little in the research literature on how creativity is produced and practised in cities, particularly the material dimensions of these activities and their mundane practices, such as socializing in bars. The work of Allen Scott (1999, 2000) has been important in exploring the ways in which the consumption activities of creative individuals create a collective creativity, where local cultures shape economic activity and economic activity becomes a key component of local social life. Places of sociability are consequently more than just meeting places, but are also a source of inspiration for creative activities, both in terms of their physicality and via the mix of people and objects within them. However, knowledge and networking remain important to creative industries, particularly tacit knowledge which is not necessarily written down yet which is needed to pick up on trends which emerge quickly. It is the intensity and multi-disciplinarity of this networking activity, and its capacity to generate new ideas and connections, which are important to creativity. However, for some the importance of proximity and face to face contact is overstated, arguing that clusters need to be seen as spaces with fluid boundaries and relations stretched out beyond the locale, increasingly globally (Hubbard 2006: 234). Krätke (2002), for example, demonstrates the complex nature of these local-global networks with regard to the clustering of the film industry in Babelsberg on the outskirts of Berlin, highlighting that firms were both closely networked within the local business area, but were also integrated into supra-regional networks of global media firms. To this end, Hubbard (2006) notes that this underpins the development of conference centres and the putting on of events such as festivals and expos which can be used to foster global links which nourish local creative milieu.

However, while ideas about fostering creativity in cities have become popular in planning circles they have been subject to some critique (Peck 2005). Some commentators have argued that pandering to the needs of a relatively well-off creative class may lead city authorities to ignore disadvantaged groups who may not experience the benefits of this economic growth and whose local cultures may be marginalized in favour of more commercially orientated cultural activities. Similarly, within discussions about the benefits of fostering the creative industries there is a tendency to glamorize the creative dimensions of this work and to play down the mundane and sometimes poorly paid nature of many of these jobs (Crewe et al. 2003). Another concern is that these bohemian, socially mixed creative spaces are short-lived, as they become incorporated into more conventional aspects of urban life and popularization leads to increasing commodification and commercialization of these areas (see case study below). Paradoxically, formal policies to enhance cultural quarters may increase rents and property prices, fostering gentrification (see also chapter eleven) and undermining the processes that created such quarters in the first place (Hubbard 2006: 225).

Case study: alternative capitalism and the creative economy, Christiania Copenhagen

The Free Town of Christiania is an autonomous community of about 1,000 inhabitants in the centre of Copenhagen. Built in the 1970s on a former military base, Christiania was set up as a community driven by the values and ideologies of the cultural revolution of the 1960s, attracted by a hippy-oriented communitarian way of life, including a liking for 'alternative' art. The Free Town is a node in a very active network of people, including many artists and celebrities, who sympathize with the Christiania experience, and it still represents, in the eyes of its many supporters around the world, a stimulating challenge to mainstream neoliberal capitalism. However, it is also now widely celebrated as a creative, 'alternative' space in the city, attracting a large number of visitors and mentioned in tourist guides and the official tourist website of the city.

The symbolism of Christiania as a creative, 'alternative' space is sustained by many of its key features: cultural activities (including cinema, theatre and music events), bars and restaurants, shops and small businesses (notably workshops for art, bicycles, ovens and furniture), public art installations scattered throughout the site, and a range of curious and colourful partially or totally self-built houses. Most things are self-constructed from recycled materials, with a strong emphasis on ecological sustainability, and there are

continued

also examples of attempts to demonetarize consumption, with stalls where people are free to leave and take second-hand goods, and a weekly 'collective kitchen' meal where people bring and share self-cooked food. However, its most (in)famous economic activity has been the selling of soft drugs in a contained space named *Pusher Street*, which in recent years has been deemed problematic by both Christiania residents and the wider city alike due to its increasing infiltration and control by organized criminal gangs.

In the paper, Vanolo argues that Christiania represents an interstitial, or in-between, economic space, where there is ongoing negotiation and hybridization between the market capitalism and the Free Town's autonomist and collective spirit and activities. A number of the institutions that are central to the Christiania experience are managed collectively, yet many people who live in the community work outside it in Copenhagen, and many businesses are run individually. The inhabitants are not self-sufficient, there are visible internal social and economic divides, and inhabitants pay fees to the Copenhagen City Hall and Ministry of Defence. The Free Town is characterized by a relaxed experience of time, with people just strolling, drinking, chatting at all times of day and night, which feeds into its creative identity but is also suggestive of subtle unemployment and social marginality. As Vanolo points out, Copenhagen's official stance towards Christiania is ambivalent, on the one hand celebrating its alternative creativity and on the other seeking to control and normalize its activities, remove its 'excesses', particularly the hash market of *Pusher Street*, and make it available for consumption in the market as a tourist attraction. In many respects, Christiania represents an 'odd' economic space that troubles mainstream conceptions of creativity and the creative city by revealing that creativity is both fluid and situated. Despite its alternative image, Christiania is still visibly situated in a capitalistic market economy, but in a place-specific and contradictory way.

Source: Vanolo (2013)

Cities and the rise of the service economy

The expansion of the service sector has been a key component of the transition to advanced capitalism in the global economy. The proportion of global output attributed to the service sector, measured as a percentage of GDP, rose steadily during the second half of the twentieth century (table 4.3).

The service economy covers a wide range of activities which take place within cities. One of the key areas of growth has been in the producer service sector, which provides a range of services to businesses (e.g. accountancy, legal services and advertising) and to people (e.g. insurance and banking). These new office functions have added to more traditional areas of urban service employment, such as retailing. While retailing has been an ever present service activity within cities, in recent years the sector has undergone

Table 4.3 The rising significance of the service sector, globally

	Structure of output				Growth of output			
	Total GDP ($ million)		Services (% of GDP)		GDP (average annual % growth)		Services (average annual % growth)	
	1995	2008	1995	2008	1990–2000	2000–2008	1990–2000	2000–2008
Low income countries	195,611	564,572	43	47	3.5	5.8	3.4	6.3
Middle income countries	4,894,312	16,722,126	51	53	3.9	6.4	4.3	6.4
Low and middle income countries	5,091,618	17,229,923	51	53	3.9	6.4	4.3	6.4
East Asia and Pacific	1,312,702	5,695,585	36	41	8.5	9.1	8.5	9.4
Europe and Central Asia	904,254	3,872,528	52	61	-0.8	6.2	0.9	6.2
Latin America and Caribbean	1,755,662	4,216,075	64	61	3.2	3.9	3.5	4.0
Middle East and North Africa	315,651	1,074,015	50	46	3.8	4.8	3.3	5.6
South Asia	476,175	1,469,613	46	53	5.5	7.3	6.9	8.7
Sub-Saharan Africa	327,684	978,062	53	55	2.5	5.2	2.5	5.3
High income countries	24,508,224	43,273,506	68	73	2.7	2.3	2.9	2.6
Euro area	7,274,360	13,566,882	68	72	2.1	1.8	2.5	2.0

Source: World Bank World Development Indicators 2010 (table 4.1: 228 and table 4.2: 232)

important transformations. In addition, many cities have also witnessed significant growth in their leisure and visitor economy sectors. The expansion of the service sector is evident in cities at all levels of the urban system, both in those cities catering for local-regional-national markets, and those servicing a global market. However, the percentage of output derived from services displays global unevenness, with the highest percentages in high income, advanced industrial economies. The scale of service sector growth has been relatively slower in the Global South, where the smaller size of the service sector is explained partly by the larger informal service sector which fulfils many of the functions provided by the formal service sector in advanced capitalist economies.

Cities and producer services

For many cities there has been a shift in their economic core from one based on manufacturing to one based on banking and service activities. At the heart of this new urban economy has been growth in the producer service sector. These services have become the fastest growing economic sectors of many large cities in Europe and North America, with significant numbers of international linkages. Producer service employment rose for a number of reasons in the early 1980s. These include the demands of businesses for specialized financial and legal services, the co-ordination required to orchestrate spatially (globally) dispersed economic activities within companies (TNCs) and the increased demands of households for services (Bryson et al. 2008).

While many of the operations of TNCs have dispersed, particularly routine production and back office functions, the command and control of functions and headquarters of major global businesses have remained within the central areas of major world financial centres (see global cities discussion above). The consolidation of corporate headquarters in large urban areas reflects their need to access national and international markets and demand for a highly skilled labour force and a range of sophisticated specialist service inputs. Only the very largest and well connected cities can viably satisfy these requirements. This has fuelled the increasing expansion of producer service operations within these cities which support the operation of these TNCs. The concentration of corporate headquarters in key world cities has provided producer service firms with a lucrative and spatially concentrated market. In common with the clustering evident in cultural industries (see above), the social environments of bars, restaurants and clubs are important to the formation of social networks which are heavily drawn on in business, and which are seen as difficult to reproduce outside these central city environments even with sophisticated telecommunication links (Sassen 2002).

Producer services have considerable impacts on the wider urban economies of which they are part. Along with certain financial and investment services, they enjoy a privileged position in their urban economies due to their ability to generate substantial profits in relation to the relative investment of time and money. Because of this, producer services are able to dominate competition for land, resources and investment at the centres of large cities. They have become the focus of new, dynamic economic quarters that have reshaped both the central areas and the wider economies of these cities (Sassen 2002). Within advanced industrial economies a great deal of optimism was placed on this rise in producer service employment. It was felt that the rise of this alternative sector might offset the losses experienced in the industrial sector in many cities of Europe, North America and Australia. However, the rise of service employment has been sectorally, socially and geographically specific. Despite growth in producer service employment, this has not fully compensated for the loss of manufacturing jobs. Equally, the places and people who have benefited from this new growth have been different from those who bore the brunt of manufacturing deindustrialization (figure 4.5).

As other sectors are squeezed out of both the spaces and lucrative economies of these city centres other activities and populations become devalued and

Figure 4.5 Docklands, London, the focus for major producer service expansion in the 1980s and 1990s with new office buildings in close proximity to poor urban communities

Source: Author's photograph

marginalized. This has social implications as the lucrative opportunities offered by the producer services sector are limited to a few highly qualified professionals. The remaining jobs in this sector have tended to be low paid, low skill jobs such as cleaning and security services. This polarization of opportunity also has negative impacts on locally orientated services as a result of their displacement by boutiques and restaurants aimed at high earning producer service workers (Short 1989; Sassen 2002).

Equally, the types of job created by the increase in producer service employment have tended to be polarized between a core of highly skilled, knowledge based jobs that are highly rewarded with good employment conditions and more routine back office jobs which are characterized by lower levels of pay, lower skill requirements, a lack of training and union representation, poor prospects and part time or temporary contracts. This has produced polarized labour markets both between and within cities, with a core of skilled, wealthy individuals in a small number of key world cities and a growing number of peripheral support workers both within these cities and within cities in the Global South where some routine office functions have relocated. Telecommunications improvements have created the possibility of 'virtual offices' which are spatially distant but which form electronically interlinked networks of offices within companies. In common with the NIDL and patterns of dispersal associated with manufacturing firms, large TNC service companies have begun to spatially disintegrate their offices according to function. This has led to some relocation of routine, back office functions and some middle management functions to peripheral locations in former manufacturing cities in deindustrialized regions and to cities in developing economies such as India (e.g Delhi, Mumbai, Bangalore) (Bryson et al. 2008).

Retailing

The trading of goods and services has been a core activity in cities from the development of the earliest ones, with market areas forming a key element of the pre-industrial city (see chapter three). Today, retail activities continue to constitute a core element of urban life and in many urban areas they are a significant land use. The geographies of urban retailing have been examined from two main perspectives: an economic perspective focusing on the physical structure of retail environments, and a cultural perspective which looks at retailing as a consumption activity.

Urban geographers have had a long-standing interest in the spatial and hierarchical organization of retailing in the city. Early research was dominated by the neo-classical economic approaches of bid rent and central place theory which sought to explain patterns of retail activity and establish hierarchical classifications of urban retail locations (Wrigley and Lowe 2002). While a range of classifications has been developed, they broadly distinguish between nucleated centres and ribbon developments, either within the city centre or in suburban locations and which are either unplanned or planned. Urban retail organization has historically been dominated by unplanned shopping areas in the core of the settlement, the central business district, specialist product areas, retail clusters at route intersections and small neighbourhood shopping areas. More recently planned shopping centres have been developed in central areas, and latterly in suburban out-of-town locations.

Recent research into the changing structure of urban retailing has utilized structural perspectives which have been concerned with connecting the changing corporate and social organization of retailing with its spatial expression, termed 'the new geography of retailing' (Wrigley and Lowe 2002). As we have seen, capital switching in search for profit is a key characteristic of the capitalist city. The changing retail environment of the city in the twentieth and twenty-first centuries is a key product of this process where constant innovation and business reorganization have precipitated shifts in retail activities from downtowns to suburbs, creating abandoned inner-city landscapes before coming back in to revitalize these to create downtown leisure-retail areas.

These changes in urban retail structure have been driven by a number of socio-economic factors, influencing both retail demand and supply, and policy frameworks, particularly local and national planning regimes. On the demand side, the most significant factor underpinning changes in the urban retail structure has been the selective suburbanization of population (Pacione 2009). Other important factors have been changing consumer attitudes and expectations, particularly demands for more convenience and comfort when shopping, the growth in female employment and increased car ownership, which has changed traditional shopping patterns to more infrequent bulk-buying of goods, particularly food. On the supply side, there have been important structural and technological changes in retailing. Most significant has been the increasing market share of large multiple retailers and distributors, largely at the expense of independent shops (Wrigley and Lowe 2002). In many cities, the number of retail establishments has decreased as smaller independent stores have closed and larger units have come to dominate. Similarly, technological changes, such as bar-coding and electronic point of sale terminals have given larger operators further competitive advantage over smaller independent stores. Further to this the increased use of the Internet for e-shopping has led to some decline in traditional city centre shopping, although city centre locations with retail environments perceived as attractive have resisted this to some extent (Weltevreden and Van Rietbergen 2007).

The demand and supply side changes have led to important changes in traditional retail structures, particularly leading to the growth of planned out-of-town suburban shopping areas, superstores and retail parks. In the US, these trends precipitated the so-called 'malling of America' from the 1960s onwards with significant growth of out-of-town shopping malls and the related decline in downtown shopping opportunities (Wrigley and Lowe 2002). The idea of the mall has proved adaptable to changing retail trends with more differentiated and niche malls opening up from the 1990s onwards and even returning to some downtowns as the result of leisure-culture based regeneration schemes, such as South Street Seaport in New York, or Pier 39 in San Francisco (figure 4.6).

These new retail patterns have affected many cities around the world, although the impact of large scale out-of-town developments has varied as a result of variations in supply side changes and particularly the actions of planning authorities. For example, in many European countries, such as the UK, the growth of out-of-town shopping has been restricted by national planning policies which have sought to protect inner city locations and neighbourhood

Figure 4.6   Pier 39, San Francisco, US – a festival retail-leisure development

Source: Author's photograph

centres as shopping locations. A key factor in all these changes is that parts
of the urban population have been left behind by these moves, mainly older,
poorer and less mobile populations with lower purchasing power. The term
the 'disadvantaged consumer' (Bromley and Thomas 1993) has been used to
describe these groups, which have less choice in shopping as a result of the
flight of shops from inner city locations, through financial exclusion or as a
result of problems in accessing out-of-town shopping locations. Some poorer
inner-city districts have been abandoned by major retailers, leaving behind
shopping 'deserts' which limit choice and compound disadvantage for
populations living in those areas (Wrigley 2002).

The other key area of research in the new geography of retailing is concerned
with consumption and consumerism. This acknowledges that increasingly
shopping is carried out not merely as a necessity, but as a social activity and
cultural pursuit, linked to identity production. Increasing interest in consumption
has been fuelled by interest in the emergence of the post-industrial city, which is
seen to be organized around consumption rather than production (Zukin 1998).
Zukin notes the plethora of spaces of mainstream and alternative consumption
(malls, multiplexes, cafés, festival market places, nightclubs, super-casinos,
heritage sites and museums) that exist within post-industrial cities. Retail
research has focused on retail spaces as sites of cultural reproduction, such as
the department store (see for example, Blomley 1996; Domosh 1996), the mall
(Goss 1999) and the street (Zukin 1995). Hannigan (1998) has charted the
emergence of new hybrid consumption activities and spaces in contemporary
cities, termed 'shoppertainment, eatertainment and edutainment'. While the
linkage of entertainment and retailing can be traced back to the emergence of
the first department stores in the nineteenth century, within contemporary cities
there is seen to be an explosion of 'experiential retail spaces' in the form of
themed malls, festival market places and global brand stores (Wrigley and
Lowe 2002).

The visitor economy

Until relatively recently urban tourism was a rather neglected area in tourism
and urban studies, as noted by Ashworth (1989) in a key review of the subject.
The visitor economy was not viewed as a distinct area of activity in most city
economies, except within recognized resort towns, and little attempt was made
to manage this provision separately from other urban management functions
(Page 1995). In part this stems from a certain difficulty in distinguishing tourist
visits from other visits to the city, for example as part of the city's service
function, contributing to the invisibility of urban tourism within urban policy
and management. However, in the context of post-industrial urban economies,

tourism is emerging as an important driver of urban development in a growing range of cities. In 2005 the United Nations World Tourism Organisation held its first global forum on urban tourism in Istanbul, Turkey, highlighting the increasing significance of tourism to city economies around the world. This increasing global significance has been accompanied by a growing range of academic work on urban tourism. Hall (2009a) notes three key themes within the diverse array of literature on urban tourism: focus on the diversity of urban tourism destinations, consideration of the relationship between the supply of, and demand for, tourist facilities and urban tourism as a characteristic of the post-industrial economy.

While visits to cities have a long history, stretching back to early pilgrimages to key religious centres, it is with the continued development and evolution of mass tourism during the late twentieth century that urban tourism has expanded as an activity within a greater range of cities. The growth in people's leisure time, particularly within advanced industrial societies, has seen people take more holidays, increasingly more frequent short breaks, to a wider variety of destinations. Many of these new niche markets are based around urban tourism, witnessed in the growing popularity of city breaks (Hall 2009a). The diversity of urban tourist niches is reflected in difficulties in classifying urban tourist destinations. Classifications have typically either been based on the supply side characteristics of urban tourism, the resources on offer such as historic assets, shops, nightlife, or on the demand side characteristics of the tourists themselves, such as local visitor, day-tripper, holiday maker, conference delegate. Burtenshaw et al. (1991) combine these characteristics together to identify some of the different functional areas of tourist cities (or types), including the historic city, the cultural city, the sport city, the shopping city, the nightlife city and the business city.

Prior to the 1980s it was rare for urban managers or commercial interests to recognize tourism as a significant economic activity. However, growing global tourism activity, coupled with widespread deindustrialization in many urban economies (see discussion above), saw an increasing number and range of cities exploring tourism as a way to address their economic and image problems. Many sought to replicate the success of tourism based urban regeneration schemes pioneered by cities such as Baltimore in the US, which in the late 1970s redeveloped large parts of its derelict Inner Harbour area with leisure and festival shopping attractions, museums, a convention centre and latterly sports stadiums (Hall 2009a). In addition, cities increasingly competed for key temporary events and festivals, such as major sporting events or cultural festivals, or developed their own events to act as catalysts for regeneration (see also chapters six and nine). The redevelopment of Birmingham's (UK) Convention Quarter in the 1990s (see figure 4.4b above) is a good example of the increasing importance of urban tourism to the economy and urban policy agenda of former manufacturing cities which underwent significant deindustrialization in the late 1970s and early 1980s. The regeneration of the city centre of Birmingham has resulted in a significant increase in tourist spending within the city and growing popularity as a conference and tourist destination, although critics have argued that these developments have done little to improve levels of social welfare among the city's most deprived communities or reduce social exclusion (Loftman and Nevin 1998).

The growing significance of the visitor economy to major provincial cities such as Birmingham highlights two important issues for urban tourism that warrant further research by urban geographers (Hall 2009a). First, the centrality of tourism to the processes of post-modern urbanization within these cities highlights the limitations of wider discussions of the post-modern city within geography. While tourism has been recognized as a characteristic of the post-modern city, there has been little discussion of the role of tourism specifically in the development of post-modern urban forms. This is largely a function of the empirical foundations of these discussions, which remain predominantly based around consideration of larger cities such as London or Los Angeles (see discussion above) where tourism is a less significant economic driver of their post-industrial economies in comparison to the financial and producer service sectors. In contrast, provincial cities have tended to lack these sorts of links to the global economy and have rather had to forge new links to international circuits of capital, largely through the development of their tourist economies. Further examination of these major tourist developments would serve to broaden consideration of processes of post-modern urbanization in cities beyond those global cities which have traditionally been the focus of such debates. Second, the growing centrality of urban tourism to the economies of cities like Birmingham raises questions about the impacts of such developments on cities. On the one hand tourist activity has a positive multiplier effect on local economies through both direct spending on activities and indirect expenditure in wages and the purchase of supplies. Conversely, there are also potential economic costs in the form of inflationary pressures on land costs and rents, resulting in tourist-led gentrification. In addition, tourism is also a vulnerable and potentially unstable industry, with tourist activity in cities threatened by competition from other destinations and new developments, shifts in patterns and tastes in tourist activity, reductions in visitor numbers and concerns over the perceived safety of destinations stemming from media reporting of crime and terrorist attacks. Tourism also has economic, social and cultural impacts on its host communities. Economically, it is argued that many jobs associated with tourism are low paid, low skilled and also highly seasonal, leading to limited and unstable economic benefits for local residents. Socially and culturally, while many residents may welcome tourists and have jobs linked to tourist activities, many others may be resentful of tourist intrusion and antagonistic to tourists, particularly where tourist activity conflicts with other aspects of local urban life, such as shopping or leisure, or where tourist behaviour conflicts with local cultural values and norms.

The informal economy in cities – alternative economic spaces

Discussion of city economies has so far focused on the formal economy. However, for all of the activities surveyed so far, the formal is either supported or replaced by the informal economic sector in many cities. In particular, within megacities in the Global South the informal sector constitutes a significant component of the urban economy and employment market. Within these cities, population growth, principally as the result of rural to urban migration, has outstripped the growth in formal employment opportunities (Drakakis-Smith 2000; Daniels 2004). In addition, within countries that have undergone recent significant political transformation, the collapse or removal of previous economic and political structures has resulted in an increase in informal activities within many urban economies, such as in South Africa (Skinner 2006) and post-Soviet states (Round et al. 2008). Equally, significant economic restructuring, need for flexibility in labour and the polarization of labour markets in many cities within advanced capitalist economies has precipitated a rise in informal employment and increased academic interest (Leonard 1998).

The concept of the informal sector is easier to grasp than to define, as it is a sector characterized by great diversity (Daniels 2004). Equally, its measurement and quantification is also difficult as it is a 'floating, kaleidoscopic phenomenon, continually changing in response to shifting circumstances and opportunities' (Dicken 2007: 506). The number of urban dwellers employed is therefore difficult to accurately measure and varies between different parts of the world, with estimates ranging from just over half of all employment in Africa to three per cent or less in high income countries (Daniels 2004). Definitions of the informal economy are varied but are usually based on either the types of activity undertaken or the characteristics of the sector.

Potter and Lloyd-Evans (1998) identify four sub-sectors within the informal economy. First, there is the subsistence sector associated with self-consumption for the household, such as urban agriculture, and services in the home, such as child and adult care. Second, there are small-scale producers involved in craft production and the retail of goods and services. These producers can either be full or part-time participants in these activities which frequently involve an extension of activities carried out in the home, the utilization of skills learnt in the home or the vending of goods also produced for home consumption. These activities are the most visible part of the informal economy as they frequently involve the trading of goods and services on the street (figure 4.7).

Figure 4.7 *Informal street trading in Marseille, France*

Source: Photograph by Alan Dixon

Third, there is the petty capitalist sector which consists of small-scale production units. Also known as the 'downgraded manufacturing sector' this informal manufacturing sector is common in cities of the Global South and is linked to the formal manufacturing sector through outsourcing to small factory units and also to workers in the home (Drakakis-Smith 2000; Sassen 2000). However, it is also present in cities of the Global North and is common in particular industries in world cities such as the clothing industry which has a history of using migrant labour in sweatshop conditions (Leonard 1998). These businesses often flout regulations related to factory or labour conditions, such as minimum wages, working hours or the use of child labour. This sub-sector is also growing in size as a result of changes to manufacturing processes, specifically the growth of small batch production and the need for flexibility and rapid changes in output, which has encouraged greater subcontracting, increased temporary working, piece work and home-working. Finally, Potter and Lloyd-Evans note the criminal sub-sector representing the undesirable face of the informal sector, including activities such as drug trafficking and prostitution. This illegal informal economy occupies a grey area between the underground, hidden or black economy, which involves paid production or sale of goods and services that are unregistered by or hidden from the state in terms of tax, social security or labour law purposes but legal in

Informal	Formal
Ease of entry	Difficult entry
Irregular hours/pay	Fixed hours and pay
Local inputs	External inputs
Family property	Corporate property
Small scale	Large scale
Labour intensive	Capital intensive
Adapted technology	Imported technology
Non-school skills	Formal taught skills
Unregulated market	Protected market

Figure 4.8   *Characteristics of the formal and informal sectors*
Source: Authors

other respects, and the underground criminal economy and organized crime (Pacione 2009).

The characteristics of the informal sector are frequently described as the mirror image of those of the formal sector (figure 4.8). For many urban dwellers, the appeal of, or inevitability of, informal sector working is the ease of entry into the sector, requiring as it does limited formal qualifications or high cost inputs which serve to limit access to formal employment opportunities for many within cities. In addition, despite the heterogeneity of the informal sector in terms of activities and workers, a common characteristic is the small scale of activities often organized around family or kinship groups, in contrast to the larger scale corporate enterprises of the formal sector.

However, while it is useful to view the formal and informal sectors as opposites in order to highlight their key characteristics, in reality the formal and the informal sectors are not separate sectors and are rightly now viewed by researchers as part of a more complex and interlinked continuum of employment. This involves complex production, supply and marketing chains and fluid employment status. Daniels (2004) notes that the global production chains for products such as foodstuffs, electronic goods, clothing and footwear span across the formal and informal sectors, with interactions within the chain controlled by a series of 'middle men' which serves to erode accountability and limit the benefits for those workers at the bottom of the chain. Equally, individuals do not solely operate within either the formal or informal economy but can operate across both sectors, holding down two or more 'jobs', or will move frequently between sectors depending on personal circumstances and the availability of employment opportunities within cities.

There is some discussion in academic and policy circles as to whether the presence of the informal economy is a benefit or a burden for cities and those within them. Within traditional economic development thinking, social progress involved modernization of urban economies and transition from informal to formal economic activities, with many city authorities seeking to formalize activities such as street trading though licensing and the provision of formal market areas (Drakakis-Smith 2000). However, more recently many development agencies have considered informal economic activities in a more positive light, viewing them as a means by which poor urban populations can help themselves by providing goods and services at an affordable price and as the provider of employment opportunities that occupy many within cities where limited formal opportunities exist, particularly the rapidly expanding megacities of the Global South (Daniels 2004). The informal economy is also viewed positively as offering an alternative to the capitalist system, either as a choice or where formal businesses are not present (see also Christiania, Copenhagen case study). For example, in some cities of the Global North, local communities have set up Local Exchange Trading Systems (or LETS) which use a local non-commodified currency which remains within the local system to exchange goods and services among scheme members (Pacione 1998). However, the reliance on informal economic activity can also be viewed as an excuse for authorities to ignore the economic problems of cities and the plight of those unable to earn a reasonable living from these activities (Drakakis-Smith 2000). Equally, growing reliance on informal or community volunteer provision can be seen to highlight the withdrawal of private business enterprises and state provision from those urban areas and populations deemed unprofitable or too expensive to provide for (Marshall 2004; Leyshon et al. 2008).

Summary

This chapter has sought to highlight the importance of examining economic processes in seeking to understand urban change both between and within cities. Cities can be seen as 'crucibles of economic change' (Short 1996: 13) and urban geographers have highlighted the importance of cities to processes of global economic change. Exploration of these economic changes has also been important to theoretical developments in urban geography, particularly the development of structuralist perspectives on the city.

Recently, the global economy has undergone a number of fundamental changes, including the increasing globalization of economic processes, a transition from an industrial to a post-industrial service-based economy for many countries in the Global North, and a new international division of labour (NIDL) with new industrial spaces emerging in newly industrializing regions around the world. This global economic change has created a new global urban order with a number of key world cities acting as the command and control centres of the global economy as a result of their role as centres of financial and producer service activities. Recent global economic change has also precipitated the decline and deindustrialization of many manufacturing cities in the Global North, which have sought to transform their economies by attempting to generate or attract new cultural or service industries. Another key concern for urban economies and populations has been the increasing polarization of their labour markets with a growing divide between those holding a small number of well paid professional jobs and a larger majority in low paid, precarious employment or active in informal economic activities.

Follow-up activities

Essay title: 'a small number of world cities … are effectively the command and control points for global capitalism' (Clark 2003: 12). Critically examine why certain cities have attained this world city status.

Commentary on essay title

An effective answer would consider the development of research into world cities and outline the attributes considered to define world city status, with particular reference to considering the location of the headquarters of global finance and producer service companies. It would then examine the hierarchy of world cities defined by these criteria and explore the global economic trends and changes that have shaped this hierarchy. An excellent answer would develop these themes and critically assess the debates surrounding the definition of world cities, including the differences between attributional and relational research, the types of data used and critiques of Robinson and others regarding the 'western centeredness' of this research. In conclusion it would explore the differing definitions of 'world cityness'.

Project idea

What evidence can you find of new economic sectors in your city or a city with which you are familiar? In what areas of the city are they located and is this in areas similar to or different from previous rounds of economic activity? Collect information on the numbers and types of new activity developing either through field observation or via the collection of local business information (for example information held by local government or business promotion organizations). You might also review local economic development policies to examine whether the patterns of growth and change you observe are driven by particular local initiatives. Think about the implications of any new economic geographies for those populations involved in previous rounds of economic activity.

Further reading

Books

- Brenner, N. and Keil, R. (2006) (eds) *The Global Cities Reader*, Abingdon: Routledge
  A comprehensive collection of key articles exploring a number of facets of global cities research. The editors provide a good overview of the research and useful introductions to each reading.

- Dicken, P. (2015) *Global Shift: Mapping the Changing Contours of the World Economy*, 7th edn, New York: Guilford Press
  Established as a classic textbook examining global economic change.

- Mukhija, V. and Loukaitou-Sideris, A. (eds) (2014) *The Informal American City: Beyond Taco Trucks and Day Labor*, Cambridge, MA: MIT Press
  An edited collection considering the growing range of informal economic spaces of the American city.

- Sassen, S. (2002) *The Global City: New York, London, Tokyo*, 2nd edn, Princeton, NJ: Princeton University Press
  Classic text on the impacts of economic globalization on world cities.

- Taylor, P.J. and Derudder, B. (2015). *World City Network: A Global Urban Analysis*, 2nd edn, New York: Routledge
  Updated edition of a key text which examines the development of the world city network and the role of service firms within this.

Journal articles

- Amin, A. and Thrift, N. (2007) 'Cultural-economy and cities', *Progress in Human Geography*, 31(2): 143–161
  A key paper arguing for the importance of considering culture to understanding the economy of cities.

- Bassens, D. and Van Meeteren, M. (2015) 'World cities under conditions of financialized globalization: Towards an augmented world city hypothesis', *Progress in Human Geography*, 39(6): 752–775
  This paper considers the enduring yet changing role of world cities as centres of capitalist 'command and control' amidst deepening uneven development in the wake of the 2008 global financial crisis.

- Brown, D. and McGranahan, G. (2016) 'The urban informal economy, local inclusion and achieving a global green transformation', *Habitat International*, 53: 97–105

This paper provides a different take on the pros and cons of the informal economy, by examining its role in the development of the 'green economy' in cities and outlining the risks that such attempts pose for vulnerable informal dwellers and workers.

- Daniels, P.W. (2004) 'Urban challenges: the formal and informal economies in mega-cities', *Cities*, 21(6): 501–511
  A useful overview of the significance of the informal economy to cities, particularly megacities in the Global South. Also critically examines the relationship between the formal and the informal economies.

- Hober, G. (2013) 'Surviving the era of deindustrialization: the new economic geography of the urban rust belt', *Journal of Urban Affairs*, 35(4): 417–434
  The article examines the differing trajectories of transformation of American 'rust belt' cities over the course of economic restructuring.

- Van Meeteren, M. (2016) 'Can the straw man speak? An engagement with postcolonial critiques of "global cities research"', *Dialogues in Human Geography*, 6(3): 247–267
  Paper which engages with recent postcolonial critiques of global cities research, and which provides a useful overview of recent debates in this area.

Websites

- Global and World Cities Research Centre (GaWC) (www.lboro.ac.uk/gawc/index.html) – a wealth of research material into world cities and the connectivity of cities. Also displays some interesting maps showing world city connectivity.

- International Labour Organization (www.ilo.org/global/lang—en/index. htm#3) – US-based organization which monitors global employment trends. Website contains useful data and research reports.

⬤5 Urban politics

The nature of urban politics

> In urban societies, towns and cities are the place of politics: of revolutions in
> Paris in 1789 and Petrograd in 1917; of long reformist struggles about
> 'collective consumption' typified by the urban politics of the UK's industrial
> cities through the first two-thirds of the twentieth century; of elite-dominated
> 'growth coalitions'; and of the new politics of sustainability with its base in
> 'new social movements'.
>
> (Byrne 2001: 167)

In many ways urban politics may seem a straightforward aspect of cities
for us to consider. It appears to be about the city hall, the local authority,
local elections and the like. However, this is actually a very restricted notion of
urban politics; it refers to what we might think of as the formal urban political
arena. This is undoubtedly extremely important and forms a considerable
proportion of the writing on the subject, from a number of disciplines, that have
emerged over the last fifty years or so. It is fascinating in itself, demonstrating
considerable variation around the world and through time in its nature and in
the activities undertaken by local authorities (Haynes 1997; Judd and
Swanstrom 1998; Herschel and Newman 2002; Savage et al. 2003; Sharpe
2005; Pacione 2009; Pasotti 2010). Urban politics, though, embraces much
more than just this formal political arena. As Sophie Watson (1999) reminds us,
there are many ways that we can think about issues of power and politics in the
city. As well as the formal political realm she, and others, cite struggles and
conflicts over the provision of services and the allocation of resources between
different groups in the city and issues of meaning, representation and identity
(referred to as cultural politics) that are embedded, in many ways, in the fabric
of everyday life, as all constituting an expanded notion of urban politics.

Perhaps the most fundamental question, and the starting point of any
exploration of urban politics, is the issue of power. What is it and who holds it?
How is it distributed between different groups? How are different forms of
power manifested and exercised? Power in the city takes many forms and is
distributed highly unevenly. There is the power bestowed on representatives
from political parties through the ballot box, the power of organizations,
perhaps from big business or occasionally other realms such as criminal
organizations, to influence the processes of urban politics in their favour.
However, there is also the power held by other, less obviously 'powerful'

groups in the city such as social groups or residents to resist or protest. It is important when thinking about power in the city, then, not to reduce it to a narrow function emerging out of formal political processes.

Questions that will run through this chapter and which follow on from the point above concern who holds power in the city. This has been the subject of a great deal of debate and has generated a huge amount of theory and counter theory within the literature of urban politics (Judge et al. 1995; Davies and Imbroscio 2009). Before exploring and thinking through the detail of these debates we would like to pose a few basic questions to guide your thinking and which encapsulate the fundamental issues that scholars and theorists of urban politics have long grappled with.

- Is urban political power centralized in the hands of a small, restricted elite or is it distributed among a wider constellation of groups?
- Is the situation above an oversimplification; can the truth lie somewhere between?
- To what extent is the state autonomous from capitalism?
- In what ways does the distribution of power within cities vary through time and across space?
- How is the realm of urban politics affected by the processes of globalization and the increasing interconnections between cities?

Much of the subsequent discussion in this chapter will concern the first three of these questions and more particularly struggles and conflicts that arise from them. The fourth is also significant, however, as we need to be careful to avoid sweeping generalizations and the trap of universalizing theories and models to contexts where they may not be appropriate (Robinson 2002, 2005a).

The final issue above opens up questions about the nature and relevance of city governments in a global world. If anything, it would appear that the city is becoming a more important political space under current conditions of globalization. A common interpretation of the impacts of globalization is that it is weakening the powers of the nation state. This would certainly appear to be the case, to some extent at least, when it comes to political power. Kiel (1998: 617), for example, has argued that globalization has witnessed the 'displacement' of the powers of national government either upwards from nation states to international political institutions such as the European Union or, alternatively, downwards to cities and regions. As globalization appears to weaken the nation state then this power does not somehow evaporate, rather, it may come to rest in places like cities, which find their political powers enhanced and which, according to Kiel again, take on some of the political characteristics that we previously associated more with the nation state. Certainly, much work on globalization puts cities, rather than nation states, as the lynchpins of its processes and networks (Sassen 1994; Clark 2003; Brenner and Kiel 2006; Kim 2008).

However, we need to be a little careful how we think about these various levels of government. We deliberately avoided using the word 'scale' in the discussion above, which suggests a neat hierarchy that ascends from the local scale, through the regional and national up to the international. Rather, we should think of all of these 'scales' as juxtaposed within sites, crucially urban sites that provide the arenas for the vast majority of the world's political power. Similarly, urban politics does not take place within cities that are hermetically sealed entities, rather local politics includes the involvement of many actors from beyond the city, such as developers and politicians from national and international institutions, who all work with local actors. It is these juxtapositions and interconnections between different scales that constitute and shape the realm of urban politics in a global world (Amin 2002a; McNeill 2009).

Before moving on it is worth pausing briefly and considering why we need city governments. Why do we have multiple levels of government within nations, typically characterized as the central and the local state, and what are the practical implications that flow from the model of multi-level government? In the first instance, it should be very apparent that nation states are far from uniform entities but rather there exists a huge amount of geographical variation within their borders. For this reason, some form of local level representation would seem appropriate (Cox and Mair 1987; Duncan and Goodwin 1988). This is especially the case where nation states are composed of formerly separate territories, perhaps with very different ethnic or cultural identities, where the 'hold' that the nation state has over its territory might be at times fragile. Local representation can be an important building block in holding the state together. Indeed, even in strong, centralized, in some cases, dictatorial regimes, some level of local government has almost always been present, and often vital to the functioning of these regimes (Byrne 2001: 169–170).

> National representation cannot always deal adequately with local differentiation, and so local electoral politics was clearly a necessary part of representative democracy.
>
> (Duncan and Goodwin 1988: 45, in Byrne 2001: 170)

Michael Pacione (2009: 419) expresses this as three principles of the local state:

1 Liberty from central authority and abuse of central power.
2 Popular participation in government, which is encouraged by the proximity of decision makers and citizens.
3 Efficiency in government, which is advanced by a scale of organization that permits locally sensitive provision of public services and functions.

However, beyond this, is there something about the city specifically that makes some form of local government important? Cities, particularly in the present period, are places of intense juxtapositions of many kinds. These might be juxtapositions of land use, social groups, levels of wealth and poverty and so on. In this context, the potential for conflict is heightened and the local state is important in the management and mediation of this situation (Watson 1999). However, historically, the reasons for the desirability of city governments and the functions they performed were somewhat different. In the rapidly expanding cities of the world's first industrial nations, for example, the speed of the growth of cities and the levels of poverty and dreadful conditions of public health that emerged under what was only crudely governed development, demanded that the state intervene and ensure that these cities became safe, clean, ordered environments for their escalating populations (Hall 2002). Despite the necessity for the local state and for urban governments, though, the relationship between these different levels has been far from harmonious. Duncan and Goodwin, cited in Byrne (2001: 170), have referred to the local state as 'both obstacle and agent for the national state' (1988: 46), something that we return to below.

The literature of urban politics has been characterized by a long-standing multi-disciplinarity of which urban geography has been only one among many contributors (McNeill 2009: 141). Each of these disciplines contributes slightly different takes on the issue of urban politics and governance. This is not to say, however, that there is not a great deal of cross-disciplinary dialogue. Indeed, this is perhaps more prevalent in the study of urban politics than in any other aspect of urban studies. This chapter will try to outline a distinctly geographical perspective on urban politics. It will draw on work from across the disciplinary range but will emphasize issues such as the importance of space, place and scale that are of particular interest to geographers and which have largely shaped their contribution to the field.

The formal political arena

The formal political arena is constituted in a number of ways. Perhaps most fundamentally it can be viewed as part of the capitalist system. What then is the relationship between the (local) state and the capitalist system? People leaning towards right of centre political views might argue for the retreat of the state and the 'freeing' of the market, characterizing local government as a bureaucratic 'drag' on the efficiency of markets. Indeed, this was the basis of much central government action in the UK in the 1980s, for example (Goodwin 1992; Ambrose 1994). By contrast, others might point to the long-standing and ongoing involvement of the state within the capitalist system, arguing that this is an inevitable condition of capitalism. David Byrne (2001), along with many others writing from a broadly Marxist perspective, most notably David Harvey (1973, 1982), are helpful here in their articulation of the seemingly unstable nature of the capitalist system. They see capitalism as a system tending periodically towards crisis as capital 'switches', seeking profitable investment opportunities (see chapter four for a fuller discussion of this). Indeed, the city is a key site of this capital switching and one where its effects are most acutely felt. The results of these periodic crises are not socially benign and, consequently, one of the purposes of the state is handling these crises and the change they produce. Marxists, therefore, would argue that this 'crisis management' by the state provides some stability and support to the capitalist system and is vital to its maintenance.

> Urban places and urban lives are not static. These systems are not close to equilibric. Their natural condition is not a steady state maintained by negative feedback mechanisms. They do change, and change in terms of transformation of form, while remaining within the general social relations of capitalism with its foundation of wage labour. The 'task' of the state and associated agents of governance is, in large part, that of handling those changes, while at the same time maintaining the social and ecological foundations of the system as a whole.
>
> (Byrne 2001: 168)

Another way in which the formal political arena is constituted, indeed the way in which it achieves the task of 'handling' capitalism's inherent instability, is through a set of practices. What is it, then, that local authorities do? Local authorities are large complex organizations (see exercise below) with many departments and many functions. Pacione (2009: 418) provides a concise overview of this complexity, identifying six key functions of local authorities:

1 Providing public services.
2 Acting as an agent of central government, as when it enforces state legislation.
3 Formulating policies and plans such as those relating to local development.
4 Representing the locality in dealings with other governments, as when seeking financial aid.
5 Resolving conflicts between competing local interests, for example, over the location of facilities.
6 Regulating private-sector activities, as in land-use zoning and building control.

Despite being a universal feature of urban societies, city governments vary a great deal internationally. This variation includes the form and structure of city government, its power and significance and its activities (Savage et al. 2003: 153; Pacione 2009: 418). This variation is well illustrated by the contrasts between the organization of city governments in the UK and the US. In the UK,

Exercise

Examine the structure of city government in an area you are familiar with. Can you find examples of all of the activities outlined above? Are there some that seem more important than others? Why do you think this is? Compare this with an authority in a contrasting urban setting – perhaps in another country. In what ways do the structures and activities of the two authorities differ? Can you think of reasons why this might be? Perhaps think about national differences in the nature of the local state and differences stemming from their contrasting urban settings. What constraints do you think the two authorities operate under (see below)?

power over city government, indeed all local government, is held by parliament, whereas in the US this power lies in the hands of states. It would appear that the power to reorganize local government is more likely to be exercised in the UK than in the US. This is evidenced by the Conservative central government's abolition of the left-wing Greater London Council in 1986, along with six other metropolitan authorities, a decision that was driven more by politics than by a desire for efficient government of large metropolitan areas. In terms of structure there are also some striking differences between the organization of local government in the two countries. While in the UK urban areas are governed by single unitary district councils, the American system is characterized by a series of overlapping levels of government topped by the state and including counties, townships, municipalities and special districts. The federal nature of American politics means that the form and structure of the organization of these levels of government varies in different parts of the country. The special district is the most widespread form of local government in the US and is responsible for the provision of specific services such as hospitals, fire protection and heritage preservation among many others. The number of special districts in the US has shown continued upward growth over the course of the twentieth and into the twenty-first centuries (Pacione 2009: 424–425).

In addition to the differences noted above, the relationships between city government and the processes of urbanization in different countries result in the emergence of very different issues internationally. For example, an issue that dominates politics in many American urban regions is the fragmentation of American urban politics linked to processes of suburbanization. While there is evidence that this exists to some extent in the UK (Pacione 2009: 430), it is considerably less of an issue. The suburbanization of the American population is a highly selective process. Generally, it is the wealthier, white, urban dwellers in the US who migrate to the suburbs, leaving behind poorer populations composed of relatively high numbers of ethnic minority groups. While this is inherently significant in socio-economic terms it becomes significant in fiscal and political terms because of the structure of American local politics. Linked closely to the 'edge city' phenomenon (see chapters three and twelve) these processes of suburbanization have seen the widespread migration of certain types of voter, and importantly, their property tax dollars, beyond the administrative boundaries of cities. The result has been the fragmentation of American urban politics and significant falls in local tax revenues to city administrations. From the point of view of these suburbanites, and the same issue applies to commercial and industrial property owners who are similarly taxed, suburbanization may be a way of avoiding contributing to the costs of the social consumption needs of poorer inner city residents. The issues around the differing needs, wants and economic resources of city and suburban residents and the possibility of defensive **incorporation** to protect the interests of suburbanites has emerged as a major feature of the social and political geographies of American cities (Pacione 2009: 425–431). The decline of the city as a centre of political power in America and the secession of power and resources to the suburbs has changed the face of American local politics in many regions, resulting in the dominance of a politics of suburban self-interest.

> The continuing suburbanisation of population and the emergence of a multi-nodal metropolis are likely to enhance the electoral power of the outer municipalities in provincial policy making and increase the social and financial disparities between inner and outer zones.
>
> (Pacione 2009: 431–432)

> Suburban politics is much more straightforward than the city variety. There is one overriding interest in suburbia – single family home ownership. Suburban politics is home owner politics and elected officials in suburban areas tend to be non-partisan managers of property taxes. The aim of their administration is to preserve the quality of suburban life.
>
> (Gottdiener and Budd 2005: 177)

From managerialism to entrepreneurialism

The discussion above has highlighted that forms of urban governance vary across space, showing strikingly different characteristics in different countries. This section considers the ways in which it has changed through time, noting one of the key shifts within urban government of recent decades. There has been widespread recognition that a new mode of urban government has emerged in cities in many parts of the world. This has been observed particularly since the early 1980s, although some commentators have recognized elements of it in earlier periods. This new phase of urban government has been characterized as entrepreneurial in nature and concerned primarily with the promotion of local economic development and the attraction of investment and jobs to localities (Logan and Molotch 1987; Cox and Mair 1988; Wolman and Goldsmith 1992; Hall and Hubbard 1996, 1998; Savage et al. 2003: 169–185). What distinguishes this phase from earlier 'managerial' phases of urban government is the diminished role of the city government as a provider of welfare services and **collective consumption** that has traditionally been seen as the key function of local authorities (Castells 1977; Pinch 1985). The implications of these changes have been felt far beyond the world of urban politics, however, and raise questions about the very production and nature of cities. It was Castells (1977) who first argued for the significance of the role of collective consumption in shaping the city, prompting many to view the city, during the managerial phase of urban government, 'as the product of bureaucratic action' (Savage et al. 2003: 161; see also Hall and Hubbard 1996: 154). By contrast, entrepreneurialism in city government has seen more piecemeal approaches to managing the city prompted, some would argue, more by economic opportunity than public welfare strategy.

This entrepreneurial phase of local government has coincided with the limitation and decrease of central government funding to local authorities in many countries (Goodwin 1992), and the restriction of the amounts that local governments are able to generate through taxation, although the differences between centre–local relations internationally have affected the ways in which this process is manifest in different places. This coincided with the severe deindustrialization of many urban economies. The result of these financial 'squeezes' from above and below was a severe drop in the amount of funding available to local government. On a more general level this reminds us that local authorities are restrained in their freedom to act and operate under a number of constraints (Pacione 2009: 419–423). This situation forced local authorities to adopt much more entrepreneurial roles, promoting development through functions that were traditionally seen as more the preserve of the private sector. These have included spectacular property development, place promotion (see chapters six and nine) and subsidy of private development. Typically, this has been achieved in alliance with private capital.

As with any attempt at neat periodization, the division of urban government into distinct managerial and entrepreneurial phases (characterized perhaps by a politics of income distribution versus a politics of growth) carries with it the risk of oversimplification. For example, while there is much discussion of urban government as falling into these two clear phases there is little by way of empirical evidence to categorically confirm its accuracy. Further, we know relatively little of the extent to which both managerial and entrepreneurial functions can co-exist within single authorities or more broadly within phases of urban government. Indeed, David Harvey (1985) has recognized that urban governors have long been key agents in the promotion of conditions favourable to economic development within cities. The question then is the extent to which entrepreneurial modes of urban government are supplanting earlier managerial modes or merely supplementing them (Hall and Hubbard 1996: 155). Despite these reservations, though, it is clear that we have moved into a phase of urban politics which is more risk-taking, proactive and prepared to work more closely with private sector interests than was the case in the past (Judd and Parkinson 1990; Keating 1991). Undoubtedly the regeneration of cities in the last thirty years in many parts of the world reflects the increasing entrepreneurialism of city governments (see chapter six).

Reflecting the points outlined earlier about geographical variations in the nature, structure, activities and outlook of local authorities it is important not to view any shift towards entrepreneurial city government as eradicating these variations. For example, a number of commentators have noted that the regimes and partnerships within urban politics under the current phase have shown enormous variation, in part reflecting local social, cultural, economic and political characteristics (Leitner and Garner 1993; Stoker and Mossberger 1994). Further, it is certainly not the case that all local authorities have been as growth obsessed as some of the entrepreneurial rhetoric would suggest. A number of case studies from parts of South West England in the 1990s highlighted that local authorities there were more concerned with a strategy of growth management than growth per se (Cowen et al. 1989; Bassett and Harloe 1990; DiGaento and Klemanski 1993; Bradley and Hall 2006). Here the concerns of these authorities revolved around issues such as the maintenance of property prices, and the selective attraction of investment from sectors such as finance and high-tech industry (Hall and Hubbard 1996: 158). Indeed, the borough council in Cheltenham, a town in the region, took the unusual step of closing its economic development unit during the late 1990s, so buoyant was the economy at the time. The fear here was more the 'overheating' of the economy causing spiralling, unsustainable property price rises, rather than recession. It appears that the more growth-obsessed local authorities tended to be found in larger deindustrialized cities within 'rust belt' regions that had suffered economically to a much greater extent than smaller urban areas in more affluent regions, which were more associated with newer waves of high-tech industrialization (Boyle and Hughes 1991; Carley 1991; Goodwin 1993; Loftman and Nevin 2003). Geography, then, it would seem, has played a significant part in the imprint of this entrepreneurial urban politics across different regions.

Theories of power and urban politics

The discussion above highlights that when we talk about urban politics we are referring to more than just the local authority. Entrepreneurialism in urban government points to the crucial role played by private capital within the political arenas of cities. However, a third group of interests that we need to consider are those of the community. Much urban political theory has sought to explore and articulate the relationships between the local state, private capital and community interests. At first glance the world of urban political theory can seem a complex terrain to negotiate. However, we can begin to understand it by returning to two of the questions outlined at the start of this chapter and mapping the main theories of urban politics (see table 5.1) along two axes. The first axis is concerned with the issue of whether power is concentrated in the hands of a relatively small political elite or whether, on the other hand, it is more distributed between a wider range of groups. All of the theories outlined in table 5.1 position themselves at some point along this axis. The second axis is concerned with the autonomy of

Table 5.1 *Different theoretical perspectives on power and urban politics*

	Key works
Elitist perspectives	
Cities are seen as being controlled by narrowly constituted, powerful elites, primarily drawn from, and supporting, financial and business interests.	Hunter (1953)
Growth coalition theory	
A development of the elitist perspective that suggests the influence of strongly pro-growth 'coalitions' who benefit from growth in local and regional economies while seeking to persuade that the benefits of this growth are widespread. A key role for the local media has been identified.	Molotch (1976); Peterson (1981); Logan and Molotch (1987)
Regime theory	
Combines aspects of elitist and pluralist perspectives. Sees power in the city as highly fragmented and the ability of individual groups to hold power as limited. Focuses on the ways that groups come together to achieve specific objectives, thus producing power.	Stone (1989, 1993); Stone et al. (1994); Peck (1995)
Pluralist perspectives	
While recognizing the presence of powerful elite interests within urban politics, argues that power is more diffuse and community interests are also represented within government.	Sayre and Kaufman (1960); Banfield (1961); Dahl (1961); Polsby (1963)
State-managerial perspectives	
The view that urban managers influence the control of social goods, for example, through their control of urban land and property markets. This reflects, and protects, their interests and class positions.	Pahl (1970); Rex and Moore (1974)

	Key works
Structuralist perspectives	
Less a specific theory or set of theories of urban politics, more a broader critical /theoretical perspective. Stressed the influence of the underlying socio-economic system on the political agenda of the local state. Provided a counter theory to earlier pluralist claims about the dispersal of power in the city.	Bachrach and Baratz (1962); Stone (1980); Manley (1983)
Neo-Marxist perspectives	
A form of structuralism primarily characterized by its empirical investigations of 'systematic, cumulative, political inequality' (Mollenkopf 2003: 238).	Castells (1977); Piven and Cloward (1977)

the state from the capitalist system. Whereas some theories of urban politics have been very economically determinist in their outlook (Peterson, 1981), others have seen the state, and city governments particularly, as less bound to capitalism and capable of exercising their own autonomy (Stone 1989). At the risk of oversimplifying somewhat, all theories of urban politics argue for different positions along these two axes. Further, there is some overlap and interdependence between the perspectives laid out in the table. For example, growth coalition theory can be seen as a development of elite theory, whereas some perspectives, such as some versions of growth coalition theory, are imbued with aspects of wider theoretical positions such as structuralism. Finally, as the subsequent discussion will highlight, and to reiterate an earlier point, we should not suppose that one theoretical position will provide an adequate explanation for all cities. Urban politics takes very different forms in different places. It may well be that some theories provide good explanations of the situations in some places but not elsewhere.

The influential studies of Floyd Hunter (1953) in Atlanta provided some of the first major theoretical insights into urban politics. Hunter argued that cities tend to be controlled by small, powerful elites. These are drawn primarily from the financial and wider business communities along with career politicians (Gottdiener and Budd 2005: 173). Although drawing criticism and inspiring an extensive tradition of empirical research, from the likes of Robert Dahl, that seemed to highlight the pluralist nature of urban politics, elite theory has enjoyed some enduring legacy in the form of growth coalition theory.

Growth coalition theory, drawing primarily on work in North American cities, has argued that it is possible to recognize elites, powerful within the politics of major cities, composed of coalitions between groups such as land owners, rentiers, state banks, utility companies and local authorities. Typically, charismatic individuals, such as mayors, councillors or other politicians, lead these coalitions, lending them an appealing 'everyman' persona. Mollenkopf discusses the ways in which private and political interests are often intertwined in the city (see also Watson 1999: 218; Savage et al. 2003: 175; Gottdiener and Budd 2005: 176; Pacione 2009: 434–436):

> Business, in particular, derives systematic power not only from its juridical status and economic resources but from its attractiveness as an ally for those who advance any policy change and from the shared subculture from which private and public officials both emerged.
>
> (Mollenkopf 2003: 238)

The aim of coalitions is the intensification of urban development and the attraction of outside investment. The outcomes of this growth are to the benefit of both private interests within these coalitions and the local authority (Logan and Molotch 1987; Cox and Mair 1988). Private interests tend to achieve their aims by developing influence within political circles, typically through the promotion of a rhetoric of growth being inherently good for all in the city (Hubbard et al. 2002: 181). This overlooks, indeed obscures, the uneven impacts and benefits of this growth that have been well documented in many cities internationally (Wolman et al. 1994; Hall and Hubbard 1996, 1998; Metraux 1999; Loftman and Nevin 2003). Given this need to persuade the population of the benefits of growth it is not surprising to see the local media playing important roles within these coalitions. Media coverage of major urban developments is typically uncritical and boosterist. Their analysis is rarely analytical and tends to favour local authority or business sources over any oppositional or critical perspective. The local media stand to gain both potential increases in circulation (or now website hits) and advertising revenue. In supporting the powerful members of coalitions, they also maintain good relations with regular and plentiful suppliers of news (Thomas 1994).

Growth coalition theory reflects the geographical milieu of the North American city within which it was developed. Applying it to cities beyond the US has drawn some criticism (Boyle and Hughes 1995). This is alongside some wider limitations that have been recognized. Two related weaknesses of growth coalition theory are its inability to account successfully for diversity between coalitions and the naïve way that it conceptualizes the issue of power within urban politics (Stoker 1995). Growth coalition theorists argue that the prime motivations behind coalition formation are economic or political (to bring more money into a city and/or to win more votes). Consequently, the picture of the urban political arena that it tends to paint is very polarized, dominated by groups who are either pro- or anti-economic development. The classic picture is of a business community that is strongly pro-growth and a public, or at least sections of the public, that are opposed to this growth. As a result of this, growth coalition theories are unable to account successfully for the public as active members of growth coalitions (Stoker 1995). However, a number of studies in the US have shown that the support of the public is a crucial factor in coalitions achieving their objectives. Thus, the motivations that have lain behind coalition formation and the groups that have taken an active part in them have been far more diverse than some growth coalition theories have acknowledged.

Much of this weakness within growth coalition theory stems from a narrow and unproblematic conception of power. Growth coalition theories regard power in the context of urban politics as the ability to control people and resources. While this might ultimately be the case, it fails to acknowledge the importance of the processes through which this power is created. An alternative set of theories emerged out of a reaction to overly economically deterministic versions of growth coalition theory, especially the work of Peterson (1981) (Hubbard et al. 2002: 182). This work, labelled regime theory and most closely associated with the work of Clarence Stone (Stone 1989, 1993; Stone et al. 1994), focuses specifically on the complex processes involved in the creation of power and, in doing so, offers much more sophisticated accounts of the operation of urban politics (Watson 1999; Byrne 2001; Hubbard et al. 2002).

Regime theory, therefore, sees power in cities as very fragmented and the ability of groups to hold power as limited. Regime theory focuses on the ways that groups within urban politics come together to overcome these limitations by forming regimes to achieve specific objectives. The crucial contribution that regime theory makes is that it demonstrates that the power to govern is not given, it is not the inherent possession of any group; rather it has to be created or produced. It is created, regime theorists argue, by different groups coming together and blending their control and resources. In doing so they are able to create the capacity to govern (Stoker and Mossberger 1994; Stoker 1995: 269–270; Hubbard et al. 2002: 182–185).

Typically, regimes are formed through some combination of a government organization that has the ability to mobilize and co-ordinate resources and private sector interests that might own resources. Often otherwise oppositional groups are able to come together within regimes by each partner offering selective incentives to the other. Sections of the public can enter regimes by offering their support for developments, either through the ballot box or expressed through community leaders (Stoker 1995). Regime theory is, therefore, able to transcend a major weakness of growth coalition theory which is able only to conceive of the relationship between certain groups in a very polarized, oppositional way (Painter 1995; Stoker 1995). Regime theory, by contrast, is able to show how, while this opposition may be constraining, it does not necessarily determine the course of urban politics. Groups may be able to transcend their divisions in specific instances, if it is in their interests to do so and if, in doing so, they are able to achieve the capacity to govern. Consequently, regime theory paints a picture of urban politics as less rigid and structurally determined than growth coalition theory and regime formation as more fluid and complex (Stoker 1995).

Regime theory recognizes that urban politics is characterized by a diversity of different types of regime that are composed of different groups often with very different aims. Again, this contrasts with the more monolithic view of the motivations of groups within growth coalition theory. Four main types of regime have been recognized from studies in American cities (Stone 1993). Maintenance regimes seek the preservation of the status quo rather than the promotion of new development. Development regimes seek to promote development or arrest or prevent decline within localities; this task requires greater resources and co-ordination than that facing maintenance regimes. Middle class progressive regimes may seek a variety of outcomes from development including social gains; they may also seek to control development to prevent or limit externalities such as environmental harm. Lower class opportunity expansion regimes seek to enhance or expand the opportunities open to disadvantaged urban groups, which typically involves major resources and substantial co-ordination. The scale of these tasks and the relatively disempowered positions of the groups in this last type of regime mean that they are often absent from American cities (Stone 1993).

Exercise

Can you detect regimes in urban areas you are familiar with? Do they fit into the categories outlined above? Who are the members of these regimes and what are their objectives? How do they seek to achieve these regimes and to what extent do you think they are successful?

Regime theory has provided an influential contribution to geographical studies of urban politics and economic development (Judge et al. 1995). Although developed in the context of North American cities it appears to be more readily applicable elsewhere, particularly to the British and European political arenas, than the cruder, more geographically specific growth coalition theories which rely too heavily on economic motivations to explain coalition formation. Regime theories also appear more versatile, being able to explain the actions of central government agencies, such as Urban Development Corporations, important players within the UK urban policy context (Imrie and Thomas 1999), as well as local public and private agencies (Stoker 1995; Hall and Hubbard 1996, 1998).

Case study: elected mayors and performative city politics

Recent years have seen the growing emergence of politically powerful territorial units and actors at the sub-national level. Within Europe this has been characterized by the increased visibility of cities and regions often associated with high profile and charismatic leaders. Some commentators, observing this, have spoken of the development of a 'Europe of the regions' or a 'Europe of the cities' that has been accompanied by a 'new mayoral class'. This process reflects increasing inter-urban and inter-regional competition and growing calls for regional cultural autonomy. Key here is the ability of mayors to promote their cities across wider political stages as well as to shape their development locally. Often these two projects are closely interlinked. Helping to attract investment from external interests may afford the mayors greater power to impose their vision on their locality, often in association with these external actors. These developments can be attributed to a number of recent factors including: re-establishing the position of Mayor of Paris in the 1970s, which Jacques Chirac used as a base from which to become French president; the high profile enjoyed by Pasqual Maragall, Mayor of Barcelona, linked to the success of that city's hosting of the 1992 Olympics; the emergence of a new, less corrupt, regional political class in Italy in the 1990s; the high profile and international influence of Mayor Rudy Giuliani's policy of zero-tolerance in New York; the creation of the position of elected Mayor of London and the associated publicity that accompanied the 2000 election victory of Ken Livingstone; and the recent creation of fora for collaboration between European city mayors.

One of the key roles that mayors play for their cities is the embodiment of the character of a city, or at least persuasive and compelling versions of that character. This represents a form of place myth that can act as an important commodity within a media dominated politics, forming impressions about cities and drawing attention to them. This emphasis on performance and embodiment is something that has been underplayed in previous studies of political power. City mayors achieve these roles in a number of ways. They often strive to be seen as the epitome of their cities, highlighting their links to the locality through birth or upbringing and through this to argue that they embody the character of the place. They may also stress elements of their appearance or accent to strengthen this association. In this sense, the local media play crucial roles in representing these individuals and their messages. Second, successful mayors attempt to narrate compelling versions of the future of their cities, often drawing on and linking this to mythical aspects of the city's history. Finally, mayors are important in the animation of city spaces. This might involve the promotion of iconic city spaces, the endorsement of mega developments, or engaging in battles with central government for the staging of events or the winning of development. Mayors, then, play important tangible and symbolic roles in the shaping of their cities.

This emphasis on the symbolic roles that mayors play is not to suggest that this is all they do. Mayors have to run administrations efficiently, marshalling and distributing resources if they are to be successful. However, this example shows that to understand how urban politics works requires an understanding of these previously neglected dimensions of political power.

Source: McNeill (2001a)

Finally, it is worth acknowledging that the political arena of cities is not above corruption, the infiltration and compromise of its integrity by a range of interests. It is difficult to assess the extent to which corruption is a major force in shaping urban politics. It is not something, understandably, that researchers have found easy to gather data on (though see Server 1996). It is unlikely that any urban areas are entirely free of scandal and corruption, although the majority of these cases will be relatively minor examples of backhanders or favours paid to local officials by both legitimate businesses and low- to mid-level criminals ('oiling the wheels' as this process is often euphemistically called), which will have only marginal impacts on political decision making. The ubiquity of political corruption as a theme within fictional drama is probably an exaggeration of both its extent and its impacts and a highly glamorized representation. It would appear though that in some parts of the world corruption is a significant aspect of the politics of cities. For example, Paoli (2005) and Saviano (2008) have discussed the long-standing and deep interrelationships between the local state in Italy and various mafia groups. Despite recent attempts to tackle the mafia in Italy it is probably true to say that it is effectively part of the local state in some parts of the country, with significant stakes in many public sector institutions. Elsewhere, although much less embedded within the state than the mafia in Italy, Tremlett (2006) discusses the infiltration of local administrations in parts of southern Spain by criminal gangs. The problem of the corruption of the local, and at times central, state is a significant issue in some parts of the Global South in which organized crime and corruption are deeply embedded (Server 1996; Glenny 2008; Hall 2010a, 2010b). Studies of the political development and activities of these regions need to acknowledge the extent to which they can and have been corrupted and the effects of this. In the context of the Global South commentators have described examples of 'soft states' where the government has only weak control over public administration and where corruption of public officials is widespread (Pacione 2009: 589). However, it should be acknowledged that this corruption is not restricted, by any means, to the Global South. Despite the recognition of corruption within urban politics, there is, to date, little by way of theory in this area and few empirically rich case studies.

Beyond the formal political arena

If we imagine urban politics conceptualized as a triangle, with local authorities at one point, private sector and business interests at another, the final point comprises the community. Or rather, it would be more accurate to say 'community groups', as urban communities are highly fragmented entities. As the discussion about suburbanization and politics in American cities above showed, while some urban residents are relatively privileged others are much less so. Consequently, different sub-groups of urban populations enjoy, or endure, different levels of power and different degrees of access to and influence on the urban political system. This is despite notionally being entitled to an equal share of democracy through the ballot box (though see Pacione's (2009: 436) discussion of systematic inequalities in electoral participation and the issue of the exclusion of many urban residents from the political process in significant numbers of cities of the Global South (2009: 588)). The involvement of community groups in urban politics is another area that has generated a great deal of debate and a number of alternative theoretical perspectives on the issue have emerged. Pluralist theorists of urban politics tended to attribute great significance to the role of community groups (urban social movements) within city politics, while **structuralist** accounts of urban politics tended to stress the exclusion of those interests that run counter to the imperatives of the capitalist system (Mollenkopf 2003; Savage et al. 2003: 192).

Without doubt the most influential theorist of this aspect of urban politics has been Manuel Castells (1977, 1978, 1983). Castells' starting point within this project was a recognition that the city should be viewed through the lens of collective consumption, the provision of welfare goods by the state (housing, transport, education and health, for example) for the reproduction of labour power, and not merely through economically determined relations of production and exchange. This was important as it shifted emphasis away from processes of competition, such as in the Chicago School models of the city (chapter three) and highlighted aspects of the city that had clearly become significant in the post-war period and which were having a significant influence over the urban forms emergent at the time (Savage et al. 2003: 162).

Castells, within initially rigidly structuralist accounts of the process, embedded his discussion of the provision of collective consumption resources by the state within the wider contradictions of the capitalist system (Savage et al. 2003: 162). Broadly speaking, Castells recognized a number of contradictions engendered by the provision of collective consumption. These resources were provided by the state not primarily through, in Castells' view, altruistic motivations but because it was unprofitable for them to be provided privately, despite their being essential for the reproduction of the labour force (we need health services, for example, to treat workers when they fall ill). These goods are, therefore, financed out of various forms of taxation that impose a cost on the profits of firms. This, along with the prospect of outcries from the public on what might be seen as excessive taxes, limits the revenues that can be raised from taxation. This in turn causes pressure on the state to cut back on the resources it invests in collective consumption which can lead to resistance from urban dwellers who either oppose cuts in collective consumption provision or else seek to maximize or maintain their access to them. Out of these ongoing struggles around collective consumption, Castells argues, emerge urban social movements.

Castells' perspective evolved considerably over the course of his work in this area. Most notably it became less deterministic. This work provided major contributions to debates across a number of urban disciplines while stimulating significant theoretical developments within a range of issues from social geography, through urban form, to urban politics. However, as one might expect for such significant contributions, they came under a great deal of critical scrutiny. It would be true to say that while Castells' original ideas have little direct influence on the study of urban politics today, they are still recognized as a major event in the history of urban studies that undoubtedly, perhaps as much through their critical reception as through the works themselves, pushed forward theoretical debates across the urban disciplines in ways that would otherwise not have been possible.

Savage et al. (2003: 163–164) recognize four major criticisms of Castells' work. First, they point out that some critics (Lojkine 1976; Harvey 1982; Massey 1984) have argued that in emphasizing the importance of collective consumption, Castells failed to appreciate the ongoing significance of production to urban relations. Second, they acknowledge feminist critiques of the concept of collective consumption that fail to recognize that women provide many services essential to the reproduction of the labour force for free (see also Watson 1999: 202–203). As well as being a significant omission, this undercuts the theoretical clarity of Castells' notion of collective consumption. Third, Castells seemed to overstate the potential of class as a basis within which alliances could be made between urban social movements and labour movements such as trade unions. These alliances he saw as essential to any transformation of the political system. However, urban politics failed to unfold in the ways that Castells envisaged, leading to some questioning the relevance of his ideas to this realm. Finally, and perhaps most damningly, Savage et al. (2003: 163–164) recognize that Castells' work was overtaken by events. By the 1980s the widespread social-democratic consensus that had developed in the post-war period in Europe and North America was breaking down. The period since the early 1980s has seen the exponential growth of privatized, rather than collective, consumption. In the case of housing, for example, we have seen the growth of owner-occupation. Further, this period witnessed the increasing market provision of services under successive iterations of **New Right** political ideology. Castells' work seemed to have little to say about this as alternative theorists emerged, most notably the sociologist Peter Saunders (1979, 1981, 1984), whose work was empirically and theoretically more in tune with the era than that of Castells.

Despite the criticisms of many aspects of Castells' work from this period, the concept of senses of identity among groups of people, that was argued to form the basis of urban social movements, has remained key within subsequent debates about the expanded realm of urban politics (Watson 1999; Byrne 2001;

Savage et al. 2003: 162; Pacione 2009). Most commentators agree, however, that the broad class based identities, within which it was argued the urban social movements that Castells identified were rooted, have been superseded, but certainly not replaced entirely, by a more fragmented, multiple politics of identity. This perspective is in part a critique of the homogeneity that some writers on urban social movements imagined characterized these groups (Watson 1999: 228; Byrne 2001: 167) and in part a recognition that more privatized forms of consumption, identified by commentators such as Peter Saunders, have contributed to the general fragmentation of social identities.

Without doubt the last few decades have seen a significant rise in political movements based around various forms of identity. Some of the more significant and widespread of these include the American civil rights movement, a range of disability rights movements internationally and women's and gay liberation movements. These have had significant impacts on politics, policy and attitudes in many countries at the national and local levels. It is important to recognize, though, that people possess multiple identities that may shift throughout their life course. They may take up various, perhaps at times contradictory, positions with regard to these movements and the politics of identity more generally.

Exercise

Can you identify any examples of politics of identity, meaning or representation in an urban area you are familiar with? What groups are involved in these struggles or conflicts? What type of power are they able to deploy within these struggles and how are they able to do this? Do issues of language and representation emerge within these examples (see case study)? Are the examples you have found purely local struggles or do they involve actors from beyond the city? What roles do the local authorities take? Do any of the theories of urban politics outlined above help to make sense of these examples?

There is an urban politics then beyond the formal political arena. This is a politics of everyday life that in recent decades has become increasingly focused around issues of identity, meaning and representation (Watson 1999). When thinking about urban politics we need to pay attention to the issues that are rooted in the social and cultural geographies of cities and the meanings and representation of urban space. They involve struggles over access to spaces, the rights of marginalized groups to use the city in ways that might be formally regulated, controlled or excluded (Amin et al. 2000; Amin and Thrift 2002). There is a perhaps more mundane, everyday dimension to this politics of space though that involves the daily, often minor, acts of transgression or resistance to attempts to inscribe authority, order and regulation across the spaces of the city (see chapter ten). Urban politics may involve a performative politics that literally takes to the streets in the form of protests, demonstrations or occupations (Watson 1999: 234–236) or the everyday transgressive acts mentioned above. It also involves struggles over the meanings that are ascribed to places and spaces through political, planning or development discourse and which often shape the ways that places develop (see chapter nine). It is at the intersections of space, meaning, representation, regulation and development that this politics of identity is often manifest. The case study below discusses one example of this, highlighting a number of key points, and these are issues that are picked up and discussed in more detail in chapters seven, nine and ten.

Case study: El Raval, Barcelona: a politics of representation

El Raval is a neighbourhood located adjacent to the Ramblas, the main pedestrian artery through the centre of Barcelona. It originally developed beyond the city's medieval walls and became associated with a number of noxious industries that were excluded from the city itself. As a result of this it tended to become associated with various groups, such as prostitutes, immigrants, criminals and political radicals, who were marginalized within the broader life of the city. To suggest, however, that it was only home to such 'outsiders' is a significant oversimplification. The 'outsider' perception of El Raval, though, remained for much of the twentieth century. The 1970s saw the influx of large amounts of heroin into Barcelona, coupled with deep industrial decline leading to severe social and economic problems in the city. El Raval, particularly, became associated with issues of drug addiction and prostitution. Again though, to define El Raval entirely through these pathologies was inaccurate. In recent years Barcelona has turned itself into one of the most popular urban tourist destinations in Europe following its successful hosting of the 1992 Olympics. This has led to pressure

for the gentrification of a number of its poorer neighbourhoods. This pressure has been particularly acute on El Raval, given its prominent central location. The flagships initiating the regeneration of El Raval have been the Museum of Contemporary Art Barcelona (MACBA) (figure 5.1) which opened in 1995 and is located on the site of a number of demolished blocks to the west of the neighbourhood, and a new boulevard (Rambla Del Raval) that cuts through an area formerly associated with prostitution. Inevitably as gentrification brings physical development and social change to an area it is met with resistance. These struggles can be summed up as involving a long-standing resident population who see their local area under threat from incoming capital and people, and associated development trajectories very different to previous ones.

The carefully managed representation of El Raval in media and political discourse became crucial in gathering support for the contested gentrification of the area. Metaphors of light and dark were liberally deployed within discussions of El Raval and its regeneration. El Raval was typically

continued

Figure 5.1 *Museum of Contemporary Art, Barcelona*

Source: Author's photograph

described by developers, planners and politicians in terms that associated it with darkness, threat, dirt and disease, while the descriptions of new developments, such as MACBA, played on metaphors of light, whiteness and purity. Talk was of opening up El Raval, bringing in light, air and order where previously there had been darkness, danger and disorder. This selective presentation of the area in these 'common sense' terms was crucial to the wider acceptance of the need to gentrify and develop El Raval and in the neutralization of opposition to this process. Alternative representations that challenged these images were unable to gain any purchase within debates about the area's future. Describing urban change in these ways is not new. Parallels can be seen, for example, in discussions of poverty in Victorian cities. However, such representations should not be treated uncritically. The use of myth and metaphor should be recognized, as should the ideological purposes of such representations. Representation and language then are crucial elements deployed within urban politics.

Sources: Degen (2003, 2004); Miles (2004)

Summary

Urban politics should not be viewed narrowly. It should be recognized that politics in cities extends far beyond the formal political arena. Crucially, everyday life in cities becomes political where struggles over resources, rights and access emerge among groups with different levels and types of power. Further, in the same way that cities vary hugely around the world, so does the nature of urban politics. This is something that should always be acknowledged, particularly when examining the claims of competing theories of urban politics. However urban politics is theorized though, it is vital that the roles of local authorities, private and community interests, and the relationships between them, are at the centre of analysis.

Follow-up activities

Essay question: 'Attempts to produce a universal theory of urban politics have proven elusive.' To what extent do you agree with this statement?

Commentary on essay question

The question demands a critical evaluation of the contributions of urban political theory. A poor answer will merely rehearse details of the main theories with little appreciation of the specific demands of the question. Such an answer, while demonstrating relevant knowledge, does not demonstrate the ability to apply this knowledge and will achieve, at best, a low pass mark. More sophisticated answers will draw attention to debates around these theories that stress both the contributions and limitations of urban theory. There are many standard texts within the study of urban politics. While good, solid answers might rely largely on these, more sophisticated answers will draw on more specialized references, particularly recent work on urban political theory. Answers that are at or near an excellent level will recognize the multi-disciplinary nature of the literature of urban politics and should seek to draw on literature from different disciplines, perhaps noting these different perspectives and the nature of debate within different disciplines. Some kind of geographical demarcation might be important here and could significantly affect the answer. For example, does the question allow you to talk about cities from across the globe or are you restricted to cities from one global region, or even a single country? If the answer demands recognition of global urban diversity, some discussion of cities in the Global South will be required and either some information about urban political theory derived from empirical

*studies of cities from this region or some commentary on the limitations of
applying urban theory derived from studies of western cities to these very
different urban contexts. The latter approach might make links to the work of
Jennifer Robinson (2002, 2005a) discussed elsewhere in this volume.*

Project idea

Undertake a study of the leader of a major urban area. Draw on the work of Donald McNeill discussed above. What roles can you recognize this political leader playing with regard to his or her urban area (promoting the city on wider political stages, embodying the city, narrating versions of the city's future, animating its spaces etc.). Do you think these roles are purely symbolic or can you detect tangible influences of this performative politics on political decisions affecting the area and on the city's development?

Further reading

Books

- Amin, A. and Thrift, N. (2002) *Cities: Reimagining the Urban*, Cambridge: Polity Press (chapter 6)
  Looks beyond conventional understandings of urban politics to explore the possibilities of an 'expanded' urban democracy (see also Amin et al. 2000, chapter four).

- Davidson, M. and Martin, D. (eds) (2014) *Urban Politics: Critical Approaches*, London: Sage
  An excellent, broadly conceived edited collection of critical essays, many by some of the leading scholars in recent iterations of urban political research.

- Davies, J.S. and Imbroscio, D. (eds) (2009) *Theories of Urban Politics*, London: Sage
  Along with Judge et al.'s (1995) earlier volume of the same name offers a comprehensive in-depth overview.

- Judd, D. and Swanstrom, T. (2009) *City Politics: The Political Economy of Urban America*, 7th edn, Harlow: Longman
  Long established as one of the classic guides to the political arenas of American cities.

- Levine, M. (2015) *Urban Politics: Cities and Suburbs in a Global Age*, 9th edn, New York: Routledge

The latest edition of a classic comprehensive overview of all aspects of urban politics. It is particularly attuned to the relationships between politics and processes of urban development and the different spaces of the city.

- Mossberger, K., Clarke, S.E. and John, P. (eds) (2012) *The Oxford Handbook of Urban Politics*, Oxford: Oxford University Press
  An extensive collection, some 32 chapters, of excellent discussions of all aspects of urban politics. Written by some key commentators and from a variety of perspectives. A key resource.

- Pacione, M. (2009) *Urban Geography: A Global Perspective*, 3rd edn, Abingdon: Routledge (chapters 17, 20, 29)
  Pacione provides some detailed, wide-ranging discussions of urban politics within the contexts of developed and developing world cities.

- Strom, E.A. and Mollenkopf, J. (eds) (2003) *The Urban Politics Reader*, London: Routledge
  A well-chosen collection of classic readings in urban politics. An essential resource for higher level study of the issue.

Journal articles

- Beal, V. and Pinson, G. (2014) 'When mayors go global: international strategies, urban governance and leadership', *International Journal of Urban and Regional Research*, 38(1): 302–317
  This article reflects on the roles of city mayors who engage in international strategies such as twinning arrangements and engaging with international city networks. It considers the roles of these activities within the political and policy arenas.

- Goldfrank, B. and Schrank, A. (2009) 'Municipal neoliberalism and municipal socialism: urban political economy in Latin America', *International Journal of Urban and Regional Research*, 33(2): 443–462
  Examines different political regimes across Latin America, challenging the myth that cities in this region are impotent in the face of powerful central governments.

- Ponzini, D. and Rossi, U. (2010) 'Becoming a creative city: the entrepreneurial mayor, network politics and the promise of an urban renaissance', *Urban Studies*, 47(5): 1037–1057
  Examines the emergence of what might be called 'creative regimes' within the context of the regeneration of the contemporary city.

- Ward, K. and McCann, E.J. (2006) '"The new path to a new city"? Introduction to a debate on urban politics, social movements and the legacies of Manuel Castells' *The City and the Grassroots*', *International Journal of Urban and Regional Research*, 30(1): 189–193

Introduction to a series of papers collected in a single issue of the journal that discuss the legacy of Manuel Castells' work on collective consumption and urban social movements.

- Wu, F. (2002) 'China's changing urban governance in the transition towards a more market-oriented economy', *Urban Studies*, 39(7): 1071–1093
  Examines the under researched realm of China's urban governance under its current economic transition. A useful counterpoint to some of the more western perspectives discussed in this chapter and the wider urban politics literature.

Websites

- www.apsanet.org/~urban/ – American Political Science Association's urban site.

  A number of city mayors are extensive users of Twitter. Some are listed below and offer good resources for understanding the issues they are grappling with as they unfold.

- https://twitter.com/SadiqKhan / https://twitter.com/MayorofLondon
  Sadiq Khan – Mayor of London.

- https://twitter.com/NYCMayorsOffice / https://twitter.com/NYCMayor
  Bill de Blasio – New York City Mayor.

- https://twitter.com/MarvinJRees
  Marvin J. Rees – Mayor of Bristol, UK.

- https://twitter.com/ericgarcetti / https://twitter.com/MayorOfLA
  Eric Garcetti – Mayor of Los Angeles.

⬤6 Planning, regeneration and urban policy

Introduction

As long as there have been cities their development has been planned and controlled in some way; all that has varied is the nature and extent of that planning and control. As Carter (1995) notes, while models of urban development based on the idea of the operation of free market economic forces proved appealing to urban theorists, the reality of city growth has always been one where these forces have been controlled and regulated in some way by those in power. Frequently, the urban planning and policy arenas have been characterized as stages on which battles between capital, the state and urban populations are played out in cities. How urban planning and policy operate within cities then has important ideological underpinnings, influenced by attitudes to the role that the state plays in influencing free market forces and to the role of the public in influencing these processes. Consequently, urban planning and policy have been viewed as either beneficial or problematic by all sides of the political spectrum from far right to far left (Pacione 2009).

This chapter continues by first providing an overview of the development of core traditions and approaches to urban planning and policy. It then moves on to consider early ideas about creating **utopian** visions of the 'good city' and how these ideas have shaped later urban developments. This is followed by two sections exploring more recent approaches to urban management: moves to conserve valued urban environments and the development of urban policy. Finally, the chapter discusses the challenges facing urban planning and policy in the early twenty-first century.

Urban planning and policy traditions and approaches

An important starting point in considering the traditions and approaches that underpin urban planning and policy is to define what we mean by urban planning and urban policy. This is no easy task!

Exercise: what do we mean by urban planning/policy?

Think about what you understand by the
terms planning and policy and what you
already know about them. In considering
these terms it is useful to think about:

● What do they do?
● How do they work?
● Who is involved?

Your responses to the exercise should hopefully have demonstrated the range of
things that could concern urban planning and policy, and the variety of purposes
that they could serve. Plans and policies can have different roles: they might
seek to restrict building, control or manage change or even look to facilitate
new development. The aims of plans and policies can also be wide ranging,
seeking, for example, environmental protection and sustainability, social
improvements, reduction in inequalities or economic regeneration and growth.
A range of agents can be involved, including planners and policy makers,
businesses, politicians and the public, all of whom often operate with different,
conflicting agendas and who possess differing levels of power and influence. It
is therefore difficult to put clear boundaries around the topic of urban planning
(Hall 2002) or to neatly define what constitutes urban policy (Cochrane 2007).
Equally, planning and policy are often considered separately within the
literature on cities, making identification of core themes and approaches tricky.

As we have seen, cities have often been viewed as disorderly places containing
unruly populations (Pile et al. 1999; Cochrane 2007). As cities have developed
across the globe and expanded in their size and complexity, the desire to
control their development and functioning and mitigate their problems has
also increased. Broadly speaking, planning can be seen to have the longest
intellectual tradition, stretching back to the city building ideas of early urban
civilizations. As we considered in chapter three, many early cities were
designed to reflect religious or cultural beliefs or followed standardized plans
when settlements were built in newly settled or conquered lands. However,
while examples of city planning can be traced back to these earliest urban
civilizations, the rise of widespread formalized attempts at planning cities has
its origins in the nineteenth century, associated with attempts to control the evils
of the rapidly expanding industrial cities of Europe and North America. It is
during this period that a number of key ideas about the planning of cities were
developed, which were the product of ideas generated by a number of planning
'visionaries' writing at the time (Hall 2002).

Here, urban planning was concerned with 'the problem of cities', managing the
growth and expansion of unruly cities and mitigating the urban social and
environmental problems associated with rapid population growth through the

regulation of land uses and building development, through the provision of key infrastructure and through the development of designs and plans to improve their aesthetic qualities and their functioning. These concerns have continued to be an important focus for urban planning in the twenty-first century, and are very much a key concern for the rapidly expanding megacities of the Global South (UN-HABITAT 2009). In its approach, planning can be viewed as a rational, modern and technical response to the unruly city, seeking to impose order through recording, quantifying, mapping and designing. These ideas and approaches developed in the early twentieth century have proved important in influencing much of what has subsequently been planned in the world's cities and have developed into the profession of planning as it functions in many countries today. Here planning ideals are communicated through plans and regulatory frameworks, ranging from grand masterplans for whole settlements to loosen controls on building heights and land use zoning (UN-HABITAT 2009). However, globally, planning systems differ as a result of their specific cultural and historical origins, particularly as a result of the link between planning and the law and the extent to which they operate within a framework of constitutionally protected property and citizen rights, the degree to which plans are flexible and planners can exercise discretion in decision making, and the balance between local and central state control (Cullingworth and Nadin 2006).

For many of the older industrial cities of Europe and North America, new challenges for planners and policy makers emerged in the late twentieth century. In particular, there was a growing disillusionment with established modernist planning traditions, particularly in terms of the top-down, comprehensive redevelopment approaches adopted and the failure of these approaches to eradicate the problems of the city and its most disadvantaged populations (Hall 2002). Equally, new problems were beginning to emerge in these cities associated with deindustrialization (see chapter four). Two important strands emerged from this crisis in modernist planning. First, there was growing public discontent at plans for comprehensive clearance and renewal, fuelling a growing popular concern for conserving urban environments valued by local populations. This was linked to concerns about what to do with urban landscapes abandoned in the wake of deindustrialization and how best to approach the regeneration of cities. Second, there was broader critique of the **Keynesian** welfarist approach to governance and the management of cities (Cochrane 2007). This precipitated the development of a range of policies aimed at stimulating the economic, social and environmental revival of depressed urban areas, from which have emerged new ways of planning and managing cities, particularly associated with entrepreneurial approaches to urban governance (see chapter five), referred to broadly as urban policy.

Urban policy then has a more recent intellectual parentage, and has been concerned with 'problems in cities', particularly concentrations of urban social disadvantage and the uneven impacts of deindustrialization and economic decline between and within cities. Although viewed as a distinct area, it is often difficult to define what is specifically urban within the broad collection of social and economic policy initiatives that have been included under this umbrella, beyond the title given to particular policies or their focus on particular cities or parts of cities, or the fact that the target populations or issues addressed are largely concentrated in urban areas (Cochrane 2007). As Cochrane notes, what is, therefore, considered urban policy is not fixed but highly fluid, linked to what is perceived as an urban problem and to changing ideas about how these problems should best be tackled.

Over the course of the twentieth century the developments outlined above have generated a significant body of theory and practice in urban policy and planning. Despite their different origins, trajectories and foci some common themes underpin a consideration of the variety of urban planning and policy approaches. Most approaches can be placed within four broad arenas, depending on their location on a continuum between state-led and market-led development and the degree to which ideas are imposed by those in power or stem from community priorities (see figure 6.1).

With the development of the planning and policy professions in many countries, and the growth in global policy and development organizations, ideas and approaches have been increasingly spread around the world

Free Market	
• Utilitarian tradition (laissez-faire) – development as required/needed • Organic development – piecemeal • Self-help/cooperative schemes	• Authoritarian tradition – policy dictated by a single person/owner/architect/organization • Grand designs and flagship developments • Entrepreneurial (New Right) – business-led approaches
• Community-led planning and policy • Public participation and community developed priorities	• Managerialist/reformist tradition – professional planners and urban managers • Utopian visions and development plans • Welfare State/Keynesian – 'big government/planning'
State Control	

(Bottom-up / Top-down axis labels on left and right)

Figure 6.1 *Common themes in urban planning and policy*

Source: Authors

(Ward 2002; Cochrane 2007). However, in many cases the translation of planning and policy ideas developed in particular economic, political and cultural contexts to other parts of the world has proved problematic. In particular, approaches developed within western modernist planning traditions and policy contexts have not always been easily or successfully translated to the planning and management of cities in the Global South (UN-HABITAT 2009). Across the world, a key concern is that urban planning and policy approaches developed in the context of western cities in the twentieth century fail to take into account the new challenges facing cities in the twenty-first century, and equally fail to acknowledge the need for the meaningful involvement of communities and other stakeholders in the planning and management of urban areas. The voices of the public have often been absent in the debates about the planning and management of cities, particularly those of the more marginalized and disadvantaged urban populations in cities of both the Global North and South, serving to further exclude them and contribute to their social and spatial marginalization.

While the popularity of planning has waxed and waned over the years, there is currently a renewed interest in planning and a desire to better manage urban areas (UN-HABITAT 2009). At the beginning of the twenty-first century urban planning and policy agendas around the world have been united in seeking to address the challenge of promoting sustainable urban development (see also chapter thirteen). The global climate change agenda has stimulated considerable discussion about how best to plan cities to make them sustainable, although what this actually means and how sustainability can be best approached and delivered are areas that are still far from clear, and which continue to pose a considerable challenge to planners and policy makers around the world. Equally, new approaches to planning and urban management have developed which broadly tend to be more strategic, flexible and action and implementation orientated. Many are also concerned with developing new institutional processes, seeking to foster new forms of community involvement and integrate planning with other local government, voluntary and private sector activities. These new approaches are broadly termed strategic spatial planning, which emerged in Europe in the 1990s and which has spread to other developed and some developing countries since then (UN-HABITAT 2009). Strategic spatial planning is seen as useful in that it focuses on a process of decision making and does not carry with it a predetermined set of urban forms or values, although many current plans promote sustainability, inclusiveness and qualities of public space. However, at present these approaches are more evident in the planning literature than in practice and it remains to be demonstrated whether these new approaches can fully address the problems facing cities in the twenty-first century.

Planning visions

In his wide-ranging history of planning, Peter Hall (2002) notes that much of the planning carried out in the world's cities during the twentieth and early twenty-first centuries has stemmed from the ideas of a few early visionaries from the nineteenth and early twentieth centuries. Many of these sought to provide visions of the 'good city' which would provide a better, more ordered environment than the nineteenth century industrial city. These visionaries are viewed as the founding fathers of modern urban planning; as Hall notes, there were very few founding mothers, a gender imbalance that has continued within planning professions (Greed 1994). These visionaries also emanated principally from Europe and North America, and their visions for the city and urban life consequently stem from these specific temporal, spatial and personal contexts. This has often created problems when their ideas were translated into quite different places and circumstances from those originally envisaged by their originators.

Exercise: what would be your vision of the good city?

Think about which elements and activities you would include in your city and how you would arrange these. When you have developed your list of attributes, or perhaps even a plan or design, share your ideas with your friends, family or fellow students. Do they agree with your vision or not? Think about which elements are agreed on and which are more contentious and why this might be – does this have anything to do with your background and values in relation to those you are discussing your ideas with?

Now read on through the rest of this section about visions. When you have read through it, go back and further reflect on whether the issues raised about other urban visions are reflected in the discussions about your vision. What issues does this raise about the problems facing those who seek to plan the future city?

While it is difficult to single out ideas and individuals from the myriad planning visions and visionaries, a short overview such as this must do so. There are perhaps two key city visions from the early twentieth century that have echoed most down the decades in urban planning and which have had the most far-reaching global impact on ideas for the reordering of cities; 'the city in the garden' and 'the city of towers', as Hall (2002) labels them.

The garden city

The idea of a garden city, proposed by Ebenezer Howard, is perhaps one of the most important in the history of urban planning (UN-HABITAT 2009). Howard was not, however, a planner but a copywriter deeply influenced by both his personal background and the broader context of the late nineteenth century. Principal influences on Howard were British land and housing reform movements and a number of utopian thinkers and model housing developments (Ward 2004). Howard's vision therefore had a number of antecedents, although he believed his combination of ideas to be unique in applying existing reformist ideas of collective ownership to the solution of urban problems.

Howard's vision was published first in 1898 under the title *Tomorrow: A Peaceful Path to Real Reform* and was reissued in 1902 with the title *Garden Cities of Tomorrow.* In Howard's vision, agricultural land would be purchased at a low price, with a self-governing community owning the land, building the homes and developing the businesses in the garden city and with the community benefiting from the increased land value arising from development, which would be used to further develop the community. The success of Howard's vision lay in both his ability to concisely convey his ideas and in his provision of both ideas and the practical means to achieve them. Howard's ideas are neatly expressed in a series of diagrams that appear in his book. The key diagrammatic expression of the vision took the form of a social city (figure 6.2), a decentralized network of garden cities of about 32,000 people surrounding a larger central city of about 58,000 people. These cities were to be separated by extensive areas of green space but also connected by transport networks. In their detailed layout, the garden cities would have abundant open space within them, wide boulevards, the functional separation of industry, good quality housing, and a range of social facilities.

Howard's ideas became the basis for a wider movement and in 1903 the Garden City Association (GCA), founded by Howard in 1899, began its first garden city in Letchworth in the UK, giving physical expression to Howard's ideas. Raymond Unwin and Barry Parker were appointed as architects at Letchworth and their designs, based on the building styles developed by the **Arts and Crafts Movement**, were particularly influential in giving the garden city its three-dimensional form (figure 6.3). Their style looked back to lessons from the past, particularly medieval town building traditions and village designs.

The garden city idea proved an important foundation for early urban planning movements and Howard's ideas were widely disseminated both in the UK and more widely across Europe, North America, Japan and Australia, with many countries creating their own garden city associations (Ward 2004). In addition,

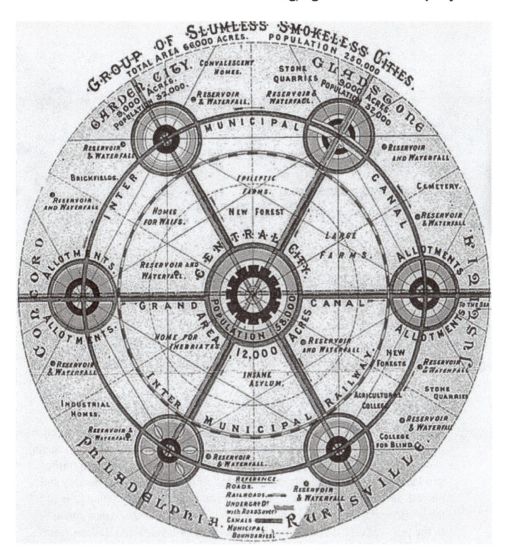

Figure 6.2 *Howard's vision of the social city*

Unwin and Parker disseminated many of their ideas through their co-operation on schemes in other countries, through the production of pattern books of their designs and through their involvement in government planning and housing committees in the UK (Hall 2002). Many of their ideas became blueprints for other planning schemes, including their holistic approach to urban design and their use of design elements such as open space, the clear zoning of functions and road layouts including pedestrian ways and cul-de-sacs. In the US, Clarence Stein and Henry Wright developed the garden city idea in their design for

Figure 6.3 *Letchworth Garden City*
Source: Topical Press Agency

Radburn, New Jersey in 1928 in collaboration with Unwin. Key ideas to emerge from this design were those of the neighbourhood unit, with groups of housing clustered around community facilities, and the separation of through traffic and local traffic, with neighbourhood areas arranged around cul-de-sacs and bicycle and pedestrian routes. This arrangement was widely copied around the world and became somewhat of a standard in the design of new towns. In the post-Second World War period, garden city layouts and designs were revived and widely utilized, for new capitals in newly independent post-colonial countries, for new towns in European countries and for speculative planned community developments, particularly in the US (Hall 2002).

Despite its widespread adoption, the garden city concept has often been criticized for promoting an anti-urban vision of low density suburban and small-town living. However, many of the developments subsequently promoted under the banner of the 'garden city' were little more than suburbs, designed as urban extensions and commuter zones linked to the main city and composed mainly of housing, far removed from the free-standing, self-governing communities envisaged by Howard (Hall 2002). These garden suburbs often lacked any commercial functions and became principally middle class commuting enclaves rather than the mixed communities envisaged. Consequently, these developments have been criticized for failing to promote vibrant mixed communities, with many new town residents experiencing exclusion, isolation and a lack of community; the so-called 'new town blues' (Attfield 1995). Equally, the approach adopted by these schemes is seen to embody a paternalistic and top-down approach to planning and urban development, where the planners know what is best for people, who appear to be mere pawns to be moved around in their orderly schemes (UN-HABITAT 2009).

Thus, while the idea of the garden city has been much discussed and emulated around the world it became, like many visions, stripped of its more fundamental ideas, particularly its more radical messages about co-operative ownership and the broader concept of the social city. Yet, while the reformist vision of the garden city might seem highly utopian and unattainable in the contemporary world, its core ideas have a clear resonance with current urban concerns and many writers still contend that significant economic and social reform is required if truly sustainable cities are to be planned in the twenty-first century (Blowers and Pain 1999).

The Radiant City

The other key vision which had a considerable impact on twentieth century urban planning was that of the 'city of towers' proposed by the Swiss-born architect Le Corbusier. In parallel to Howard, Le Corbusier proposed an ideal for the 'good city' which would address the perceived problems of nineteenth-century urbanization. Hall (2002) argues that, like Howard, Le Corbusier's vision was very much a product of both his background and the wider context within which he was working. His family background in watchmaking is seen to have influenced his desire for order and his conception of the house as a machine for living in, where the city can be compared to a watch or engine, consisting of various components that need to be carefully placed in the correct, rational order, so creating a harmonious and functioning whole. His work can also be seen as a reaction to the problems of congestion and slum housing in early twentieth-century Paris, the city he lived and worked in for most of his life.

Le Corbusier developed his planning principles in two key visions: the *Contemporary City* (1922) and the *Radiant City* (1933). His paradoxical solution to the problems of the crowded, disorganized city was to decongest cities by increasing their density, and also to improve circulation by increasing the amount of open space and by careful geometrical design. This could be achieved, he argued, by demolition of the existing city and rebuilding on a cleared site with modern high-rise building using the latest building technology

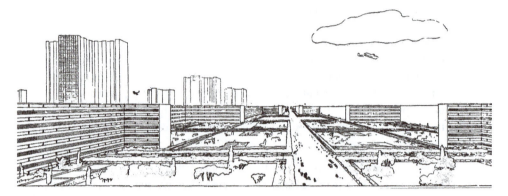

Figure 6.4 *Le Corbusier's* **Radiant City**
Source: FLC/ADAGP Paris and DACS

and wide boulevards to accommodate traffic (figure 6.4). Le Corbusier's ideal was therefore the epitome of modernist design thinking, decontextualizing the city and working with a clean slate without any sense of historical context. However, Le Corbusier's vision did not just encompass the physical design of the city but also sought to order life within the city. Everyone was to live in giant collective apartment blocks or vertical streets called *Units*, with apartments built to the same space standards and which would benefit from the collective provision of services, such as cooking, cleaning and childcare. Le Corbusier's views on urban planning were therefore quite authoritarian; he believed that the design of cities was too important to be left to the people and that schemes designed by experts should be imposed from the top-down. These views were influential in the development of the planning and architecture professions and the idea of design being undertaken by experts who knew what was best for urban populations (see also chapter eight).

Yet, despite developing plans for a number of major cities in his writings none of his broader urban visions was ever built and he struggled to find people to support his grand schemes. Beyond some single *unités* that were built in a number of cities, particularly in France (figure 6.5), his only planning achievement on the ground was Chandigarh, a new capital for the Punjab built following Indian Independence in 1947 by a number of prestigious modern architects, including Le Corbusier. Chandigarh was therefore not solely Le Corbusier's, although he did take a lead in the development of the scheme and he was principally responsible for the design of the commercial centre. A key criticism of the scheme is that the architect's focus on its visual form and symbolism resulted in a disregard for the realities of the local environment and the lives of its Indian population, imposing a western future onto an Indian present (Hall 2002; Shaw 2009).

Figure 6.5 *Corbusian* unité, *Nantes, France*

Source: Author's photograph

Similar problems were also evident in the Corbusier-inspired design of Brasilia as the new capital of Brazil, where the planned city, the *Plano Piloto*, proved inflexible and consequently unable to accommodate the majority of the city's working population (see case study). The problems associated with the development of cities like Chandigarh and Brasilia have come to epitomize the failure of Le Corbusier's modernist planning ideas and the negative consequences of imposing design ideas top-down onto local populations, particularly onto cultures quite different from the European origins of those designs (UN-HABITAT 2009).

The take-up of Le Corbusier's ideas in the planning of cities in Europe and North America was also problematic. For Hall (2002) one of the most problematic legacies of Le Corbusier's city vision arose from its misapplication to the planning of working class housing in the 1950s and 1960s, which he notes was 'at best questionable, at worst catastrophic' (p. 219). Le Corbusier's argument that every great city needed to rebuild its centre in order to survive was to be enthusiastically endorsed in the replanning of many cities in the period after the Second World War. City authorities embraced the idea of a

Case study: the problems of modernist planning legacies, Brasilia, Brazil

The development of a new capital for Brazil was a bold and extreme example of modernist planning ideals in action. The city was designed by Lucio Costa and Oscar Niemeyer with a ground plan in the shape of an airplane. It was inaugurated as Brazil's capital in the 1960s. From early in its existence the original layout, known as the *Plano Piloto*, has been protected by strict zoning to preserve its original layout and in 1987 it was placed on the UNESCO World Heritage List (an ironic end for a modernist vision?), creating further barriers to change. A number of academic studies of Brasilia have noted that the city's plan did not create the utopia it sought as it ignored the social and economic realities of Brazil and has in fact exacerbated many of these problems. In their 2007 paper, Dowell and Monkkonen argue that Brasilia now has a distorted urban form and abnormal land market as a result of its earlier planned legacies. A key problem is the inflexibility of the original urban layout and the distortions to the urban land market caused by planning restrictions and government ownership of land.

The authors use population and land market data to compare the socio-economic structure of Brasilia to the more 'traditional' cities of Recife and Curitiba, also in Brazil. The restrictions on Brasilia's urban development have had a particular impact on the spatial distribution of the city's inhabitants. Brasilia has a low density centre in contrast to other cities in Brazil, as this area is functionally segregated and dominated by government and commercial functions, with population concentrated outside the central fifteen kilometres of the city region. Due to its restricted urban land market Brasilia has been unable to meet the demand for housing from its urban population, which has pushed up prices in the central city and forced poorer residents to seek accommodation beyond the city core in 'satellite cities'. Thus, the planned central city, occupied by the elite, has become surrounded by unplanned urban developments far beyond the edge of the planned city. This has created a decentralized and sprawling urban pattern which has an inverse population density to other Brazilian cities.

This social and spatial division of Brasilia has been detrimental to the quality of life of its residents, particularly those on lower incomes. As most formal jobs are located in the *Plano Piloto* many have to endure high transport costs and long commutes from surrounding 'satellite cities' to access employment. The authors argue that this has created an unsustainable urban form which promotes sprawl and creates high costs for infrastructure provision. In this respect the lessons from Brasilia highlight the failures of modernist planning visions to provide flexibility and adaptability and to deal with changing urban realities. Like many urban planning visions of the 'good city' it has exacerbated the problems of many urban residents rather than alleviating them.

Source: Dowell and Monkkonen (2007)

clean sweep of bomb damaged areas and dilapidated nineteenth century townscapes and developed comprehensive modernist plans for both commercial centres and public housing projects. For example, within the UK, a combination of factors precipitated the widespread adoption of Corbusier-style designs for public housing schemes in the 1950s and 1960s. However, beyond a few architectural show pieces, the majority of people found themselves living in hurriedly constructed system-built flats lacking amenities, environment and community (Hall 2002). While initially they provided better accommodation than many families had known, problems with many of these developments soon emerged. It is not clear exactly why they became labelled a failure, but it stemmed from a combination of poor design, poor maintenance, poor management and widespread publicity for building failures which fuelled scepticism about high-rise living and labelled many of the developments 'problem estates'. Similar developments also took place in other European countries and in North America. For example, in France large Corbusier-style public housing estates were built on the edge of most major cities in the 1960s as a response to its post-war population boom. Many were poorly constructed and lacked key amenities, and in recent times these estates have emerged as problem areas, becoming places with high concentrations of social and economic disadvantage, particularly youth unemployment (Dikec 2006).

Critiques of urban visions

By the end of the 1960s the tide was turning against clearance, modernism and grand planning. Within many cities in Europe and North America, the masterplans and comprehensive clearance schemes proposed by planners were increasingly opposed by local populations and local pressure groups seeking to defend their neighbourhoods. Significantly, in the US, the threat of clearance to Jane Jacobs' neighbourhood in New York prompted her to embark on a successful campaign to defend the neighbourhood and also resulted in the publication of the book *The Death and Life of Great American Cities* (1961), which offered the 'right message at the right time' (Hall 2002: 254) and became one of the most influential books in city planning. Jacobs criticized the great planning orthodoxies of the time, both the egotism of the Corbusian designers and the perceived paternalism and anti-urbanism of the garden city movement. The critiques stimulated by Jacobs undermined the certainty (she would have said arrogance), of ideas stemming from these urban visions, questioning the 'state bulldozer' approach to planning and the idea of planners and designers as the all-knowing experts.

Jacobs' view of the city was one from below, from the street, rather than the Corbusian architect-planners view from above. Her urbanism celebrated the traditional inner-city neighbourhood, arguing that there was nothing wrong with high densities as long as there was no overcrowding. Jacobs' four principles for good neighbourhoods included mixed uses, mixed blocks with buildings of different age and condition, conventional streets with short blocks and numerous corners and a dense concentration of people, all to encourage a lively and diverse community. These ideas have been influential in the development of 'new urbanism', an approach to planning that is opposed to the growth patterns of cities exemplified by urban sprawl and restrictive residential enclaves (Gottdiener and Budd 2005). However, the urbanism proposed by Jacobs and the new urbanists has perhaps proved just as susceptible as modernism to having its egalitarian impulses subordinated to capitalist development interests. In hindsight, calls to abandon clean-sweep planning and preserve the diverse qualities of older established neighbourhoods have underpinned the subsequent transformation of many inner-city neighbourhoods through gentrification. Equally, many are critical of ideas that seem to suggest that idealized visions of community life can be easily mapped on to larger cities, to solve urban problems and create harmonious environments merely through design rather than through addressing more fundamental socio-economic problems (Gottdiener and Budd 2005).

Overall, the real criticism of the legacy of these urban visions is of design solutions laid down on people without regard to their preferences and ways of life. Professional planners and designers had no real feel for the way ordinary working families lived and the problems resulting from the application of abstract design ideas on various urban populations provides an important lesson for future generations of planners. However, this is not solely the problem of the planners and designers but is also linked to the wider realities of urban development processes. As Cochrane (1999: 313) notes, 'the grand visions of the utopian urbanists were translated into urban practice in a series of grey, day-to-day piecemeal decisions, driven by the priorities of developers and the exigencies of budgetary constraint'.

Finally, these critiques highlight the fundamental issue of whether the unruliness of the city can or should be ordered by rational planning interventions. A key question is whether the disorderly city is necessarily bad, or whether disorder might enhance a city's character. For Jacobs, 'good' disorder was an important part of the vitality and creativity of urban life. This links to the development of new attitudes to planning and conflict, exemplified in the writings of Leonie Sandercock (1998; see also chapters seven and eight). She argues that planning needs to be conscious of the politics of difference and diversity and that planning should be a 'messy' discourse of contestation and discussion rather than an imposed blueprint. Again these ideas exist more on paper than in the reality of planning cities today, although they have been developed through a number of radical planning approaches (Miles 2008). The key point here, however, is how order and disorder are defined and by whom, acknowledging that there will be many different ideas about what these might be, and that the power to impose these ideas is unequally distributed in the city.

Conserving urban landscapes

The critiques of modernist urban planning that developed in the 1960s, principally in western Europe and North America, created a climate of discussion into which new ideas about managing the city emerged. In many respects this was a popular reaction to existing planning orthodoxies, driven by urban pressure groups rather than by key shifts in the intellectual traditions from within the planning and architecture professions. However, these protests tapped into, and were galvanized by, longer standing intellectual traditions underpinning the development of conserver societies in the west.

The idea of conserving the past has a long intellectual tradition, stretching back particularly to the nineteenth century and even before this, linked to the development of the **Enlightenment** (Harvey 2001). However, it is within Europe in the nineteenth century that it became expressed in a more widespread desire to preserve particular monuments and buildings, with an increasing role for the state in this process. In its early phases it was a reaction to the increasing pace of change brought about by industrialization and urbanization, drawing on the same intellectual traditions as early urban planning visionaries such as Howard. Across Europe, there was a similar trajectory in the development of conservation concerns (Larkham 1996). Initial concern was frequently driven by wealthy, educated elites who were involved in the collection of artefacts from the past. These concerns were then taken up and further popularized by key publicists, particularly artists and architects who advocated a return to the design traditions of the past, particularly pre-industrial, medieval traditions. From this, pressure groups were formed which campaigned for the protection of particular buildings. For example, the artist and designer William Morris founded the Society for the Protection of Ancient Buildings in 1877 in Britain, which initially focused its campaigning on the protection of medieval buildings, particularly churches. In France, the architect Viollet-le-Duc was active in the restoration of medieval walled towns such as Carcassone. These developments then led to the development of national inventories of those buildings considered important to preserve and the passing of legislation to protect them.

From their origins within western Europe, ideas and frameworks concerning the conservation of urban landscapes have spread to many other countries around the world, through similar mechanisms to those underpinning the translation of other planning traditions, particularly through the activities of supra-regional and global organizations. Within Europe, the Granada Convention (1985) has provided a framework for the development of a consistent approach to conservation law and policy (Pickard 2002). On a global scale, the UNESCO World Heritage Convention (1972), to which 193 countries (state parties) belong (UNESCO 2017), provides a forum for the discussion of heritage issues and maintains a list of cultural World Heritage Sites, which contains a number of urban areas, such as the historic centres of Quebec, Canada, Liverpool, UK, Havana, Cuba and Macao, China. In recent years, UNESCO has been active, through its *Global Strategy for a Representative, Balanced and Credible World Heritage List*, launched in 1994, in encouraging heritage protection in countries beyond the Global North, highlighting the culturally specific nature of certain terminology and approaches used and seeking to redress the global imbalance in the designation of World Heritage Sites.

As a consequence of these developments, whether urban conservation systems are long established or relatively recent, they have some commonality in their practical and theoretical concerns, particularly in the selection criteria used (age or architectural or historical significance), the methods of control and protection employed (protection for single structures and/or areas, one level of control or different levels) and in the forces for change which underpin the protection of architectural heritage (economic and cultural). However, important differences exist in approaches to the repair and restoration of buildings (full restoration to the original or ongoing repair), attitudes to the social consequences of conservation (e.g. whether gentrification is an accepted consequence of building conservation), the financial incentives offered to conserve buildings (private money or government grants, loans or tax relief) and the degree of central government involvement in local activity (centralized or devolved control) (Larkham 1996; Pickard 2002).

There are a range of justifications for the conservation of buildings; the principal ones being intellectual, psychological and financial. The intellectual tradition is the longest standing rationale for conservation, where buildings are retained because of their artistic, architectural or historical qualities and therefore for the role which these then serve in illuminating the cultural achievements of a society. This rationale underpins much of the early effort in urban conservation where particular structures, such as religious structures, important state buildings, monuments and the homes of the ruling elite, were retained because they were seen to embody important cultural traditions of a society or nation. It is through this tradition that the role of the state in urban conservation has developed, although this can create biases in those buildings afforded official protection, devaluing those buildings and landscapes valued by local populations and minority groups. A key concern here is the politics of heritage, where the use of the past in the present is a highly contested process bound up with the production of individual and collective identities (Graham et al. 2000; see also chapter seven).

The psychological rationale for conservation is more recent, linked to reactions to the increasing pace and scale of urban and industrial change in the twentieth century. Here the loss of urban environments through comprehensive redevelopment leads to 'future shock' (Toffler 1970) where residents feel a sense of loss and dislocation. This can be seen to be one of the principal drivers behind more recent conservation campaigns by local groups. This wider popular involvement in the conservation of locally important urban environments has served to widen the scope of urban conservation beyond the designation and protection of a few important structures to more extensive protection of areas of cities including more 'everyday and mundane' buildings, or the 'familiar and cherished local scene' to use words from recent British urban conservation guidance. This has heralded a move to designate and afford protection to larger parts of cities rather than just individual structures. In addition, it has broadened the scope of conservation beyond the protection of the historic gems, cities with pre-industrial urban legacies, to include some nineteenth and twentieth century urban landscapes (Ashworth and Tunbridge 2000).

Finally, the financial rationale for conservation is again more recent in origin, and has further extended the scope of building conservation activities. With the rise of environmental concerns from the late 1960s, the conservation of buildings was promoted in terms of the economic and environmental benefits of conserving the embodied energy in a building through its reuse and in terms of conserving resources by not building new structures. However, as more buildings have been conserved and sustainable building demands have increased, the fit between conservation concerns and sustainability objectives have been increasingly debated. In particular, there are concerns about the energy efficiency of older buildings in comparison to modern structures, with old not being necessarily good (Strange 1997). More recently economic arguments around urban conservation have focused on whether conserving old buildings is a financial benefit or a burden. Much has been written on the burdens associated with building conservation, including the costs of maintaining buildings, which often require specialist materials and expertise to repair, and the potential economic cost of preserving or adapting a, perhaps, out-dated structure as opposed to replacing it with a more profitable structure. It is in this context that development and profit considerations often outweigh conservation concerns, particularly in economically dynamic city centres, although this is far from being a straightforward equation, with decision making processes highly dependent on the local planning context (Larkham and Barrett 1998). However, increasingly urban conservation activities have been supported by the potential economic benefits that can be gained from using conserved buildings in contemporary developments, adding a so-called 'heritage premium' (English Heritage 2010).

In the commercial context, certain businesses, particularly services such as banking, professional services or entertainment/leisure activities, can derive a certain cachet from occupying conserved buildings, and prestige developments with a conservation element can outperform more mundane modern schemes in terms of rental income. Equally, within residential areas, those areas that are designated as valued or homes that occupy conserved buildings are seen to enjoy a premium in terms of real estate value (Gibson and Pendlebury 2009). Increasingly, old buildings are being reused and incorporated into regeneration schemes, particularly in those cities seeking to reinvigorate their economic fortunes by boosting their service and creative economies (Healey et al. 2002). Here, old buildings and historic urban quarters are revitalized in order to provide the unique selling point for redevelopment schemes to attract new business investment, or to encourage urban tourism and leisure pursuits in an increasingly competitive global economy (see also chapter four). Thus the conservation, reuse and promotion of historic buildings and urban quarters for business and tourism purposes has expanded beyond the so-called tourist-historic gems to include many more towns and cities around the world (Ashworth and Tunbridge 2000; see case study below). However, this potential for a heritage premium has led to concerns about gentrification and exclusion within protected urban heritage areas, with existing businesses and residents who previously enjoyed the lower costs of these older buildings and areas being replaced by wealthier groups with the capacity to meet strict building controls as these areas are transformed into prestige areas.

Case study: urban heritage in Singapore

Complex issues surround the protection of urban heritage in the multi-cultural and fast changing city-state of Singapore. Singapore is better known for its rapid modernization, industrialization and rapidly changing urban landscape, resulting from its key role in the economically dynamic Asia-Pacific region. It has only recently adopted policies to encourage building conservation and the protection and use of its cultural heritage. In 1989 a conservation masterplan was developed which designated a number of ethnic enclaves as Historic Districts. For the government, protection of this ethnic heritage through the conservation and refurbishment of buildings of historical and architectural significance serves to promote ideas of harmony and multi-culturalism within the racially heterogeneous state of Singapore, both to local citizens and to outside visitors who are told through brochures and other tourist media to expect a kaleidoscope of different vibrant cultures that are united as Singaporean.

Henderson's (2008) article focuses on the district known as 'Little India' within Singapore which has traditionally been a centre for the city's Indian ethnic communities. The enclave was designated an Historic District because of its significance as a centre representing the long term presence of these Indian ethnic communities within Singapore, and it remains a leisure and commercial space for these groups. However, a range of other

continued

groups have a 'stake' in the area, including migrant workers from the Indian sub-continent, who use it as a meeting place at weekends and a source of traditional products, tourists, government agencies and private businesses. The author notes that these ethnic urban enclaves are therefore complex entities which serve multiple purposes and which are used in various ways by different groups. These multiple meanings and uses serve to highlight the contested nature of urban heritage management. A key concern has been the impact of alterations to the townscape and commercial and residential life resulting from tourist-led developments. Here there is a concern that the authenticity of the district is being eroded, and its meaning lost for both locals and visitors. While redevelopment activities seek to promote an Indian theme, many developments can be seen to have undermined the original character of the area. Rising rents have seen many original businesses replaced by themed businesses operated by groups beyond the local Indian community, although this has not been as extensive as in some of the other Historic Districts, particularly Chinatown.

A key tension therefore is the pressure for change against that for preservation and continuity which remains a fundamental issue for urban heritage management to address. However, within this, activities such as tourist development do not sit easily on one side or the other but rather can be seen as acting as instruments for both change and conservation, and which can therefore create mixed reactions from local stakeholders to proposed plans.

Source: Henderson (2008)

The rise of urban conservation and heritage concerns has thus had an important impact on the development trajectory of many of the world's cities. Many cities now have large parts of their urban areas, the cores in particular, protected in some way. This has generated considerable debate over the style of new development within these areas. The desire to preserve and enhance their character has been seen to have encouraged the use of contextual designs for new developments and the use of revival styles and pastiche architecture (see also chapter eight). While many architects and planners applaud the use of these styles, others claim that the overuse of bland conservation-area-architecture and absence of modern architecture in these areas is stifling urban change and the development of cities. Property owners, developers and architects can view conservation controls as an unnecessary burden and an infringement of property development rights. Equally, the need for new buildings to meet modern business needs juxtaposed against the desire for conservation often produces uneasy compromises through the planning process, most graphically illustrated by façadism where the fronts of old buildings are retained with new development behind (Larkham and Barrett 1998). The need for conservation within the planning and management of cities has clearly placed new demands on both the planning and development professions. In particular, the ability to negotiate good schemes that harmoniously and profitably blend the old and the new requires specialist skills and knowledge that are often lacking in urban planning teams, particularly in cities of the rapidly urbanizing Global South (Steinberg 1996). However, for many cities, protecting urban landscapes is often a luxury which ranks low on the list of priorities in comparison to economic development concerns and the need to provide basic infrastructure. Here, the lack of funding to support conservation activities is a key concern. In addition, for many cities in post-colonial countries the urban past has been something to be escaped from, with a desire to replace the problematic urban landscapes of the colonial period with new development reflecting independence and modernity (Tunbridge and Ashworth 1996). Therefore, while the conservation of urban landscapes has become an increasingly important part of urban planning around the world, its impact has been far from even on a global scale.

Urban policy and regeneration

Urban policy can be viewed as another area of urban management activity that emerged into the hiatus created by the crisis in modernist planning. As we noted earlier in this chapter, it is difficult both to define urban policy exactly and to pinpoint its exact emergence as a discrete field of professional activity (Cochrane 2007). However, Cochrane notes that some of the first urban policy measures emerged in the 1960s, initially in the US, at the same time as the shine was fading from modernist planning as an arm of the Keynesian welfare state. In its early incarnations urban policy can be seen as an attempt to deal with the problems that had not been solved by previous planning interventions, focusing on social concerns within a welfarist paradigm.

Despite the huge variety of urban regeneration initiatives and policies enacted around the world it is probably true to say that they have all aimed to achieve one or more of the following four goals:

- Improvements to the physical environment (which have more recently come to focus on the promotion of environmental sustainability);
- Improvements to the quality of life of certain populations (for example through improvements to their living conditions or by improving local cultural activities or facilities);
- Improvements to the social welfare of certain populations (by improving the provision of basic welfare services);
- Enhancement of the economic prospects of certain populations (either through job creation or through education or reskilling programmes).

Often regeneration has pursued more than one of these goals within any one policy, programme or project. In many cases policies have been pushed forward on the sometimes questionable basis that causal links exist between some of these goals. For example, at times policies whose explicit aim is economic development, have been advocated on the basis that they will also lead to improvements in the quality of life within localities undergoing regeneration (as in the case of the Urban Development Corporations in the UK in the 1980s and 1990s). Finally, the relative emphasis on the goals of regeneration has shifted over time. For example, it is only since the early 1990s that environmental sustainability has risen substantially up many urban policy agendas. Below we outline a framework to guide analysis of either specific projects of urban regeneration or the more general policies or programmes of which they are part. Depending on the focus of any study and upon the nature of the case being examined, not all questions will be relevant in every case. However, they highlight a number of key themes and issues to consider in the interrogation of urban policy and regeneration schemes.

A framework for analysing urban regeneration: twelve key questions

Urban problems

- What urban problem, or problems, have been identified?
- What has been identified as the cause of the problem or problems?

Policy contexts

- What is the origin of the policy/ programme/project?
- What is the relationship to earlier approaches or those being implemented elsewhere?

Funding

- Where does the funding for the policy/ programme/project come from?
- In what way is funding allocated?

Stakeholders

- Who are the stakeholders involved?
- What are the relationships between the stakeholders?

The nature of regeneration

- In what ways does the policy/ programme/project seek to achieve its aims?
- What are the outcomes of the policy/ programme/project?

Impacts of regeneration

- What are the impacts of the policy/ programme/project?
- In what ways has the policy/programme/ project been evaluated?

Urban problems

Urban policies are designed to address urban problems. However, urban problems do not just present themselves to policy makers; rather they are defined in various ways. The definition of urban problems has an important role to play in the legitimization of urban regeneration and policy. While acknowledging that urban problems are real, as well as rhetorical, their articulation may be thought of as playing crucial roles within urban regeneration discourses. Commentators have recognized that urban problems include various combinations of environmental problems (such as derelict land, redundant industrial capital, inadequate housing stock, pollution and contaminated land), social/cultural problems (a lack of social cohesion within communities, crime, antisocial behaviour, poor schools and other public facilities) and economic problems (long term structural unemployment, a lack of indigenous economic dynamism). Yet, while urban policy makers have long recognized that urban problems are multi-faceted, urban regeneration initiatives have tended to define a particular problem as the root cause of their woes.

In addition, inherent in all urban regeneration programmes is some articulation of the causes of the problems they are seeking to address. It is this articulation of causation that shapes regeneration. Put simply, urban regeneration is designed to address whatever policy makers or practitioners think, or want to believe, is causing the problems they observe. However, as students of urban geography you will know that the causes of urban problems are the subject of some debate! Early urban policies located the problems of urban areas in the failings of the people who lived in them, often referred to as a social pathology approach (Cochrane 2007). Subsequent interpretations problematized the notion that you could draw a line around 'problem areas' and explain their problems by simply looking within them. In the 1970s, structural explanations emerged which suggested that the problems of such areas stemmed from the consequences of economic structural adjustment and a lack of economic dynamism within certain localities. More recently structural explanations have suggested that the persistence of urban problems stems from the failure of these localities to compete within a globalized, post-industrial world economy (Cochrane 2007). Equally, there has also been a recent focus on social exclusion, the lack of social and economic relations between certain localities (for example inner city areas or peripheral housing estates) and the rest of society. However, while a shift towards community regeneration can be seen as a positive move, in recognition of problems with some earlier more economically driven policies, these more recent community focused policies have drawn criticism for their emphasis on tackling the operational failures of local institutions, rather than focusing on more radical interpretations that might see urban problems as stemming from unequal distributions of resources and opportunities (Imrie and Raco 2003).

Policy contexts

Urban regeneration does not emerge within a vacuum; rather it is reflective of a number of contexts. Most obviously it is influenced by initiatives active in earlier periods and other places. One aspect of the analysis of urban regeneration is to situate individual policies or initiatives historically, as part of an ongoing process of policy development through time; geographically, by tracing the influence of policies and practices from elsewhere; and ideologically, as reflective of prevailing political ideologies. However, while it is tempting to characterize urban policy in a series of distinct periods of development, this tends to underemphasize the degree of continuity between periods and overemphasize a clear sequential trajectory to policy development. Cochrane (2007) highlights the complex and interwoven nature of the contexts within which policy develops, offering an alternative approach to the examination of policy development, which traces the key themes and ideas, such as community and entrepreneurialism, which have shaped urban policy. Nevertheless, as a pedagogic device the periodization of urban policy does have some merit when viewed in broad terms, particularly in highlighting the impact of **New Right** ideologies on urban policy specifically in the UK and the US.

In the late 1970s, a new political ideology was beginning to take hold within central government in the UK and the US. Termed the New Right, this so-called neo-liberal ideology challenged the prevailing consensus of the central role of the state in the provision of welfare and other key services such as regeneration. This can be seen as part of wider changes in the **mode of regulation** within cities and shifts towards post-industrialization (see chapter four). Existing urban policy was viewed as an impediment to economic development and the New Right vision was for a market-led pathway to a post-industrial society (Cochrane 2007: 89). The position of the New Right was cemented with election victories for neo-liberal governments in both the UK and the US in the 1980s. This shift in urban policy to adopt a neo-liberal stance emphasized economic development through incentives to, and partnerships with, the private sector. The model of regeneration which emerged during this period drew on developments that had taken place in a number of North American cities dealing with the impacts of deindustrialization, the most notable example being Baltimore, which had pioneered the use of spectacular flagship regeneration projects around its run-down Inner Harbour area since the 1970s (Bianchini et al. 1992: 246). The influence of American developments on urban regeneration in the UK, and subsequently in other countries, is readily apparent with an extensive exchange of ideas and examples from the 1980s onwards.

The promotion of more commercial aims of urban regeneration through central government policy and in local authority practice has been characterized as a shift away from managerialism towards entrepreneurialism. While this characterization does tend to gloss over some continuities in policies between the two periods, it is true to say that the 1980s did see local governments become more concerned with risk-taking and growth-orientated activities in the name of urban regeneration (Hall and Hubbard 1996, 1998; see also chapter five). Subsequent periods have seen both continuities and changes from the 1980s. The private sector is still an important partner in regeneration, although greater access has been granted to the community and voluntary sectors. While further priorities have emerged, there remains an emphasis on the promotion of economic growth and urban competitiveness, increasingly within a global context. Neo-liberal approaches continue to dominate urban policy agendas and strategies, championed as the strategies of choice by global organizations such as the IMF and the World Bank (Cochrane 2007).

Funding and stakeholders

The majority of funding for urban regeneration stems from central governments, although it is also possible to find examples of local authority funds, charity funds, private sector funds and those stemming from supra-national organizations such as the European Union or the World Bank. Whatever the origin of regeneration funding, funders have tended to use one of two models of allocation to decide where their money and resources should go. Traditionally, policy funds were allocated on the basis of the demonstration of need, typically measured by looking at the scale and extent of social or economic deprivation within localities. However, more recently, and linked to the emergence of neo-liberal agendas, the allocation of funding has moved to a competitive bidding process. Under this mechanism funding is allocated on the quality of bids, the economic opportunities in run-down localities and the capacity of local stakeholder coalitions to manage the funding and policy. This was evident in the UK in the 1990s in the Challenge Fund Programme, which represented an attempt by central government to use regeneration funds to foster innovation within deprived localities rather than to simply alleviate need. However, this competitive funding allocation model proved controversial, with questions raised about the amount of time and money spent by local stakeholder coalitions in preparing successive bids, creating 'bidding fatigue', and the loss of funding to severely deprived areas that were unable or unwilling to prepare acceptable bids (Oatley 1998).

The range of stakeholders involved in or affected by urban regeneration varies from case to case, as does the ways in which these stakeholders are involved. Healey et al. (2002), in a study of regeneration in the North of England, recognized a wide range of stakeholders with different relationships to regeneration. Those from the state sector include central and local government officers involved in a range of urban management activities, including transport management, housing provision, regulation, property development, conservation and business development. Within the private business sector there are a number of stakeholders with economic links to those areas undergoing regeneration, including providers of consumer and product services, manufacturers, retailers, leisure suppliers, property developers, landowners and consultants. Finally, they identified a range of stakeholders from within civic society including residents, workers, shoppers, service users, leisure users and civic associations (Healey et al. 2002). To this can also be added a variety of stakeholders from the third sector of charity and voluntary organizations who play an increasingly important role in service delivery and regeneration initiatives. These stakeholders produce complex webs of linkages that include local, regional, national and international links of various types. Mapping the impacts of regeneration through such networks of stakeholders is a complex yet important task.

What is important to note is that there have been shifts in the relationships between key stakeholders over time, particularly between central and local governments, the private sector, community/voluntary organizations and local residents. A key shift has been the increasing influence of business interests within urban regeneration, as a consequence of the emphasis on economic aims and neo-liberal ideologies. Often this has shifted power away from local governments as control has passed to non-elected boards of development agencies. Since the 1980s, policy makers (US and UK) have increasingly advocated the partnership model of urban regeneration, specifically a three-way partnership between local authorities, private business and the community and voluntary sectors. This has sought to address concerns about reduced local accountability. However, the extent to which a genuine partnership of equals exists in policy initiatives has been frequently questioned (see case study below). It is argued that urban regeneration initiatives remain extensively top-down in approach, driven and controlled by international organizations, central governments and big business concerns. While communities have become involved more extensively in regeneration programmes, this involvement is often strictly prescribed and overseen by those 'in power' at the top.

Case study: inner-city regeneration, Salford, UK

Henderson et al.'s (2007) study examines the role of the local state in the regeneration in the English city of Salford. They consider the idea of 'actually occurring regeneration' where broader central government urban policies are mediated by specific local contexts. They suggest that rather than wiping the slate clean, new regeneration initiatives are layered on to geographical areas that are made up of

continued

existing actors, who will exhibit different responses to policies, historical relations and past layers of policy and that as a consequence the implementation of new policies is never straightforward. In particular they note the influence of past relations between local government and central government, the market and the community in shaping the ways in which new initiatives are received and implemented (what they term path dependencies).

They examine these ideas in a case study of the role of Salford City Council in the redevelopment of the inner area of the city over a twenty-five year period from the early 1980s. The city is seen as exemplifying many of the persistent problems associated with the deindustrialization of former manufacturing cities (see chapter four). However, the local city council was viewed as unusual in the English context in its willingness and ability to attract national urban policy expenditure. The most prestigious development was the transformation of the former docks into the Salford Quays development, although this was only one part of the council's wider regeneration programme.

In seeking funding for the regeneration of the docks the local council initially adopted a 'pragmatic' approach, being prepared to engage with neo-liberal market orientated central government initiatives, despite their desire for local-state led development. Despite this central-local tension, the council

were awarded funding as they were viewed as able to deliver initiatives, as a result of past successes, and because they adopted an entrepreneurial stance and aspirational masterplan for the docks which chimed with national government regeneration ideas. Development was driven by the 'vision' of key city council officers and consultants and can therefore be viewed as top-down, with little meaningful input from the local business community or local residents. The authors argue that this created a particular path dependency which shaped subsequent city council responses to the changing urban policy context and the development of new initiatives. This fed through into the use of similar approaches in the development of larger regeneration schemes, specifically the Lowry Arts Centre in the 1990s and the approval for an Urban Regeneration Company (URC) in 2005 seeking the regeneration of a larger part of inner Salford. However, lack of earlier community engagement has created negative path dependencies where local communities have become sceptical about the benefits regeneration schemes will bring to their lives, particularly in terms of the job opportunities created. Consequently, there is a difficulty in engaging the local community despite the requirement for more community involvement specified in recent central government urban policy initiatives such as the URC.

Source: Henderson, S. Bowlby, S. and Raco, M. (2007)

The nature and impacts of regeneration

Urban policy and regeneration initiatives can achieve their aims through interventions in a number of dimensions of localities. These include their natural environment, built environment, local social networks, economy, regulatory framework (including local planning regulations and taxes) and externally perceived image. It is common for regeneration initiatives to address more than one of these dimensions. Having said this it is possible to discern broad shifts in the dimensions of localities that urban policy has sought to intervene in over time. Historically, urban policy, drawing on earlier urban planning roots, was overwhelmingly concerned, although not exclusively, with intervening in the built environment, particularly housing. This occurred either through the provision of new housing or through the refurbishment of elements of the housing stock, bringing them up to certain standards. As urban policy developed in the 1970s and 1980s, so the aspects of localities within which it sought to intervene expanded. Beyond continued interventions in the built environment, urban regeneration became concerned with regulatory frameworks of localities, their externally perceived images and, in a limited way, their natural environment.

Policies drawing on neo-liberal ideology sought to improve the economic dynamism of localities by attracting private sector investments and relocations. This has been achieved in a variety of ways, but particularly by relaxing local planning restrictions and local tax burdens within specified locations, the development of publicly subsidized flagship commercial developments, such as major office developments or convention centres, and addressing the image of run-down localities. In addition, initiatives to clean up contaminated land in former industrial areas and aesthetic improvements through landscaping schemes have also been a key component in efforts to attract economic investment to depressed areas. More recently, the range of interventions has broadened again to include enhanced engagement with the natural environment, linked to efforts to promote environmental sustainability. In addition, policy has also moved beyond a focus on material outcomes to seek to enhance local capacity, defined as the structures and networks stretching between a range of stakeholders within localities. Here, developing this enhanced capacity is deemed likely to sustain regeneration beyond the life of specific projects or funding streams.

So what have been the impacts of over forty years of formal urban policy and regeneration interventions in towns and cities? To what extent have they had positive impacts upon cities, or specific areas within cities? To what extent do pressing physical, environmental, economic, social and cultural issues remain? To some extent, urban policy is part of the various strands of the ongoing processes of urban change and development. It would be unfair to imagine that a successful urban programme would necessarily produce long term, unchanging conditions of stability and prosperity for an area and all its urban populations. Cities are dynamic and ever changing. The external processes within which they are implicated are similarly dynamic. Consequently once an urban policy has run its course it is likely that cities will be facing new sets of local and global challenges. Having said this, it is possible to recognize both areas in which the impact of urban policy has been significant, and also the persistence of long term seemingly deeply embedded urban problems where urban policy has, on a large scale at least, appeared to make less of an impression. Michael Carley, writing about the impacts of urban policy on deindustrialized cities in the UK, argues that their regeneration has been partial: 'while the nation has become better at property-led regeneration, it has not cracked the hard nut of helping households disadvantaged by long-term unemployment or the inability to work' (Carley 2000: 273). A similar view could be taken of policy initiatives undertaken in urban areas in other European countries, the US and other parts of the world. As Cochrane (2007: 4) notes, acknowledgement of the lack of success of urban policies in achieving their stated or implicit aims is an unfortunate area of seeming consensus among academics and professions.

Future challenges

It is perhaps too harsh a statement to say that urban planning and policy is in crisis at the beginning of the twenty-first century, although it is perhaps fair to say that there is some debate and disagreement around the directions we should be going in the future. The twentieth century ended with a certain amount of disillusionment over the abilities of urban planning and policy to deal with the ills of the city. Despite over one hundred years of intervention, in the case of some of the older industrial cities of the west, problems of urban poverty have proved impossible to eradicate. Equally, the growth of the new megacities of the Global South has thrown up new urban challenges. Hall (2002) contends that we are perhaps back where we started in trying to deal with the concentration of problems in the giant city. However, the global context for addressing these problems is clearly different from the early twentieth century.

There is, however, a certain agreement around the urban challenges facing us in the twenty-first century. Equally, despite the range of caveats considered above, there appears to be a continued commitment to a role for urban planning and urban policy in addressing these challenges. UN-HABITAT (2009: xxii–xxiii) identifies five main challenges for urban planning and policy to deal with. First, there are demographic challenges, particularly the problems associated with rapid urban population growth within the world's megacities. Second, there are environmental challenges, specifically issues associated with climate change and the declining availability of resources, particularly fossil fuels (see also chapter thirteen). Third, there are economic challenges resulting from the impacts of global economic restructuring and economic crises on city economies and urban labour markets. Fourth, there are socio-spatial challenges, specifically the increasing social and spatial polarization of urban populations and the increasing marginalization of poorer groups. Finally, there are institutional challenges and a recognition of the need for new and more responsive and inclusive forms of urban governance.

The overall conclusion of the UN-HABITAT (2009) report into urban planning for sustainable cities was that planning systems in many parts of the world were not up to meeting the challenges of the twenty-first century and needed to be revisited. As noted above, in many cases planning had contributed to problems of exclusion and marginalization rather than helping to resolve these issues. Yet, the report notes that there is no one planning vision that addresses the problems for all cities in the world and consequently revised planning systems need to be embedded within, and be responsive to, their local contexts (UN-HABITAT 2009). However, there are some common themes that urban planning systems need to address in rising to meet the challenges of the twenty-first century. A key issue, raised at the beginning of this chapter, is the relationship between planning and the market. Recent economic crises have resulted in less certainty in the growth and strength of the private sector and there is a certain consensus that governments have a key role to play in leading development initiatives and ensuring basic needs are met through some regulation of market forces. However, to be effective urban planning needs to be clearly embedded within urban governance functions, and able to play an integrating role in terms of coordinating different functions, levels of government and stakeholders in a flexible and responsive way. Yet, fully integrated spatial planning systems remain more in theory than in the practice of urban planning around the world. Equally, planning systems need to be able to accommodate the plurality of views and agendas that exist within cities and particularly to incorporate the views of those groups often marginalized by the planning process.

Just as there is some consensus over the challenges facing cities and the broad planning frameworks needed to address these, there is also an emerging global consensus in terms of the overall policy approach to managing cities in the twenty-first century. As Cochrane (2007) notes, the language and politics of neo-liberalism have been actively promoted by global development agencies and through global policy networks. He notes that global agencies such as the World Bank, UN-HABITAT and OECD have played a key part in this process, adopting policy stances that assert the key role that cities play in enabling further development as growth engines, seeing cities as having a positive role, rather than merely being viewed as a burden. Equally they have emphasized the goal of urban sustainability and the importance of social capital and its development. For these agencies the task is then to remove obstacles to this growth through good governance and the empowerment of urban populations. Policies such as slum upgrading in some cities of the Global South reflect this 'capacity building' stance (Cochrane 2007; see also chapter eleven). However, for some the capturing of the urban by the agenda of neo-liberalism and competitiveness is equally problematic, offering a further straitjacket that is neither locally responsive nor inclusive (Peck and Tickell 2002). Alternative policy visions based on the potential of cities to provide a platform for social and political engagement and to foster participation and the negotiation of issues among urban societies remain mainly the preserve of academic writing (see Sandercock 2000; Amin and Thrift 2002; Healey 2004). The challenge remains to illustrate whether small scale experimental initiatives adopting these approaches can be developed into mechanisms and practices that can empower urban populations to meet the challenges facing cities in the twenty-first century.

Summary

This chapter has sought to provide an overview of the role of urban planning and policy in shaping the city and in mitigating the problems within cities. Both urban planning and urban policy are difficult terms to define and in practice they encompass a broad range of issues, approaches, agents and activities which have varied over both time and space. Cities have been viewed as unruly places that need to be managed, and in the nineteenth century the rise of large cities presented problems and challenges that urban managers sought to mitigate. Urban visions of the 'good city' developed by a number of key thinkers drove the development of the modern urban planning profession in the twentieth century. However, by the end of the twentieth century the rise of new economic and social challenges facing cities, and disillusionment with the approaches employed by planning and in its ability to solve urban problems, led to the reconsideration of planning and policy agendas. In the late twentieth century, key changes have been a move from comprehensive redevelopment to greater conservation and rehabilitation of urban environments and a shift from social problems to economic competition as a focus for urban policy. In the twenty-first century the key challenges facing urban managers are sustainability, economic competition, social polarization and demographic shifts. The challenge for urban planning and policy organizations is to develop approaches to tackling these issues that are both locally responsive and socially inclusive.

Follow-up activities

Essay title: 'Making cities competitive internationally is the main priority facing urban planning and policy makers in the twenty-first century.' How far do you agree with this statement?

Commentary on essay title

An effective answer would highlight the problems facing cities in the twenty-first century and provide an overview of the different planning and policy approaches applied to address these issues over time. In particular it would highlight the increasing significance of policies aimed at making cities competitive for urban planning and policy agendas and comment on whether this is a necessary and desirable approach for urban managers to adopt. An excellent answer would move beyond this to consider the deeply contested nature of urban planning and policy approaches and would critically consider the issues surrounding the application and impacts of policies, the wider structural constraints on policy makers and the influence of local 'path dependencies' in shaping 'actually occurring regeneration', highlighting the significance of local contexts.

Project idea

Look at a range of planning documents or urban policies over the last forty years for a city with which you are familiar. In what ways have the identified urban issues/problems and the approaches to planning and urban management changed over this time? Use prompts from the discussion of the development of urban planning and the framework for the analysis of urban policy to help you.

Further reading

Books

- Bandarin, F. and van Oers, R. (2012) *The Historic Urban Landscape: Managing Heritage in an Urban Century*, Chichester: Wiley Blackwell
  Offers a comprehensive overview of the intellectual developments in urban conservation and management of historic landscapes, drawing on examples from around the world.

- Couch, C. (2016) *Urban Planning: An Introduction*, London: Palgrave Macmillan
  Student friendly introductory text on urban planning. Wide-ranging and internationally-focused, it addresses a range of key urban issues currently facing planners.

- Edwards, C. and Imrie, R. (2015) *The Short Guide to Urban Policy*, Bristol: Policy Press

Concise, but wide-ranging book which critically examines the multiple ways in which urban problems have been defined and addressed in different places at different times.

- Fainstein, S.S. and DeFilippis, J. (eds) (2016) *Readings in Planning Theory*, 4th edn, Chichester: John Wiley & Sons
  A classic edited collection providing a range of key readings about planning ideas and approaches.

- Hall, P. (2014) *Cities of Tomorrow: An Intellectual History of Urban Planning and Design since 1880*, 4th edn, Chichester: John Wiley & Sons
  A classic overview of the development of modern planning traditions and the people behind urban visions of 'the good city'. Reflective and critical in considering these legacies.

- UN-HABITAT (2009) *Planning Sustainable Cities*, London: Earthscan
  Provides a comprehensive review of the development and practice of urban planning around the world and reviews some of the key challenges for cities and urban planning as a profession in the twenty-first century.

Journal article

- Albrechts, L. (2013) 'Reframing strategic spatial planning by using a coproduction perspective', *Planning Theory*, 12(1): 46–63
  Considers how spatial planning might be reshaped to deal with the problems and challenges societies are facing in an innovative/emancipatory and transformative way.

- Chu, C.L. (2015) 'Spectacular Macau: visioning futures for a World Heritage City', *Geoforum,* 65: 440–450
  Considers the tensions in redeveloping Macau between conservation of its colonial heritage and the development of spectacular fantasy landscapes associated with the growing casino industry.

- Craggs, R. and Neate, H. (2017) 'Post-colonial careering and urban policy mobility: between Britain and Nigeria, 1945-1990', *Transactions of the Institute of British Geographers* 42(1): 44–57
  Provides an in-depth insight into urban policy mobility between the Global North and South through the concept of 'careering', tracing the career path of one colonial administrator and his influence on the transfer of ideas.

- Harris, A. and Moore, S. (2013) 'Planning histories and practices of circulating urban knowledge', *International Journal of Urban and Regional Research*, 37(5): 1499–1509
  Paper providing a useful overview of some of the recent research into the transfer of urban policy and planning models, ideas and techniques.

- Ripp, M. and Rodwell, D. (2015) 'The geography of urban heritage', *The Historic Environment: Policy & Practice*, 6(3): 240–276
  Comprehensive overview of developing interest in urban heritage management. Principally focusing on the European context, but also examines recent UNESCO approaches and policies. A companion paper in the following issue of the journal addresses the governance of urban heritage.

- Theodore, N. and Peck, J. (2012) 'Framing neoliberal urbanism: translating "common sense" urban policy across the OECD zone', *European Urban and Regional Studies*, 19(1): 20–41
  Explores the evolution of urban policy discourses among advanced industrial nations in the period since the early 1980s focusing on a case study of the Organisation for Economic Co-operation and Development as a 'transfer agent'.

- Wensing, E. and Porter, L. (2016) 'Unsettling planning's paradigms: towards a just accommodation of Indigenous rights and interests in Australian urban planning?', *Australian Planner*, 53(2): 91–102
  Building on the need for urban planning to deal more effectively with difference and diversity, the paper explores the contentious issue of Aboriginal land rights in the planning of Australia's cities.

Websites

- www.planning.org/ – Website of the American Planning Association, the American professional planning association.

- www.gdrc.org/ index.html – The Global Development Research Centre, an online network of researchers. The website contains many useful links to other sites.

- www.rtpi.org.uk/ – Website of the Royal Town Planning Institute, the British professional planning association.

- http://whc.unesco.org/ – Website of the UNESCO World Heritage Centre, which contains lots of useful information about heritage conservation in cities.

- www.unhabitat.org/ – UN-HABITAT website, key UN organization concerned with urban settlements and their planning and development.

⬤7 Cities and culture

Introduction: the cultural turn and urban geography

Human geography, like other social science disciplines, is dynamic, characterized by fundamental shifts in theoretical orientations, methodological practices and subjects of enquiry (chapter two). A recent widespread change that has affected disciplines across the social sciences has been the 'cultural turn'. This has seen an engagement with a variety of cultural issues, an interest in a range of theories that we might broadly term post-modern that have helped to shift the prevailing analytical framework of the social sciences from their Marxist-structuralist orientation, and the increasing employment of qualitative research methods. Human geography has certainly been influenced by this change and we can detect two effects of this cultural turn within the discipline. First is the rise and growing influence of cultural geography from the early 1980s onwards (Jackson 1989; Crang 1999; Anderson 2010). Second is a transformation, to varying degrees, of the other sub-disciplines of human geography whose engagements with cultural issues had, in some cases, been patchy, often marginal previously. This is perhaps best exemplified by the changes witnessed in economic geography since the mid-1990s (Coe et al. 2007).

Urban geography's involvement within the cultural turn is a little less emphatic. While there has certainly been a growth in the number of cultural studies of cities and studies of the cultures of cities, these scholars and researchers have tended not to identify themselves primarily as urban geographers. Many would see themselves as more cultural than urban geographers (see for example, Crang 2000; Pinder 2005, 2008). Further, some of the most influential contributions to the studies of the cultures of cities come from beyond human geography entirely, including sociology (Zukin 1995) and planning (Sandercock 1998, 2003). What urban geography seems to lack, despite the timeliness of much of this work, is a sense of what exactly constitutes, what we might call, a cultural urban geography. Recent work on cities and culture, then, has something of a peripatetic quality. Given this, this chapter aims to provide a framework that will help you think through, organize and understand these diverse literatures. In doing so it will also aim to provide a guide to the many and varied spaces of culture within the city. It aims to provide a framework that will allow you to conceptually map these spaces. Hopefully, it will allow you to both navigate the literatures on cities and culture and to go out and undertake your own field research in this area. Before we pursue this though it is worth pausing and considering the reasons for the growth in interest in the cultural aspects of cities. Broadly we can attribute this to a series of changes that were occurring across cities and within academic disciplines that can be gathered together in their different ways under the banner of post-modernism.

Culture seemed to become much more central to understanding processes of urban change than had previously been the case. This centrality of culture to urban change has been manifest in a number of ways. First are apparent changes to the social geographies of cities and the implications that these changes have for the cultural landscape of urban areas. It is an almost universal characteristic of contemporary cities that their social geographies are marked by greater cultural diversity and an intensification of juxtapositions of difference than was the case in the past (Zukin 1995; Watson 1999; see chapter five). The collages of diversity that characterize contemporary cities are composed of different cultural groups each with differing combinations of norms, values and lifestyles. This is a reflection of the fact that 'publics have become more mobile and diverse' (Zukin 1995: 3) than has been the case in the past. This increased cultural diversity brings with it the potential for change, the emergence of new ideas, identities and ways of living but, at the same time, for new forms of conflict to emerge in the city. The challenges posed by and potentials of the greater cultural diversity of contemporary cities have risen sharply up both political and academic agendas (Sandercock 1998, 2003; Binnie et al. 2006).

On another level culture appears to be becoming increasingly significant to the development and planning of cities and to their economies. We have seen, for example, the revalorization of many urban spaces through various forms of cultural 'labelling'. These have been crucial dimensions of the extensive physical redevelopment of cities in the last thirty years. We can recognize, for example, the increasing importance of image to processes of urban change, most notably through the process of place promotion (which we discuss at length in chapter nine). This has typically involved the reimagining of whole cities through the composition of spectacular collages drawn from the inventory of cities' internal spaces and recast to fit in with prevailing fashions. Often these are explicitly constructed around the supposed cultural resources of cities, drawing on both aspects of high culture (opera, ballet, art galleries) and sanitized aspects of popular culture. In addition to this, many physical changes to urban landscapes have sought to emphasize or enhance their cultural value. This includes the increasing deployment of spectacular architecture and flagship developments across urban landscapes. Often the focus of these developments is cultural, as in Barcelona's Museum of Contemporary Art (see chapter five) or Bilbao's Guggenheim Museum (see chapter eight). Heritage, the appropriation of aspects of the cultures of the past, has also emerged as a significant and much debated force shaping the landscapes of contemporary cities (Jacobs 1992, 1996, 1999; Samuel 1994; Wright 2009). Urban landscapes then have been re-dressed to emphasize marketable aspects of their histories either to paying customers or the urban citizenry more generally. Often, where it is thought that urban landscapes lack enough marketable culture it has been injected or emphasized through extensive programmes of public art (Hall 1997, 2003, 2007; Miles 1997). These initiatives have been facilitated through shifts internationally in planning practice and policy. Policies such as the designation of cultural quarters have been designed to explicitly emphasize the cultural qualities and distinctiveness of cities or spaces therein (Bell and Jayne 2004).

Culture seems also to have become increasingly central to processes of economic change in the city. The cultural and creative industries have come to play ever more significant roles in the economic development of cities recently. Richard Florida (2002), an urban theorist who has undertaken extensive research into the role of creativity in urban change, has labelled this a transition between an industrial age and a creative age. It is estimated, for example, that approximately forty per cent of people employed within the UK and US economies now work in what might be classed as 'creative' industries (cited in McEwan 2008: 274). The definition of creative industries is broad, encompassing the arts, media and science, but this emphasizes the significance of knowledge, intelligence and cultural production to the economies of urban spaces. Whereas once urban and economic geographers looked at location and change in manufacturing industry and assessed its significance to cities, now they are as likely, if not more so, to look at these issues in relation to advertising, design, film making or the music industry (Scott 1999, 2002; see also chapter four).

Recent changes in the social sciences have been in tune with the growing importance of culture to understanding contemporary urban and social change noted above. Broadly, grouped under the banner of the post-modern challenge or critique, these disciplinary changes have highlighted the potential of culture as a lens through which we might examine and understand the world. This post-modern challenge to the social sciences has encompassed issues around the social construction of knowledge, the plurality of voices, knowledges and experiences characteristic of contemporary society and their cultural groundedness. This has necessitated engagement with the cultural plurality of the contemporary urban scene. This has been partly an acknowledgement of the tendency of earlier accounts of the city to erase or ignore this, partly a celebration and partly recognition of the potential of alternative voices and perspectives to destabilize the meta-narratives and totalizing accounts that had dominated earlier accounts of the city (see Robinson 2002, 2005a).

Mapping the spaces of culture in the city

The discussion above should have highlighted that culture is complex and multi-faceted, meaning different things in different contexts. This complexity has long been recognized by cultural theorists and noted, for example, by Raymond Williams (1953, see also Deffner 2005). If, therefore, culture, in the context of the city, can encompass everything from everyday practices such as taking a pet dog for a walk to the construction of a new opera house, how can we begin to make sense of it conceptually? Our starting point is the axiom that has emerged from a long tradition of cultural analysis and research that culture can be equated with ways of life. It is what people do, how different groups of people do different, or perhaps similar, things. It is the norms and values that people hold and the ways they are expressed through their lifestyles. The cultural geographer James Duncan provides a useful definition:

> Most people consider culture something that we simply 'have' because we are born into a particular culture … that existed before we came into the world and will continue long after we die. It is precisely this collective quality of culture that makes it appear to be something external to us as individuals. It can be argued, however, that it is analytically more useful to think of culture as something that we actively (re)produce rather than something external to us.
>
> (Duncan 1999: 54)

We want to move now to look inside the Pandora's Box of culture a little and try to impose a little order on this seemingly amorphous realm. In the remainder of this chapter we aim to consider the range of literature that has been produced on urban culture but to do so by categorizing it into a number of themes. In doing this we hope to outline the main dimensions of culture as manifest in an urban context. This will provide you with a basis either to explore individual aspects of urban culture or to examine the cultural geographies, in all their diversity, of particular urban areas. We want to do this by considering five domains of difference. These encompass the ways in which different cultural groups are viewed or valued within society, the nature and development of different cultures and cultural groups within cities, interaction between different cultural groups, the relationship between cultural groups and the materiality of the city and finally the relationship between cultural groups and issues of space, scale and globalization.

The conceptual framework that we use to discuss culture is summarized in table 7.1. Each of the domains of difference that it sketches out is discussed in the subsequent sections in some detail. They all represent continuums between two poles rather than mutually exclusive categories. Post-colonial critiques of culture remind us that we should be wary of binary classifications of culture

Table 7.1 *Domains of cultural difference*

Cultural aura	–	Cultural other
Manufactured cultures	–	Organic cultures
Cultural diversity	–	Cultural homogeneity
Cultural hybridity	–	Cultural exclusivity
Material cultures	–	Immaterial cultures
Local cultures	–	Global cultures

(Valins 2003: 160). We cannot reduce something as complex as culture to a simple either/or. Even attempting to locate the urban cultures that you encounter along the continuums outlined below is too crude a measure of culture. We would ask that you recognize that the cultures that you encounter typically straddle categories or span continuums. The spaces of cultures in the city contain complex mixtures of the different domains outlined in this framework. In fact, the most interesting work that you could do would be to explore the 'betweenness' of urban culture, the ways in which it contains and blends aspects from either end of any particular domain of difference. Try not to be too mechanistic in your thinking about urban cultures but rather be flexible and responsive. We illustrate how these domains of difference might be used to negotiate the complexity of urban cultural geographies in the subsequent discussions.

Valuing culture: cultural aura/cultural other

Traditionally analysis of culture has drawn a distinction between supposed 'high' and 'popular' cultures, the night at the opera or ballet versus the night in a bar. This is a distinction less employed now because of its elitist overtones, but one that we would not like to dismiss entirely. For us this distinction reminds us that different ways of life, and the activities associated with them, are valued very differently within society. Analytically we find attitudes towards different aspects of culture a useful starting point in understanding the geography of culture within cities. Rather than high and popular culture we would like to make a distinction between cultural 'aura' and cultural 'others'. The former consists of aspects of culture to which great value is attached within societies. Obviously, these will vary greatly between different societies. These, in the context of the Global North, might include classical music, ballet, gallery art, fine dining, as well as elements of contemporary glamour such as fashion and designer cultures. They might also include more general lifestyles that are deemed desirable or the objects of aspiration, such as those of middle and upper class bohemia. These, for example, are epitomized by media personalities, models, actors and musicians such as Alexa Chung, Sienna Miller, Kate Moss and Zooey Deschanel (at the time of writing the ultimate embodiments of the fashionable 'boho-chic' lifestyle/look). Cultural 'others', by contrast include groups to whom this cachet is not attached or who are routinely demonized in public, media or political discourse. Cultural studies of the city, particularly those within sociology and British cultural studies, are replete with examples of **sub-cultures** that have become the objects of moral panics, such as muggers, mods and rockers, punks, as well as numerous ethnic minorities (Hebdige 1979; Gelder 2005, 2007, see also Jackson 1989 for a distinctly geographical take on this issue).

Exercise

Can you recognize contemporary cultural groups, subcultures or lifestyles that are demonized? Monitor the media for a short period and try to collect examples of this, noting the characteristics of these groups and attitudes towards them. Why do you think these groups are demonized in this way? Does it translate into actions against them, for example, specific policies or policing measures? How are these groups manifest in an urban setting? Do you think these examples are unique to the current period or part of more historically deep seated attitudes towards different groups?

The case of subcultural groups reminds us of the spatial and temporal dynamism and contingency of cultural attitudes. Gay men, for example, while until recently demonized, targeted and repressed within cities are now frequently a celebrated element of a city's cultural landscape. This attitude, however, is far from universal. In many societies elements of this repressive, sometimes violent attitude towards gay men remains (Aldrich 2004).

Manufactured and organic cultures

When thinking about culture in the city we can also make a distinction between the manufacture and manipulation of culture for some external economic or social developmental end, and organic cultures. In some cases the distinction between the two is clear cut and unproblematic. The building of a new opera house, for example, is a case of the manufacture, or at least manipulation of culture, the provision of finance or resources to promote the development of that cultural activity. The motivation for the development of cultural facilities such as this is often not just altruism or the belief that classical music is inherently a good thing. Rather it might be bound up with the cultural economy of the city, part of its cultural policy or strategy, whose aims are in part to boost the city's economy through visitor spend and perhaps to enhance its externally perceived image. On the other hand, groups of Latin American immigrants who meet in public parks in cities such as Los Angeles to play soccer (Hamilton and Chinchilla 2001), which over time become formalized into leagues and associations is an example of a more organic cultural formation. Its development is primarily not driven by any external agency (although clearly, we should not be innocent of the globalization of soccer and the commercial interests that are bound up in the promotion and spread of the game around the world). In other cases though, the distinction is less clear, less easy to make. Taking the case of gay men again, communities may have grown up organically within specific city districts such as The Castro in San Francisco, Chelsea in New York and around Canal Street in Manchester, developing their own independent facilities, institutions and events. However, at the point when a local authority recognizes the contribution of its gay community to its economy and its externally perceived image, and supports, subsidizes, promotes and expands its Pride festival, the distinction between organic and manufactured culture blurs somewhat.

Organic cultures are those that tend to develop without the stimulus of any external organizing institution, initiative or policy. The processes of migration and the clustering of groups based upon some common characteristic such as ethnicity or lifestyle can underpin to some extent the organic development of cultures within cities. Essentially, these cultures arise from the bottom-up. However, one should avoid the danger of romanticizing the autonomy and agency of cultural groups. Few such groups, with the possible exception of the super-rich (Beaverstock et al. 2004) are simply free to choose where they might settle. The geography of cultural groups in the city is often the product of the operation of power through the combined economy and bureaucracy of the city along with wider social attitudes towards different groups of people. The ethnic enclaves that grew up in British cities from the 1950s onwards, for example, were not simply a reflection of the operation of mass choice by these groups, rather they reflected the post-war boom in the industrial economy that necessitated a supply of cheap manual labour. In addition, this was complemented by institutional racism among a range of urban gatekeepers, exemplified by the notorious practice of 'redlining' by mortgage lenders (see chapter eleven). Similar racist attitudes were reflected in the practices of local authorities and private landlords (Rex and Moore 1974; Rex and Tomlinson 1979; Sarre 1986; see also chapters two and eleven). However, the case study below of ultra-orthodox Jews living in Britain reminds us that groups can also choose to draw boundaries around themselves as a form of resistance or defence. Here again, then, it is important to look beyond the surface of seemingly organic cultural formations and to excavate the processes that gave rise to them.

The manufacture or deliberate development of culture, typically the highly-valorized cultures discussed in the preceding section, has emerged as a key strategy for cities attempting to ensure their ongoing economic dynamism. Cities have enacted numerous policies and practices designed to sustain or enhance the vitality of their cultural, symbolic or creative economies (Zukin 1995; see also chapters four, six and eight). This has included creating spaces for culture, for example the provision of major cultural facilities such as art galleries and concert halls, and policies, plans and strategies that include the designation of spaces and cultural quarters. Typically, these are a characteristic of the most entrepreneurial or aspirational cities and the prime motivation for them, notwithstanding that they may also contribute to the quality of life for citizens, is economic. However, caution is needed again here so as not to miss some of the complexity within these seemingly manufactured spaces of culture. Sharon Zukin's (1989) seminal discussion of the gentrification of SoHo in New York City reminds us of this. What started off as an organic clustering of artists using the cheaply available former industrial buildings as studio and living spaces became ever more institutionalized through city policy and the real estate industry as more wealthy professionals were attracted in increasing numbers to the fashionable bohemian buzz created in the area. This model of urban upgrading has subsequently proven the inspiration for policies aimed at the rehabilitation of former industrial spaces across the globe. Here again, what might appear at first glance to be clearly manufactured cultural spaces, on closer inspection reveal richer, more complex local and global histories.

Mongrel cities: cultural diversity/homogeneity and cultural hybridity/exclusivity

We want to consider the next two dimensions in our framework (diversity/ homogeneity and hybridity/exclusivity) together as we believe they are very closely related. As we noted earlier, increasing cultural diversity is a defining characteristic of contemporary cities, acknowledging of course that processes of exclusion and segregation are still very much part of the urban scene. Leonie Sandercock, who has written extensively around issues of planning in, and for, multi-cultural cities, has described the contemporary urban world as characterized by 'mongrel cities', cities in which 'difference, otherness, multiplicity, heterogeneity, diversity and plurality prevail' (2006: 38; see also Sandercock 1998, 2003). This cultural diversity brings with it the potential for innovation, mixing and intercultural dialogue. This is perhaps best demonstrated through the example of popular music. New forms of popular music almost always emerge out of a specific location. Examples include modern jazz from the black inner city neighbourhoods of Chicago in the 1920s, reggae music from the Jamaican townships and hip-hop and rap from neighbourhoods such as Harlem in upper Manhattan and the Bronx in New York City in the 1970s. Often these innovations are the product of a mixing of different musical and cultural traditions within specific localities. Bristol, a large city in the South West of England, gave birth in the late 1980s and 1990s to a globally popular brand of music known generally as 'trip-hop'. It was most associated with groups and performers such as Massive Attack, Portishead, Tricky and Roni Size. This unique sound developed out of the mixing of musicians from the city's active reggae scene and its white working-class punk music scene. The latter was characterized by a more melodic, multi-instrumental form of punk than that which was developing, primarily in London, at the time. Other influences that were brought together in Bristol's musical melting pot included Bollywood film music, jazz and hip-hop. This cultural milieu gave rise to a form of slow, dub-heavy hip-hop initially unique to the city but which became global in its popularity following the success of various members of this scene. Hybridity, then, the emergence of new cultural forms, that sometimes challenge or destabilize existing norms and authorities, is a widely-recognized potential of the diverse cultural landscapes of contemporary cities (Knox and Pinch 2010: 45).

This diversity, however, can also give rise to challenges and more negative outcomes. The nature of these challenges and possible directions towards their resolution have been most comprehensively explored in the previously cited work of Leonie Sandercock (1998, 2003, 2006). Her work starts from the point that mongrel cities, as she terms them, are an undeniable reality and that it is an obligation of both the city building professionals and citizens to respond to the challenges those cities present. The problems associated with these mongrel cities that we might be most aware of are those that appear regularly in the media. These include civil disturbances, or 'race riots' as they are typically presented, the election of extreme right wing local politicians, protests or violence against immigrant communities, institutions or buildings, or media campaigns against policies or initiatives designed to foster multiculturalism or integration.

However, these incidents and attitudes do not arise within a vacuum. Rather, they emerge out of deeper, widespread social attitudes. Drawing on the work of a range of key social theorists, Leonie Sandercock highlights the more subtle problems of mongrel cities that underpin their more visible manifestations. Richard Sennett (1994), for example, has argued that mongrel cities tend to lack civic culture. Despite the impression that might emerge from the media, mongrel cities tend to be overwhelmingly characterized by tolerance, rather than hostility, between different cultural groups. This is not to deny that intercultural hostility does exist, but rather to argue that this is not the defining characteristic of contemporary cities. While tolerance is undeniably preferable to hostility, it tends, in contemporary cities of the Global North, to be a kind of indifferent tolerance rather than an active engagement with cultural diversity. This is a form of indifference across which very little intercultural dialogue takes place. This represents a form of stasis in which cultural groups are predominantly inward looking, little common cultural ground emerges between groups, few meaningful cross-cultural institutions develop and little cross-cultural political capacity emerges through which cities can be shaped in desirable ways. The challenge facing these mongrel cities then is less concerned with stamping out intercultural violence and hostility, desirable as this may be, and more with finding ways to foster solidarity between different cultural groups (Calhoun 2002: 108, cited in Sandercock 2006: 39). Richard Sennett's work argues that this involves engaging with 'the challenge of living together not simply in tolerant indifference to each other, but in active engagement' (Sandercock 2006: 40).

Sandercock recognizes that some of the most useful pointers towards more convivial futures for mongrel cities come from the work of Ash Amin (2002b). She argues, for example, that Amin demolishes several popular policy responses to the problem of fostering multi-culturalism. One such response involves the promotion of intercultural encounters in the public spaces of the city. This is typically manifest in design-led initiatives which aim to provide public spaces within which the sense of the city as a shared resource can be fostered through encounters between different cultural groups. In reality though, Sandercock argues, drawing on Amin's observations, this utopian potential is rarely realized:

> The depressing reality, Amin counters, is that far from being spaces where diversity is being negotiated, these spaces tend either to be territorialized by particular groups (whites, youths, skateboarders, Asian families) or they are spaces of transit, with very little contact between strangers.
>
> (Sandercock 2006: 44)

Exercise

Do you think Sandercock is right to characterize the public spaces of contemporary cities in this way? Undertake a period of observation of a prominent urban public space. To what extent do your observations mirror those of Sandercock and Amin, above? Is the space dominated by a single group or a number of groups with little or no contact between them? Is it used as a space of transit, and if so, what are the implications of this for the groups who use it? On the basis of your observations, to what extent do you agree with Amin's view that 'the city's public spaces are not natural servants of multicultural engagement' (2002: 11, cited in Sandercock 2006: 44)?

Amin's observations were made in the context of British cities. To what extent do they apply to the public spaces of cities in other countries? If you get the opportunity, undertake a comparative exercise in a non-British city. What differences do you observe from your original study? Do Amin's observations apply here? How can you account for any differences that you observe?

Rather than pursuing the conventional design-led policies that have characterized governments in countries such as the UK in recent years (see Rogers 1999), Amin proposed an alternative policy focus around the potentials of 'micro-publics'. These are settings such as workplaces, schools and colleges, community centres and sports clubs. They are contexts in which it is sometimes difficult to avoid intercultural dialogue and negotiation, albeit often around mundane matters or 'prosaic negotiations' (Sandercock 2006: 44). These settings, Amin argues, offer contexts within which cultural differences can be bridged around a recognition of shared interests and presences and common goals. These micro-publics at least would seem to offer the capacity for the types of dialogue necessary to produce more convivial cities.

Amin and his supporters, such as Sandercock, are keen not to idealize these micro-publics or their potentials, however. They recognize that desirable dialogues will not just happen in these settings, rather this is something that needs to be shaped and fostered. While there are undoubtedly examples of micro-public settings that have been successful in this regard, it is easy to be sceptical about their potential. Sports clubs, for example, are equally likely to be settings in which reactionary attitudes are fostered, incubated in cultural milieux within which difference is rarely encountered, except perhaps in the form of opposing teams. For example, in some English cities such as Birmingham, it is common to see Asian cricket teams and leagues running in parallel to the official league structure. These are formed partly as a result of perceptions of exclusion of Asian players from some clubs. Similarly, the existence of long-standing 'ethnic minority' cricket and rugby teams such as Bristol West Indians, Nottingham Cavaliers, London Welsh and London Irish, reflects the historical role of some sports clubs as centres for the consolidation of ethnic identities rather than cross-cultural dialogue. While these teams are generally much more mixed now than was the case in the past, the result of various national sports policies, the reality of many sports clubs seems to undercut somewhat the potentials that Amin and others outline for them. Equally, empirical research has uncovered instances of social segregation and territoriality that enhance senses of difference among student bodies, as typical of many education settings (Nairn 1999; Buckley 2010). This is the staple of many high-school film and television dramas of which the most original and extreme was the 1989 film *Heathers*. This sceptical assessment is not to dismiss the potential of these micro-publics entirely. Given the failure of more conventional policies aimed at, for example, housing and public space, these settings offer valuable alternatives through which to develop policy. However, it would be naïve to imagine that they are any more immune to the politics of indifference and hostility than any of the mongrel city's other spaces and settings.

While the norm within contemporary cities might be a tolerant indifference to other cultural groups we can also find examples of cultural exclusivity, the active rejection of diversity, multi-cultural contact and dialogue and hybridity. Commentators have referred to this as the presence of stubborn identities: 'individuals and groups who are actively seeking to resist the effects of, as they see it, global homogenization and the loss of distinctive and traditional ways of life' (Valins 2003: 160).

Examples include political extremists and ultra-orthodox religious groups (see case study). While not 'stubborn' in the same way as these groups, or for the same reasons, a desire for cultural exclusivity is commonly found among the very wealthy and super-rich (Beaverstock et al. 2004). Although mobility can be an important component in maintaining the exclusivity of the super-rich, typically the identities of many groups are strongly grounded in specific spaces. They are apparent in the residential concentration typical of some orthodox religious groups (Valins 2003) or the landscapes and security infrastructures of gated residential communities of the wealthy (Brunn 2006). Despite their differences these groups are concerned with security, the rejection of difference and the preservation of their lifestyles. The case of the wealthy, however, does not involve a total rejection of cultural otherness. Wealthy residents routinely emerge from their gated communities to engage with cultural otherness through business, travel, the arts and cuisine, for example. However, these are engagements of inequality and very much on the terms of the consumers where cultural otherness is rendered non-threatening, often highly sanitized. Chance encounters with unknown cultural others tend to be planned out of the lives of the very wealthy and are seen as something to be feared and avoided rather than embraced. This was a theme central to Tom Wolfe's highly successful book and subsequent film *The Bonfire of the Vanities* (1987). The contemporary desire for cultural exclusivity among the very wealthy is also a recurrent theme in the science fiction of J.G. Ballard and forms a central concern of novels such as *Cocaine Nights* (1996) and *Super-Cannes* (2000).

Case study: resisting diversity: ultra-orthodox Jews in Broughton Park, Manchester

Discussion of cultural diversity, hybridity, flow and interconnection runs the danger of blinding us to the fact that there are pockets in many, if not most, cities of stubborn identities. These refer to groups who wish, for a variety of reasons, to resist mixing with other cultural groups. While many geographers have identified discrimination by wider society as a key reason why these groups become so concentrated and segregated, it must be remembered that boundaries can also be drawn from within. Groups may inscribe and maintain boundaries around themselves in a number of ways as

continued

acts of resistance, a rejection of the outside world. One such example is ultra-orthodox Jews living in contemporary Britain.

Roughly ten to fifteen per cent of British Jews are ultra-orthodox and they constitute the fastest growing component of Jewry in both Britain and worldwide. Valins (2003: 159) describes these groups:

> These Jews are known for their traditional style of dress, their stringent interpretations of ways of life inscribed in ancient, sacred texts and their clear construction of socio-spatial boundaries to separate themselves from those they consider to be 'other'.

Ultra-orthodox Jews in Broughton Park, Manchester show distinctive residential clustering with some streets at the heart of the neighbourhood being entirely Jewish, while many others have over eighty per cent Jewish occupation. This residential distribution is mirrored in the locations of Jewish institutions within and around the area.

For these residents, the neighbourhood has a very clear identity linked to views of it as middle class, well to do, and distinctly Jewish. This is in strong contrast to the outside which is perceived as both dangerous and morally decaying. When asked, Jewish residents recognized clear boundaries separating them from supposed 'others' outside and often articulated the identity of the area in terms of it being an 'island of decency' (Valins 2003: 168). Within the neighbourhood strict orthodox practices are adhered to by the majority of residents and instances where these are breached are strongly disapproved of. Non-orthodox Jews when interviewed mentioned the practice of being stared at when driving their car on the Sabbath as one

way in which this disapproval was manifest as well as the social distance they felt between themselves and their ultra-orthodox neighbours. The activity patterns of ultra-orthodox residents reflect the inward orientation of this community, with trips to facilities beyond the neighbourhood kept to a minimum or carefully managed. A number of ultra-orthodox Jews interviewed articulated the importance to them of imposing socio-spatial boundaries between themselves and wider society and resisting mixing, integration and what they see as the dilution of their identity. The recognition by this community of a patch of ground that they see as distinctly 'theirs' is of vital importance to the preservation of their identity. One respondent said:

> in a sense we've got to enclose ourselves within our own boundaries in order to protect, as far as our children are concerned, in order to protect our children from the influences of the rotten society, of the immoral society within which we live.
>
> (Valins 2003: 168)

This case highlights the importance of boundaries that still exist at many scales despite some of the rhetoric attached to the impacts of globalization. In the case of ultra-orthodox Jews in Manchester, their boundary building and maintaining practices can be seen as a form of resistance to modernity and what they see as its pernicious influences. For this community the rejection of diversity, mixing and hybridity are important ways in which their identities are stabilized and the processes of modernity slowed down. All of this is crucially grounded and brought together within a specific, identifiable location.

Source: Valins (2003)

Material and immaterial cultures

The example above of the gated residential enclaves of some very wealthy groups remind us of two things about the relationship between culture and the materiality of the city. The first is that the urban landscape is a reflection of the cultural norms, values and sometimes fears of the groups who produce, occupy and use the city. This idea of the 'city as text' is one that we have explored in some detail elsewhere (chapter eight; see also Bender 1993; Robertson and Richards 2003; Wylie 2007). However, more specifically it reminds us that the cultures of some groups are more 'artefactual' than others. In other words, the lifestyles of some groups are more reflected in the urban landscape, or in other material artefacts of the city, than those of others. We might think of the former as cultures that are relatively material. By contrast some cultures are much less associated with tangible, material artefacts, more with ways of life or social practices. Such cultures are reflected much less in the urban landscape. This is not to deny that they make an impression on the materiality of the city but that this impression might be through relatively minor modifications to the landscape such as posters, graffiti, signs and decoration rather than through plans, buildings, monuments and so on. These differences have a great deal to do with the relative power of these different groups. The latter groups we might refer to as less material cultures or as more immaterial. We do not mean this in the sense that these groups are not important, far from it, rather, that they are cultures less embedded in the material landscapes of the city (see case study).

Case study: landscape, politics and heritage: the Spitalfields Market controversy

The relationship between culture and materiality, and some of the conflicts that are bound up around this, are clearly demonstrated in the case of heritage. Jane M. Jacobs (1992, 1996, 1999) has provided detailed commentaries on conflicts involving a range of cultural groups around the redevelopment of the Spitalfields area of East London during the 1990s. Spitalfields was a deprived part of the East End of London where pressure for development came from its proximity to the expanding financial centre in the City of London. It is home to a range of different cultural groups linked to waves of immigration that came into the area from the early eighteenth century onwards from, variously, France, Ireland, Poland, Russia and more recently Bangladesh. The plans for the ongoing redevelopment of the area in the 1990s selectively incorporated distinctive architectural forms that existed in the area and which were closely associated with some aspects of the area's rich history. However, some cultural groups, despite having a long-standing presence in Spitalfields, whose histories were less artefactual, tended to become marginalized within the ongoing redevelopment of the area. For example, while Spitalfields'

continued

Figure 7.1 *Terraced housing near Spitalfields showing weavers' lofts*

Source: Author's photograph

heritage of Georgian elegance, embodied in the distinctive terraced housing originally occupied by Huguenots weavers (figure 7.1), were prominent aspects of redevelopment proposals, the histories associated with Spitalfields' Bangladeshi and radical left-wing communities were largely excluded. These groups were less able to embody their histories in artefacts in the landscape and hence could gain little purchase within this conflict and were disempowered within the processes of conservation and development, having little chance to influence them.

The Spitalfields Market controversy was in part a conflict of differently empowered pasts and discourses. In the current state of British planning, histories which are embodied in the built environment and which are less challenging to redevelopment objectives are clearly privileged. Pasts with more deeply oppositional potential and which are present in forms and practices less readily appropriated into redevelopment objectives, are marginalized ... Much of the contemporary city may appear to have histories but increasingly they are histories of artefacts, not ways of life.

(Jacobs 1992: 209)

Sources: Jacobs (1992, 1996, 1999)

The tendency for some cultural groups, whose histories are not so artefactual, to be written out of the urban landscape has prompted responses from artists and artist groups internationally. There are many examples where artists have attempted to address this by making visible, through the creation of various artefacts, the histories of relatively disempowered groups whose cultures leave little material trace on the urban landscape. The 'Power of Place', a multi-media group who have worked mainly in Los Angeles, have sought to create such alternative monuments, not to the powerful individuals who tend to dominate the monumental landscape of cities, but rather to more humble groups from the city's past. One example is the monument they created in the Little Tokyo area of Los Angeles to Biddy Mason, a black ex-slave and midwife (Hayden 1995;

Miles 1997). The monument commemorates the life of Biddy Mason but also provides a material artefact for an otherwise invisible cultural group. This and other monuments have subsequently become incorporated into the conservation of this historic district. This is in sharp contrast to the more exclusionary heritage based redevelopment of Spitalfields. A similar example is Suzanne Lacy's 'Full Circle' project in Chicago (Lacy 1994). This involved the installation of one hundred rock monuments in central Chicago, each one commemorating the life of a Chicago area woman. It provided both a counterpoint to the masculinity of the city's existing monuments and again represented an attempt to memorialize in material form a group whose cultures tend to be more domestic than public, more associated with a set of social practices than material public objects.

Local/global

Culture appears to exist at a range of scales. It is common to hear discussion of local, regional, national or global cultures. This is helpful in that it indicates that one aspect of cultural diversity is related to the **spatiality** of culture. Like many aspects of culture, while at first glance distinctions between different scales of cultures seem clear, closer inspection reveals a deeper complexity and ambiguity. A dialect that emerged historically within a tightly defined spatial scale, that is now being diluted or dying out as migration breaches this cultural isolation seems unambiguously to be an example of a local culture. By contrast Hollywood movies, watched in virtually all countries of the world, are an example of cultural products global in their reach. However, we need to treat this scalar classification of culture with some caution. It tends to simplify and overemphasize the separateness of cultures. It fails to capture the ways in which cultures at different scales are actually interrelated. There is a danger also that it presents cultures at different scales as being in conflict. The example of the Hollywood movie, for example, ignores the ways in which these cultural products are differently consumed and made meaningful within local cultural settings around the world. It is common, for example, to hear discussion of the homogenizing effects of global cultures and their threat to local cultural diversity. Global cultures, at times, are presented in a number of discourses as bulldozing over local cultural distinctiveness and producing a kind of international cultural 'blandscape'. We aim to demonstrate that this represents, at best, an oversimplification.

We do not wish to spend time here discussing the nature of cultural globalization. There are a number of excellent discussions of this already available (Steger 2003; Murray 2006; McEwan 2008). Rather, we want to demonstrate that this view of the inevitability of global cultural homogenization is somewhat overstated, we want to talk about the ways in which cultures at different scales are not necessarily in conflict but are actually interrelated in multiple and complex ways within local contexts. Finally, we want to talk about the ways in which this is manifest in the spaces of the city.

There are many arguments and examples that run counter to the myth that cultural homogenization is an inevitable outcome of globalization (Hall 2007). Globalization, for example, is not as complete a process as is often supposed. The penetration of multi-national companies, while being very extensive globally, is in reality uneven and incomplete. Further, globalization, although it may sometimes seem so, is not a monolithic process stemming solely from the Global North. There are a number of contemporary globalizations associated, for example, with major world religions such as Judaism, Christianity and Islam and those linked to international diasporas such as that of the global Chinese community (McEwan 2008: 278). It is wrong, therefore, to equate globalization solely with economic processes exclusively from the Global North. Processes of globalization then are more complex and diverse than is sometimes supposed.

There can be a tendency to represent global and local cultures and cultural processes in very simplistic terms. There is a tendency to portray global cultures as powerful, dominant and destructive of local cultures. By contrast local cultures can be seen as weak or vulnerable. As with all mythologies there is some truth in these characterizations but they are a significant oversimplification and miss some of the key ways in which the global and local mix in the spaces of the city. The interactions of global and local cultures within these spaces are complex and multiple. We would not suggest that this interaction should be studied by attempting to weigh up the balance of global or local or to determine which is dominant. We would argue that in all spaces of the city there will be elements of global and local cultures present. The key is to understand these presences and their interactions within local contexts.

The approach that we advocate here is best demonstrated with reference to a brief example. Elsewhere we have discussed Barcelona's Museum of Contemporary Art (MACBA) and the impacts its development had on the neighbourhood of El Raval within which it is situated (see chapter five). In that discussion, it could be argued that we are guilty of solely presenting MACBA as an agent of cultural globalization, arguing that its development was a catalyst to the gentrification of the neighbourhood. In turn, we suggested, this has lead to the partial eradication of El Raval's cultural diversity and architectural distinctiveness and its replacement with a form of international style modernism that was, primarily, a space within the global circuit of commerce of art. While we are not arguing anything contrary to this interpretation here, we would like to extend this discussion a little. A closer reading of the development of MACBA reveals a more complex mixture of a range of global, regional and local cultures within the space of MACBA and its surroundings.

El Raval's recent cultural mix is a product of numerous global flows of immigration, most notably since the 1970s from Bangladesh and Pakistan. The flat, smooth spaces of the Plaça Ángels, the square built around the site of MACBA, created an unexpected, and unplanned opportunity for the children of these communities. It is common, especially during sunny evenings, for large parts of the Plaça Ángels to be appropriated by gangs of boys from families of Bangladeshi and Pakistani immigrants who use the square as an ideal surface on which to play games of cricket (figures 7.2a and 7.2b). This inserts another set of global flows into this space which speak of colonialism, appropriation and resistance. Cricket is a game originating in England which was internationalized through the globalizations associated with the British Empire, reaching and becoming popular in the Indian sub-continent during the nineteenth century. The game of cricket is now a key dimension in the assertion of South Asian identities on a global stage. In emulating their cricketing heroes, it could be argued that the children of these immigrant communities create spaces for the exploration of aspects of their cultural identities within an otherwise Catalan/Spanish context. Often this practice leads to minor conflicts with locals, tourists and security personnel as well-struck tennis balls, often bound with tape to produce movement off the pitch, rain down on the unsuspecting users of the square and bounce off the windows of nearby buildings (figure 7.2b). The Plaça Ángels is also very popular with skateboarders, practitioners of an activity that had its origins in California but is now almost global in its reach and popularity (Borden 2001). As well as being appropriated by these surprising activities, which can be interpreted as subcultural in the context in which they are performed here, the square is also an important space for local cultural activities such as evening promenading and as a location for stalls selling books and roses, traditional gifts on St George's day, the Catalan region's patron saint. MACBA then offers a space for the consolidation and demonstration of certain local-global and regional cultures as well as their repression through gentrification and development.

MACBA itself is a major space for the display of the work of regional artists, granting them a stage of international standing. These include artists such as Antoni Tàpies whose work has frequently taken the question of Catalonian cultural identity as its theme. Tàpies' work has been important to the cultural revival of Catalonian identity. It has been part of the questioning of the region's political identity as part of Spain in the Franco and post-Franco eras (figures 7.3a and 7.3b). The presence of such a major cultural institution within the city was part of Barcelona's extensive regeneration in the 1990s based around the hosting of the 1992 Olympics, which itself acted as an important platform for the display of Catalan identity (Dodd 2003). This points to the importance of a resurgent civic culture under its charismatic mayor, Pasqual Maragall (McNeill 2001). Unravelling the cultural geographies of this space, which at first seems

Figures 7.2a and 7.2b *An improvised game of cricket in the Plaça Àngels adjacent to MACBA*

Source: Author's photographs

Figure 7.3a *Antoni Tàpies exhibition, MACBA (2004)*

Source: Author's photograph

Figure 7.3b L'esperit català, *Antoni Tàpies (1971) based on the Catalan flag*

Source: © VEGAP, Madrid and DACS, London (2011)

simply a question of the role of art in alliance with regeneration and capital, actually involves an appreciation of the interactions of numerous local, civic, regional, national and global cultures.

Summary

As we have seen, culture is an increasingly important lens employed by geographers and others to understand the city. Undoubtedly culture, in its many forms, is becoming more central to the shaping of the city economically, politically, socially and materially. While geographers have engaged enthusiastically with these issues we cannot recognize a coherent 'cultural urban geography' that has emerged from this work. Engaging with the literatures on the cities and cultures is rewarding but requires an awareness of their diversity. The framework outlined in this chapter will guide you through both these literatures and the many and diverse spaces of culture in the city that you will encounter.

Follow-up activities

Essay title: 'The cultural geographies of the contemporary city are ones of tolerant indifference.' Discuss with reference to examples.

Commentary on essay title

The question draws on the assertion made within the work of Sandercock and Amin, discussed above, that the contemporary, culturally diverse, 'mongrel city' is one lacking in collective civic culture. A successful response to this question requires an evaluation of this statement drawing on evidence from a variety of cities internationally. The answer might begin by exploring examples of cities where the statement appears to hold true. These might include the cities of contemporary Britain, from which Amin's view derives, and car-based cities, such as Los Angeles and Perth in Australia, where the spaces and opportunities for public culture have been significantly undercut by the dependence on car-based mobility. It might then consider evidence from cities that seems to challenge the assertion in the question. These might include cities such as New York and cities of Mediterranean Europe which appear to have more vibrant public cultures. In evaluating the evidence from the latter cases, the answer would ponder the extent to which the appearance of a lively public culture in these cities equates to greater cross-cultural dialogue or the extent to which the public cultures of these cities are still culturally segmented. An excellent answer would extend this by critically considering policy routes to promoting civic culture in contemporary cities around the world.

Project idea

Select an urban area you have easy access to. Spend time in your chosen location and observe it as closely as you can. Think about the dimensions of culture that are explored within this chapter. Which of these can you detect in your selected location? In what ways are these manifest (grounded) in your selected location?

Select one of these dimensions of culture that emerge from your initial survey and explore it in more detail within your selected location using photographic research methods. The aim is to use photography in a critical, investigative way rather than

descriptively. For example, you may wish to record in depth the ways that the issues you investigate are manifest in the area you are investigating. Produce a photographic report to represent the findings of your research.

When you have done this reflect on the strengths and weaknesses of using photography as a method of research compared to other non-visual research methods.

For discussions of photographic and other visual research methods see Pink (2001); Knowles and Sweetman (2004); Rose (2006); and Hall (2009b).

Further reading

Books

- Anderson, J. (2015) *Understanding Cultural Geography: Places and Traces*, Abingdon: Routledge
  Not specifically urban in its focus but an excellent guide to the sub-discipline of cultural geography. Contains a wealth of material, theoretical and empirical, that will inform the student of urban cultural geography.

- Binnie, J., Holloway, J., Millington, S. and Young, C. (eds) (2006) *Cosmopolitan Urbanism*, London: Routledge
  Wide-ranging examination of the nature and implications of cosmopolitanism as the defining characteristic of contemporary cities.

- Miles, M. (2007) *Cities and Cultures*, Abingdon: Routledge
  Excellent introduction to a diversity of cultural forms and their roles in shaping, and in turn being shaped by, the city.

- Miles, M., Hall, T. and Borden, I. (2003) *The City Cultures Reader*, 2nd edn, London: Routledge

Extensive collection including classic key readings and lesser known works encompassing a number of aspects of city cultures.

- Sandercock, L. (2003) *Cosmopolis 2: Mongrel Cities of the 21st Century*, London: Continuum
An important contribution to contemporary debates about the nature and potentials of cosmopolitan urbanism. See also Sandercock's *Towards Cosmopolis* (1998). A key author on issues around cultural planning in the contemporary city.

- Stevenson, D. (2014) *Cities of Culture: A Global Perspective*, Abingdon: Routledge
Broad in its scope, covering theories, issues, themes, cultural institutions and a wealth of international case studies from a diversity of city locations. While it appears in a series of sociological texts it is very geographically attuned, both through its reflections on the role of place and in its comparative dimension which emerges through its contrasting case studies.

- Zukin, S. (1995) *The Cultures of Cities*, Oxford: Blackwell
Although published some time ago this (and Zukin's other work) remains a key text. It explores the nature of urban cultures and particularly their relationships to economics and the ways in which cultures are harnessed in reshaping the contemporary city.

Journal articles

- Barnes, T. (2003) 'The '90s show: culture leaves the farm and hits the streets', *Urban Geography*, 24(6): 479–492
Discusses the initially reluctant but eventually enthusiastic embrace of culture by urban geography. A good starting point for investigating urban cultural geography.

- Bell, D. (2007) 'The hospitable city: social relations in commercial spaces', *Progress in Human Geography*, 31(1): 7–22
Considers the cultures and social relations that take place within the increasingly commercialized spaces of the contemporary city, an often overlooked aspect of the cultural landscapes of cities.

- Boland, P., Murtagh, B. and Shirlow, P. (2016) 'Fashioning a city of culture: "life and place changing", or "12-month party"', *International Journal of Cultural Policy*, http://dx.doi.org/10.1080/10286632.2016.1231181
A critical perspective on the potentials of cultural events to transform the lives and fortunes of cities. This is rooted in an analysis of the Derry/Londonderry's roles as UK City of Culture, 2013.

- Borer, M.I. (2006) 'The location of culture: the urban culturalist perspective', *City and Community*, 5(2): 173–197

Reviews research in urban sociology that has attempted to understand urban culture. Uses this to outline a theoretical perspective and framework through which to investigate the dimensions of urban culture.

- Jayne, M., Holloway, S. and Valentine, G. (2006) 'Drunk and disorderly: alcohol, urban life and public space', *Progress in Human Geography*, 30(4): 451–468
  Examines the roles that alcohol and the practice of drinking have come to play within the cultural (and economic and material) geographies of the contemporary city. Considers the regulation of drinking and the geographies of different drinking spaces and practices.

- Ottaviano, G.I.P. and Peri, G. (2006) 'The economic value of cultural diversity: evidence from US cities', *Journal of Economic Geography*, 6(1): 9–44
  Assesses the value of cultural diversity to cities from the perspective of economic geography. Finds evidence that cultural diversity brings with it positive economic impacts on cities.

- Sheringham, M. and Wentworth, R. (2016) 'City as archive: a dialogue between theory and practice', *Cultural Geographies*, 23(3): 517–523
  A dialogue and a reflection on the potentials of understanding urban experience by considering the city an archive and the contributions offered by artistic and visual methods. Part of a collection of articles on the urban archive.

Section 3

Issues

8 Architecture

Introduction

It is rare to find a chapter on architecture in an introductory text on urban geography. However, this is perhaps an odd omission given that cities are composed of collections of buildings and structures and that landmark buildings are often key markers of a city's identity. This is not to say that architecture has been ignored by urban geographers, but that its consideration was rather tangential to the core of the discipline in the past. Indeed, as we saw in chapter three the consideration of urban form is one of the longest areas of enquiry in urban geography. However, detailed consideration of architecture as both form and practice has remained outside the mainstream of urban geographical research and discussion until relatively recently. It is with the resurgence of culture as an area of enquiry in geography (see chapter seven) that discussion of architecture has increased. In particular, there has been a focus on examining the symbolic quality of architecture in reflecting the ideologies and values of those producing the building or structure. Equally, interest has been stimulated by the increasing promotion of landmark, spectacular architecture for urban development schemes by city authorities, often designed by globally recognized architects, in order to enhance the city's image in an increasingly competitive global urban network (see chapter nine).

The chapter continues by asking the question 'what is architecture?', a question that appears simple to address but which on further consideration reveals the complexity of what architecture is, is for and represents. The chapter then continues by examining the ways in which urban geography has approached the study of architecture, outlining early approaches to study but focusing more fully on more recent discussions considering the symbolism of architecture and also people's reaction to architecture and their use of buildings. The chapter then goes on to explore three topics in the study of architecture: gender issues in architecture and housing design, the rise of spectacular iconic architecture and the architecture of fear.

What is architecture?

Whether we are aware of it or not architecture is part of everybody's life. Many aspects of our lives, our work, rest or play, take place in a building of one sort or another that has been designed and constructed by someone with a particular purpose in mind. But are all buildings classed as 'architecture'? For Nikolaus Pevsner (1943), the great chronicler of British architectural traditions, architecture was an art and was associated with the design of great buildings (in Europe) rather than ordinary ones. This is a quite traditional, restricted and Eurocentric view of what constitutes architecture which ignores many everyday buildings and indigenous building traditions, or vernacular architectural traditions as they are known. Although Pevsner expressed this view over seventy years ago, many have pointed to a continuing preoccupation in architectural writing with special or iconic buildings rather than those that are part of most people's everyday environments (Samuels 2010). However, some more recent chroniclers of the development of architecture have adopted broader definitions and include everything which is built (see, for example, Kostof 1995). Indeed, some have even suggested that the definition of what constitutes architecture could be extended to include structures from the non-human world such as ant hills (Hersey 1999).

The development of architecture

Architecture first evolved out of the dynamics between human needs (shelter, security, worship, etc.) and means (available building materials and skills). Until the technological developments of the twentieth century, there have been two principal ways of building, either employing a frame or skeleton covered with a skin or by putting one block on top of another. These techniques are used around the world for building, with variations in building forms developing from these basic principles, linked to the differing needs of users. Additionally, variations in early building styles resulted from the use of different materials. Around the world a wide variety of materials have been, and are still, used for building, including stone, clay, wood, skins, grass, leaves, sand and water. Those materials that were readily available have therefore had a profound effect on early architectural forms, producing a wide variety of building traditions around the world which have played an important role in the development of place identities. Through processes of trial and error, improvisation or the replication of successful building types, local architectural traditions developed. These traditions continue to be produced in many parts of the world, and vernacular buildings continue to make up an important proportion of the built world that people experience every day (Oliver 2006). However, as Abel (2004) notes, increasingly in the modern urban age there are relatively few examples of truly indigenous architecture with many styles influenced by cross-cultural contact.

Gradually, societies developed and formalized their building knowledge, initially through oral traditions and then via written codes of architecture and design. It is with the development of urban societies that architecture developed through the processes of city building, technological development and the regulation of building through codes and laws (Chant 2008). As we noted in chapter three, within ancient civilizations architecture and urbanism reflected either an engagement with the divine and the supernatural or power and the state. It is from the demands for religious buildings and through the patronage of the powerful and wealthy that the great architectural traditions of the world have developed with the construction of key buildings and the production of canonical texts on architecture. For example, within the European context, the traditions of cathedral building and the revival of the architectural traditions of ancient Greece and Rome, through the rediscovery in the Renaissance of the writings of the Roman architect Vitruvius, produced the important gothic and classical styles respectively (figure 8.1). Within Asia and the Arab world, the different religious contexts produced quite different architectural traditions (Cruickshank 1996; Ching et al. 2006). From this, individuals began to be identified as architects and the developers and champions of particular styles.

From the eighteenth century onwards, increasing economic and cultural global connectivity led to the wider spread of architectural ideas and practices. The development of pattern books of architectural ideas, the growth of formal professional training through architectural schools and professional societies and the industrial production of building materials facilitated the increasingly

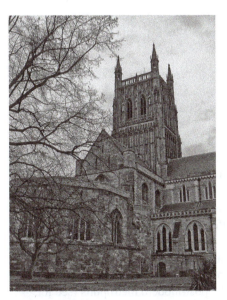

Figure 8.1 *Worcester Cathedral (early English Gothic) (left) and Law Courts (classical Palladian) (above)*

Source: Author's photographs

widespread adoption of particular architectural styles. Colonial expansion by European nations led to the spread of key classical and gothic styles to their colonies, although as Anthony King (1984) notes this traffic of ideas was not one way and architectural traditions from the colonies also filtered back to the design of buildings in Europe, something that is all too easily overlooked in accounts of architectural history.

From the early twentieth century onwards two broad architectural trends have emerged: the modern and the revivalist. These stylistic trends can be seen as part of a broader ideological struggle between instrumental (rational) and expressive (romantic) ideals that has been evident in western thinking since the eighteenth century (Ley 1987). Developing from instrumental, technocratic traditions, the modern functional style developed in Europe in the early twentieth century as a reaction to the established architectural styles of the time. It was a rational, technologically-inspired style where the form of the building stems from its function, with no unnecessary ornament or detail. Its early proponents were radical architects who wanted to transform urban societies (see, for example, Le Corbusier discussed in chapter six). However, the style's versatility and practicality, coupled with its popularity among key architectural practices, led to its widespread adoption for commercial developments and mass housing schemes. It has developed as an international style in cities around the globe as processes of urban change and development have become increasingly connected (Knox and Pinch 2010) (figure 8.2). This has led to some claims that many urban landscapes are becoming very similar, particularly in the commercial centres of cities, or 'placeless' (Relph 1976, 1987) (see chapter ten).

However, running alongside the promotion of modern architecture has been a growing architectural movement that looks back to past styles for its inspiration, as part of the expressive, romantic tradition. The continued popularity of past styles has latterly been given great impetus by the urban conservation and new urbanism movements (see chapter six). Here, the quest for buildings that have a more human scale and which respect the architectural context provided by retained historic buildings has spawned a range of historicist styles for new developments in cities (Larkham 1996). These styles are part of the broader architectural movement known as postmodernism (Jencks 1991). Buildings in the post-modern tradition appear to offer greater detail and interest than buildings in the modern tradition (see figure 8.2). The emergence of these post-modern architectural styles within cities has stimulated renewed urban geographical interest in architecture, with these new forms being seen as a key manifestation of the move towards the post-modern metropolis (see for example Knox 1991, Soja 1995). However, many of these commentators have been critical of these new styles, arguing that they are nothing more

Figure 8.2 *Modern and post-modern office buildings, Birmingham UK*
Source: Author's photographs

than the new cultural clothing of advanced capitalism designed to enhance, or, for the more cynical commentators, mask large corporate urban redevelopment projects (see Harvey 1989a). A key point about architecture that emerges from this discussion and the quotations below (figure 8.3) is that architecture is very much a product of a particular time, place and people and that architectural works can be perceived as potent cultural and political symbols. Indeed, societies are often defined by their surviving architectural achievements, and ages or epochs become characterized by particular architectural styles (Knox and Pinch 2010). Critically, this point also highlights the controversial and contested nature of architecture, in that buildings can arouse great passions both positive and negative, celebratory or destructive. These cultural meanings embodied in buildings have been a key area of investigation for urban geographers (see below).

"(An) Institution for the general advancement of Civil Architecture, and for promoting and facilitating the acquirement of the knowledge of the various arts and sciences connected therewith; it being an art esteemed and encouraged in all enlightened nations, as tending greatly to promote the domestic convenience of citizens, and the public improvement and embellishment of towns and cities..." (Charter of the Royal Institute of British Architects 1837)

"(Architecture) provides a key to the habits , thoughts and aspirations of the people and without a knowledge of this art the history of any period lacks that human interest with which it should be invested". (Bannister-Fletcher first published 1896 (20th edition 1996, xxv)

"...I know that architecture is life; or at least it is life itself taking form and therefore it is the truest record of life as it was lived in the world yesterday, as it is lived today or ever will be lived. So architecture I know to be a Great Spirit..." (Frank Lloyd Wright (1939) in Brooks Pfeiffer and Nordland 1987: 7)

"Architecture is the will of the epoch translated into space" (Mies van der Rohe 1926)

Figure 8.3 What is architecture?

Exercise

Either individually, or with a small group, imagine that you are part of a particular social or cultural group who typically wish to express their power in the city. Imagine that your group is contemplating a major building project in a large city. The building project has to act as a symbol of your group, its interests and values. Design a building, perhaps with other landscape elements such as monuments, statues and so on where appropriate, that act as symbols of the power of your group.

Groups:

- A dominant religious group – you want to build a major place of worship in a modern urban centre

- The managers of a worldwide multi-national company planning a new headquarters
- A new democratic government in a formerly totalitarian state – you wish to remove the memory of the previous government and fill the capital with symbols of democracy

Issues you may like to consider:

- The location of your building
- The design (physical layout) of your building
- The ornamentation or style of your building.

What constraints do you think you will be under in the construction of your building – lack of land, planning constraints, economic restraints, local opposition, internal group tensions?

Architecture as a profession

In the contemporary world, architecture can be broadly defined as the process and product of planning, designing and constructing environments for people that reflect a range of functional, technical, social and aesthetic considerations. It is a specialist profession that designs buildings and cityscapes and helps realize the requirements of its clients and communities. Architects transform people's needs (e.g. to eat, work, live, play) into concepts and then develop these concepts into building images that can be constructed. A wide definition of the profession of architecture can include all design activity, from the macro scale urban design and landscape architecture of city areas, to the micro scale construction details of buildings, building interiors and furniture. These broad characteristics of architecture have a long history, exemplified by the founding charter of one of the world's oldest professional architectural organizations, the Royal Institute of British Architects which was founded in the early nineteenth century (see figure 8.3). Another key point about architecture embodied in this definition is that it is both an art and a science. This important dual role for architecture is reflected in some of the earliest written principles about architecture set out by Vitruvius in the first century in that a building should have the attributes of durability, utility and beauty. It is this need for practicality and robustness that is seen to differentiate architecture from other arts such as sculpture and painting.

These definitions of architecture highlight a key way of thinking about what architecture is, in that it is a profession underpinned by a set of principles and ideas that are communicated to others and then utilized in the design of further structures. It is this aspect that defines the development of architecture as a profession as distinct from non-written indigenous building traditions. However, in constructing architecture as a distinctive profession, those researching current architectural practices have pointed to the ways in which a particular view of architecture and architects has developed over time. Jones (2009) highlights how architecture defines itself as a distinct field of activity into which other architects are socialized by learning the codes of practice – it is then an exclusive field that admits/accepts only those people and practices that embody the dominant ideas of the profession. Similarly, Habraken (2005) has highlighted the ways in which architecture defines itself as the story of gifted and successful individuals, driven by particular architectural self-publicists. He and others, note that this promotes the view of the architect as the expert imposing order on the world and acting in an autonomous and distanced way. Here buildings arrive from the architect's pen (or these days CAD package!) as perfect structures with little concern for their life beyond being built. Equally, it promotes the view of architects as acting autonomously, when in fact they are part of the wider urban development process, working within the broader parameters imposed by client demands and building codes.

Those critical of current architectural practice highlight the problems stemming from this view of the role of the architect, including a focus on style and form rather than the needs of users; that those in the profession can seem distant and out of touch with the needs of 'ordinary' people; a lack of consideration that buildings need to adapt and learn rather than being seen as totally finished structures; and a view of the building development process as an autonomous artistic practice rather than as the more messy and complex process that it actually is (Brand 1995; Hill 1998; Habraken 2005; Till 2009). Architects therefore play a key role in 'their transcription of economic, social, cultural and political dynamics into the evolving physical settings of the city' (Knox and Pinch 2010: 134). They are both the products and carriers of the broader socio-cultural ideas and power relationships inherent in urbanization, holding a key role in the socio-spatial dialectic. Therefore, who the architects are and how they engage in architectural practices are important areas of study for researchers in urban geography and in other related disciplines.

Architecture and urban geography

The roots of geography's engagement with architecture can be found in the rural landscape tradition of American cultural geography. In this tradition, natural landscapes were considered to be transformed into cultural landscapes through human practices and traditions indigenous to an area (Hubbard 2006). Much of this early work was focused on examining vernacular architecture as a way of exploring the uniqueness of landscapes and their peoples (see, for example, Kniffen 1965 on folk housing). Buildings were viewed as artefacts that reflected cultural values and a society's level of technological development, with studies mapping geographical patterns of architectural styles. However, much of this work was rurally focused and ignored cities, failing to build on key early work in the urban context such as Lewis Mumford's *The Culture of Cities* (1938). Latterly this work has been criticized for its naïve correlations between architectural types and cultural ones and its lack of theoretical engagement with architectural interpretation (Goss 1988).

In its early engagement with architecture, urban morphology adopted a similar atheoretical, descriptive perspective to that of early cultural geography (Larkham 2006). In some early studies, architecture (as building form) was considered one of the key elements contributing to urban morphology, which reflected the spirit of the society that produced it (Conzen 1966). However, this early work did not entail studies of architectural style per se, but were explorations of style as a manifestation of the processes of creating form, namely the interaction of the agents and the processes of change. Latterly, research on architectural form has been linked to the types of agent responsible for its creation (organizations and individuals), developing as a key strand of research in British urban morphology from the 1980s onwards (Larkham 2006). Using building records, studies of both commercial development (Whitehand 1992) and residential development (Whitehand and Carr 2001) have examined the diffusion of stylistic innovations and have highlighted the geographical links between agents, places and the nature of the physical changes planned and implemented in cities (Larkham 2006).

In a key review of early geographical work on architecture, Jon Goss (1988) noted that geography had generally failed to come to terms with the complexity of architectural form and meaning. His paper called for the development of a new architectural geography which explored four theoretical categories of buildings: as cultural artefacts, as objects of value, as signs and as a spatial system. More recent work on architecture within urban geography has sought to respond to the challenges set out by Goss, particularly viewing the city as a 'text' which is both written (inscribed with meaning by the various agents involved in the urban development process) and read (understood by the users of the urban environment).

In thinking about the city as written, urban geographers have examined architecture and urban design as elements in the political economy of urbanization, linked into the dynamics of urban change. This work builds on the ideas from urban morphology of architectural style as a manifestation of the processes of creating form, namely the interaction of the agents and processes of change, linking this to political economy perspectives on the dynamics of capitalism in creating and destroying urban environments (see chapter four). Here architecture promotes the circulation of capital and helps stimulate consumption and the extraction of surplus value by providing new products and designs for different market segments, such as new office blocks, shopping centres, or housing developments. It can also add an aura to developments which serves to legitimize existing economic and social relations and suggest stability and permanence (Pacione 2009; Knox and Pinch 2010). As noted above, geographers such as Harvey (1989a) and Knox (1991) have highlighted the relationship between the development of global advanced capitalism and the emergence of new post-modern architectural forms. The property industry can be seen to have adopted post-modern spectacular styles to promote product differentiation in an increasingly competitive global market and to both create and supply demand for an increasingly consumerist society. Similarly, the dystopian side to post-modern architecture has also been evident in new development in the creation of fortress architecture of security and surveillance systems designed to exclude 'undesirables' such as beggars or rough sleepers from the increasingly privatized urban public realm (Pacione 2009). Here architects play an important role in the internal survival mechanisms which have evolved to meet the needs of urbanized capital, particularly as part of the processes of mediating crisis and change (Jones 2009; Knox and Pinch 2010; Kaika 2010).

Since Goss's call for further work, there has been a considerable amount of research by urban cultural geographers in reading urban landscapes, interpreting the meanings of the built environment as sign and symbol. Indeed, it has been argued that the cultural interpretation of urban landscapes has been a key part in the broader cultural turn (or turns) in geography (Hubbard 2006). The adoption of terminology associated with **semiotics** (science of signs) became common in geographical writing on landscape as 'text'. The idea that landscapes as texts contain signifiers that send particularly ideologically charged messages to different social and cultural groups was influential to those geographers seeking a more rigorous methodological and theoretical framework for considering the symbolic qualities of the urban environment (Hubbard 2006). Lees (2001) identifies two broad strands to this cultural geographical work: those inspired by Marxist cultural materialist perspectives, such as the work of Raymond Williams, who understood architecture as a social product that both reflected and legitimized underlying social relations (see for example Cosgrove and Daniels 1988); and those inspired by semiotics and cultural anthropology, specifically the work of Clifford Geertz, who sought to read the urban landscape as a text in which social relations were inscribed, offering 'thick descriptions' of buildings (see for example Domosh 1989). However, as Lees notes, these two strands were not mutually exclusive and much of the subsequent work in this vein combined elements and ideas from both areas. This geography of architecture informed by social theory has focused both on macro scale urban landscapes, such as housing areas (see for example Domosh 1996) and individual buildings such as skyscrapers (Domosh 1989) and shopping malls (Goss 1993) (see chapter seven).

In reviewing this work, Lees (2001) notes that it was in danger of running out of steam with a continuing search for novel sites and signs offering diminishing returns. She also suggests that geographies of architecture had been too focused on discursive meanings of landscape and had had relatively little to say about the non-representational dimension of architecture as performance, both as a practice and a product. This builds on Goss's call for the consideration of architecture as a spatial system which offers both opportunities and constraints in the reproduction of meaning and lifestyles. As he notes:

Architectural form and style are not simply the concretization of cultural values and ideology, nor simply the reflection of material function and social relations of production, nor equally matters of individual perception and interpretation. Human life is multiple-sided and complex, and the meaning of a building cannot simply be read without considering the interaction of the subjects who are ultimately the sources of all its functions and meanings.

(Goss 1988: 400)

Lees highlights two key problems to be addressed for the continuing development of a critical geography of architecture: the problem of reading the meaning of buildings which are multiple and contested, and also the absence of consideration of the consumption of architecture. A key issue is how ordinary people engage with and inhabit the spaces architects design and how urban meaning is not inherent in architectural form and space but rather changes according to the social interaction of city dwellers. She develops her ideas for a new critical geography of architecture that considers both the representational and non-representational aspects of architecture through a consideration of the new public library building in Vancouver, Canada (Lees 2001, see case study below).

Case study: critical geographies of architecture – reading Vancouver's public library

In her 2001 paper, Loretta Lees uses the example of the building of the new public library in Vancouver, Canada to develop a critical geography of architecture. The paper is interesting as it charts her developing approach to examining the library building from a 'traditional' political semiotic reading of the building, seeking to understand its symbolic meanings, to a more critical approach seeking to understand the building through reflection on her active and embodied engagement with its spaces.

The library opened in 1995. It was designed by Moshe Safdie, an Israeli-born architect educated in Canada and working in the US, and its external appearance was seen to resemble the Colosseum in Rome. As a symbol and civic landmark for Vancouver the public library sparked intensive discussion and debate. Lees' analysis of the public library initially explores the debates over the symbolism of this new civic landmark within the multi-cultural city of Vancouver and the politics surrounding its architectural interpretation. Through examination of media coverage and interviews with public officials, she highlights the cultural political debates around race and identity reflected in the symbolism of the building. First, the colosseum design was criticized by some as racist and Eurocentric, ignoring the multicultural nature of the city, particularly the culture of recent Asian migrants and First

continued

Nation Canadians. Second, some criticized its placelessness and inauthenticity, seeing the design as reflecting a globalized postmodern corporate architectural style used for developments such as shopping malls and symbolizing a 'Disneyfied' American-style built environment.

These varying views of what the library symbolized highlight the problems in providing a definitive reading of the building. In attempting to provide a 'thicker' reading, Lees examined the writings of the architect for clues to what it might represent. However, Safdie's idiosyncratic and contradictory views on architecture, the development of the building and what it might represent, served to further complicate Lees' reading. Her conclusion is that the design of the colosseum is ambiguous and can be read as multi-perspectival where users can locate their own places through the different memories and images the design throws up for them.

In seeking to examine the concrete social process through which this multiple meaning develops, Lees draws on the work of Sandercock (1998) to think more fully about the public consultation process and the contested meanings within it. Reviewing her research notes, Lees suggests that the public architectural vote and the critical discussion surrounding the selection of the design represent a successful example of Sandercock's idea of cosmopolis, where development is messy and design solutions are always contested and which offers a positive way of engaging with difference in the urban realm. The public consultation represents a medium through which the multiple meanings of the library were enacted.

The idea of meaning being enacted leads onto Lees' final exploration of the library's meaning. She suggests that if meaning is enacted then it cannot be finished and merely represented and that the public consumption of architecture is also important. The key question is then not what the library means but also what it does, focusing on the use of the library and how meaning is embodied, performed and enacted. Reviewing field notes on the use of the library by various users, she argues that the meanings produced through the consultation process are always in the process of being enacted and subverted (such as by schoolchildren playing on the library escalators). Identity and its formation then become a central theme of a critical geography of architecture that does not merely focus on form and producer but also considers the everyday users and consumption of architecture.

Source: Lees (2001)

Gender and architecture

Knox and Pinch (2010: 134) suggest that the **patriarchal** qualities of the built environment are one of the most important, if often overlooked, dimensions of the socio-spatial dialectic. However, work by feminist theorists has been important in examining the ways in which the built environment embodies fundamental gender divisions and conflicts (see WGSG 1997). Here architects, planners and designers play an influential role in transcribing gender roles and relations into the physical fabric of the city. Researchers note that around the world the urban environment is largely 'man-made' (WGSG 1997). The built environment professions remain largely male, particularly at the most senior levels, and consequently patriarchal values have become embodied in the professional ideologies and praxis of built environment professions such as architecture, planning and construction (see for example Greed 1994, 2000). Authors such as Greed have argued that although more women are now entering built environment professions, the underlying principles, codes and conventions of these professions still embody traditional assumptions about the socio-economic roles that women and men will undertake.

A clichéd theme in architectural theory has been identification of 'masculine' and 'feminine' elements in design, usually focusing on anatomical references to tall towers (masculine power) and curving structures (feminine softness) (Knox and Pinch 2010). However, these discussions have tended to trivialize consideration of gender issues in architecture and essentialize masculine and feminine characteristics in the consideration of building design. As some feminist interpretations of architectural history have shown, it is the issues on which architecture is silent that reveal more about the **masculinist** practices of architecture rather than crude anatomical metaphors (Wilson 1991). For example, in developing so-called radical new visions for buildings and urban environments modern architects and urban designers have rarely had anything to say about divisions between the sexes and gender divisions of labour in particular, taking these for granted.

These assumptions about gender roles and relations are embodied in the design of the urban environment at all scales, from that of the whole city down to that of building interiors or pieces of public art. Mackenzie (1988) has argued that the evolution of urban structure can be interpreted as a series of solutions to gender conflicts bound up in the separation of home and work which was central to nineteenth century industrialization. This resulted in the division of urban structure into masculine centres of production and feminine suburbs of reproduction. The ideas embodied in the design of suburbs solely as spaces of social reproduction are reflected in building society adverts from the 1920s and 1930s studied by Gold and Gold (1994). Here homes were depicted in green settings, removed from spaces of production, with a woman and child often welcoming the 'man of the house' home from work, invoking a traditional view of gender roles (which some suggest continues in advertising associated with the home today (WGSG 1997)). These taken-for-granted gender roles were also reflected in the internal structure of suburban domestic buildings. This strong gender coding of domestic architecture has been revealed in analyses of housing designs from the nineteenth century onwards (see for example Roberts 1991). Ideals of domesticity and nuclear family living are embodied in the floor layouts, décor and design of single family dwellings centred on functional kitchens and a series of gendered domestic spaces (see figure 8.4). The significance of these codings lies in the ways that they universalize and legitimize a particular form of gender differentiation and domestic division of labour, presenting these differences as natural rather than socially constructed and contested (Knox and Pinch 2010).

Exercise

Undertake an audit of the design of your own home environment. Think about the floor plan of this space – what rooms are provided, for what purposes and where are they positioned within the house? Does this suggest any particular assumptions about gender roles and the domestic division of labour? Look also at the interior décor and the use of these spaces – are there particular 'masculine' or 'feminine' spaces? Again, consider whether this reflects any assumptions about gender roles or whether they demonstrate different ways of living and a resistance to traditional gender roles and relations.

If you have undertaken the exercise above, then your observations may well demonstrate that traditional assumptions about domestic roles and the gender division of labour are not universal or uncritically adopted but that roles are multiple, fluid and contested. Again, researchers have highlighted the problems encountered in the built environment by those whose needs appear not to have been considered in the design of those spaces, which highlights the gap between assumptions about roles and the actual activities that people

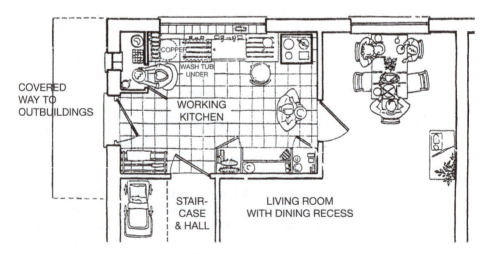

Figure 8.4 *Gender roles and housing design*

Source: Roberts (1991: 85)

perform (see chapter ten). Feminist researchers have highlighted the problems encountered by women with responsibilities for caring for children in negotiating public city centre space where few facilities are provided for those involved in childcare (those facilities either having limited provision or being confined to suburban 'reproductive' environments) (Greed 2006). Similarly, the isolation felt by women in suburban environments, distanced from access to employment opportunities and public facilities, as a consequence of zoning, has also been noted (WGSG 1997).

As well as identifying these problems, many feminist architects and designers have also attempted to challenge conventional ways of designing built environments, so questioning dominant ideas about masculinity and femininity and ensuring that the particular needs of women and other groups disadvantaged by design are catered for. Dolores Hayden (1981) has highlighted nineteenth century housing designs which did not place the nuclear family as central, offering alternative shared cooking and laundry spaces, so liberating women with families from domestic tasks (although in most cases this was liberation for middle-class women based on the communal domestic work being undertaken by working-class women). In the contemporary context, some architectural practices continue to produce designs which offer alternative forms of domestic living or community spaces, often working closely with their particular client groups in developing the design to meet their needs (WGSG 1997). Women have also been active in remaking existing environments, such as through involvement in safe city projects where women come together to change the design of their neighbourhoods, addressing issues such as poor street layout or poor lighting and signage (Wekerle and Whitzman 1995). Liz Bondi (1999) also highlights the role of women in the gentrification of housing in inner Edinburgh, where refurbishment of these houses offered the opportunity to create different domestic environments and an escape from traditional suburban housing structures.

Spectacular architecture and global architects

Increasingly, in an era of competitive globalization, architecture is being used as a tool for economic development within cities, particularly through the development of spectacular developments and iconic buildings (Harvey 1989a; Beriatos and Gospodini 2004; Sklair 2005). Urban geography's recent examination of the development of these iconic buildings has been part of the subject's interest in the political-economic processes underpinning the emerging forms of the post-modern city and also its increasing engagement with matters concerning consumption (Fyfe and Kenny 2005). In an early study focusing on these emerging architectural forms, Paul Knox (1993) examined the changing, or 'restless', urban landscape of Washington DC. He outlines the connections between the economic and social changes associated with the move to a post-industrial capitalist economy, the culture of consumption influencing the new urban elite and the emergence of post-modern, spectacular architectural styles. He argues that these new post-modern landscapes are linked to the rise of global capitalism and serve to legitimize the activities and position of new social, professional, commercial and financial elites. Building on this line of enquiry, a number of researchers have examined what has been termed 'the Guggenheim effect' (Plaza et al. 2009), named after the Guggenheim Museum development in Bilbao, northern Spain. Designed by the internationally renowned architect Frank Gehry, the Guggenheim Museum is seen to epitomize this new wave of iconic urban buildings, which have been designed to reposition cities on the global stage and to act as a focus for economic regeneration activities (figure 8.5).

In the case of the Guggenheim Museum Bilbao, as in other development projects of this kind, the image-making and tourism-attracting capacities of this iconic building have been viewed as successful in transforming the city's image

Figure 8.5 *Iconic architecture: the Guggenheim Museum Bilbao*

Source: Author's photograph

from declining and deindustrializing to being a global tourist destination and cultural centre. However, academic work on the Guggenheim effect has been largely critical of these types of development as manifestations of new post-industrial flexible accumulation strategies, arguing that they represent a cultural 'mask' for the real estate speculation, profit-making and urban marketing activities of global businesses, yet offer little benefit for local residents or the long-term development of cities (Harvey 1989a; Mouleart et al. 2003; Plaza et al. 2009).

While iconic buildings have long been used by urban elites to symbolize power and generate surplus value from space (Domosh 1996; Sudjic 2005), their current manifestation is seen as new and linked to the emergence of a transnational capitalist class who have become the new drivers of the production and representation of architectural icons (Sklair 2005). In the wake of the perceived success of projects like the Guggenheim Museum Bilbao, many entrepreneurial city authorities have commissioned so-called 'starchitects' to design eye-catching buildings with the explicit aim of positioning their city favourably in this competitive global environment, and less explicitly with the aim of embedding broader processes of economic and social restructuring in a more socially significant and resonant form for local populations (Jones 2009). Recent work has sought to move beyond consideration of these iconic buildings as merely masks for new urban developments to examine the specific social and political conflicts and economic dynamics involved in the commissioning and construction of iconic buildings in an era of globalization (Jones 2009; Kaika 2010).

Jones (2009), for example, has sought to link together an examination of the creative practices of architects with the political and economic context in which this takes place. He explores how the recent development of iconic buildings reveals the ways in which architectural practice is conditioned and normalized through its relationship with the dominant political-economic context, where 'architecture is both configured by power and is a resource for power' (Jones 2009: 2531). Jones uses Bob Jessop's work on cultural political economy to understand how corporate and state actors and institutions mobilize architecture as one way of making new political and economic strategies socially meaningful. Due to its tendency towards recurrent crisis, capitalism is in need of 'social repair work' from time to time where new economic reconfigurations resulting from crisis need to be adopted as normal practice by societies. In the context of recent economic crises, the use of iconic architecture is not merely a strategy for legitimization but is an integral part of economic projects, generating discourses and semiotic elements, or 'economic imaginaries', through which the reconfigurations associated with global capitalist change become real and embedded. Thus, architectural production is more than just the objectification of the desires of a tycoon, corporation or a state but is part of a new radical imagery and tells society what to desire and how to desire it (Kaika 2010, see case study below).

If these new economic imaginaries are to be persuasive and resonate then they must chime across different cultural practices. Jones (2009) uses Bourdieu's idea of the 'field' to highlight how architectural practice, as a cultural field, recasts the economic imaginaries associated with global capitalism within its architectural discourses. Of significance is the way in which internationally renowned starchitects are able to cast themselves as artists and distant from commercial practice through an emphasis on the aesthetic and semiotic in architectural design discussions (such as through the architectural press and awards). Starchitects, by virtue of the prestige and mystique socially accorded to creativity, add value to iconic buildings through their decisions about design, conferring prestige and a presumption of quality even if this is not apparent to the observer. Thus, the iconic building becomes seen as part of the long-standing canon of great architecture rather than merely a commercial venture.

This links to Sklair's (2005) observation that in acting as agents of global capitalism, starchitects play a key role in mediating the demands of the transnational capitalist class, the demands for local place identity and the aesthetic rules of the architectural field in the design of iconic buildings. In seeking to appeal to the internationally mobile capitalist class, these iconic structures are designed to be visually consumed in a touristic way which leads to an emphasis on the spectacular and out of the ordinary. This encourages distinctive and memorable designs which are particularly successful if they are media friendly and can be reduced to a logo image (Sklair 2005). The skyline becomes something of an obsession and cities become a 'backdrop to a display of curious architectural objects' (Kaika 2010: 471). Thus, for Sklair (2005) one of the defining features of an iconic building is that it generates wide discussion and debate, with a key tension being between the production of these spectacular and unusual transnational architectural spaces, that seem to float free from the city itself and could be almost anywhere, and desires for these structures to exhibit a certain place rootedness and act as local markers. The connection between these iconic structures and the cities in which they sit, and their use by urban populations, are key issues highlighted both in Lees' (2001) case study of the new public library in Vancouver (see above) and Kaika's (2010) examination of London's changing commercial skyline (see below).

Case study: iconic buildings and the changing skyline of the City of London, UK

In her 2010 paper, Maria Kaika examines the development of recent iconic architecture in the City of London as a response to a period of crisis and change in the institutional structures governing the City (this being the central historical and financial core of London rather than the wider metropolitan area). Here iconic architecture produced under moments of restructuring can be seen as an active part of the 'radical imagery' produced by elites or institutions whose identity is in need of reinvention. They act as 'totems of the city in the making'. Her analysis differs from more standard analyses that see iconic buildings such as these as signifiers of economic success.

From 2000 onwards, the skyline of London has been significantly altered by the development of new, iconic, tall buildings, of which the Swiss-Re Tower (or The Gherkin) was the first to be built and is the most widely recognized. Kaika argues that the emergence of these iconic structures is linked to the recent institutional reconfiguration of the Corporation of London, the authority that runs the City and that has significant real estate holdings in the City and planning powers. Prior to the 1970s the Corporation, an ancient institution with its origins in the twelfth century, had a fairly traditional and insular 'English' character and outlook. This was challenged from the 1970s onwards by the rapid liberalization and internationalization of London's economy, particularly its financial markets. Resistance by the Corporation to this 'foreign' invasion of companies, people, architectural styles and new technologies and practices weakened the City's global financial status as key businesses lost global ranking or were taken over, and also led to threats from Prime Minister Tony Blair's Labour Government, keen to promote London as a global business hub, to abolish the Corporation.

In response to this challenge, the Corporation rebranded itself in the early twenty-first century as an outward-looking institution, opening up to link with the surrounding localities (through a range of charity initiatives) and to London's new transnational business elites. This new image was reflected in the new architectural language of the 2002 Development Plan for the City which favoured tall buildings over the conservation orientated ideas of earlier plans and which resonated with the pro-tall-building stance of the Mayor of London. Whereas the City's earlier iconic buildings, such as the classical style Bank of England building, identified with the City's traditional institutions, these new icons operate more as branding objects for transnational corporations or as speculative objects for real estate developers, reflecting a new type of footloose architectural patronage.

Those transnational corporations and elites commissioning the buildings display little place loyalty, with these buildings frequently sold on not long after they have been built (in the case of Swiss-Re) or in some cases before they have been built (e.g. Bishopsgate Tower). Similarly, urban life and the buildings erected are viewed as relatively transient, with the role of urban governors to facilitate the development of 'acupuncture points' in the city – an urbanism of negotiation with transnational business elites. Like the

continued

organizations and people commissioning them, these iconic buildings display little attachment to the city that surrounds them. Based on interviews with city workers, Kaika argues that these buildings create little unsurveilled public space that is actually used and serve to further remove the City of London from the cognitive maps of urban dwellers, who while recognizing these buildings as London icons find it hard to locate them on a map of the city!

Source: Kaika (2010)

Architecture of fear

If iconic buildings and spectacular developments represent the playful and carnivalesque side to post-modern urbanism then fortress architecture represents its dystopian side. Urban scholars have been increasingly engaged in examining the privatization of the public realm, the proliferation of security and surveillance systems and gated defensive communities (Fyfe and Kenny 2005; Minton 2006, 2009; Pacione 2009). The public realm is a contested terrain within the city where attempts by authority to impose and maintain order meet, and sometimes clash with, people either knowingly resisting this authority and/or unknowingly transgressing accepted social, legal and spatial norms (see also chapter ten). Architecture, planning and urban design have a long history of designing urban landscapes with the aim of controlling bodies of people and fostering particular modes of behaviour and urban living, either explicitly in the form of orders and instructions inserted into the urban landscape or more subtly through the design of elements of it (see figures 10.4 and 10.5 in chapter ten).

Coaffee and Murakami Wood (2008) note that from the 1960s onwards, cities in the Global North have increasingly used defensive architecture and urban design as a response to urban riots and perceptions of increasing crime and disorder within cities. This was reinforced by behavioural research which indicated a relationship between certain types of environmental design and reduced violence and crime. Oscar Newman's discussion of defensible space (1972), where he advocated the use of a range of real and symbolic barriers and enhanced opportunities for surveillance to combat crime, stimulated intense debate on the relationship between crime and the built environment. However, his message of employing considered design of the built environment to promote resident control of neighbourhoods became somewhat misrepresented in later discussions of situational crime prevention and ideas about the target hardening of buildings and areas against criminal and anti-social activities.

Recently, defensible space approaches and pro-security discourses have resurfaced to again influence the design and management of the urban landscape (Coaffee and Murakami Wood 2008). In response to increasing socio-economic fragmentation and urban insecurity, city authorities have employed a range of crime displacement measures and surveillance strategies, seen to be most extreme in American cities, particularly within Los Angeles (see case study below). Los Angeles' overemphasis on its militarization and fortress urbanism has seen the city portrayed as 'an urban laboratory for anti-crime measures' (Coaffee and Murakami Wood 2008: 359). Mike Davis is one of the most cited authors on 'Fortress LA' where he paints a future vision for the city as technologically and physically segregated into zones of protection and surveillance (Davis 2000b). While some have questioned his vision of Los Angeles as extreme, it is nevertheless clear that more people within cities live, work or play in some form of secured access environment, segregated from those 'others' perceived as threatening or undesirable, which has potentially dangerous implications for the equality and conviviality of urban life (see chapter seven).

Case study: security in public space: New York, Los Angeles and San Francisco

The paper provides a rare empirical examination of the much-discussed loss of publicness in cities due to the increased presence of anti-terror security zones and related behavioural and access controls. The paper highlights how security landscapes have shifted from the hard, intense, militarized architecture of the late 1990s and early 2000s, explored in much of the writing on Los Angeles as a post-modern city, to a more camouflaged, less obtrusive, approach in recent years. Additionally, Németh points out that security policy and planning have become more decentralized, currently undertaken less by singular, public entities and increasingly by networks of public and private actors. The paper moves beyond consideration of security measures for individual buildings and structures to consider the district-wide security apparatus in the civic centres (largely public buildings) and financial districts (largely private buildings) of the selected 'high risk' cities of New York City, Los Angeles, and San Francisco. While of

similarly high risk, the three cities and the districts studied in each differ in size, population and urban density, and exhibit very different configurations of public space.

In each neighbourhood Németh evaluated security landscapes using a simple set of criteria to assess the intensity, permanence, and location of individual security measures. Areas were scored on their access restrictions, including bollards, planters, gates, or fences located at entry points to a space or building, behavioural controls, including posted signs prohibiting activities or design features to discourage actions like sitting or gathering, and surveillance measures, including security guards and other human surveillance, but not CCTV cameras given their ubiquitous nature in cities. Geo-tagged photographs were taken of the security measures and uploaded into a GIS system which was then able to calculate the extent and intensity of the security zones in the study areas. Variables relating to the security policy landscape in each district were also added

continued

into the assessment. Statistical analysis reveals that the security landscape across all six districts covers a significant 20.4 per cent of the total public space footprint. This varies between cities, with the security landscape only covering 3.4 per cent of San Francisco's Civic Centre, but 35.7 per cent of New York's Civic Centre. Additionally, the security landscape is most prevalent and intense in New York City, linked principally to policy responses following recent terrorist attacks. The research also reveals that security zones governed by multi-stakeholder networks are more intense and militarized than zones managed by a single entity, and that these networks of actors are playing an increasingly significant role in mediating and regulating the governance of urban space and creating new geographies, or landscapes, of security in cities.

Source: Németh (2010)

Summary

This chapter has explored urban geography's relationship with architecture. Architecture is the art and science of designing buildings, and as a practice and profession it has a profound impact on the look and functioning of a city and the lives of people within it. Urban geography has moved from studying architecture purely as a cultural artefact representing human achievement to considering it as a commodity as part of capitalist processes of urban development and also examining its symbolic qualities in reflecting economic, cultural and political power. More recently it has also begun to explore the everyday, embodied consumption of architecture by the people who use it.

As a profession architecture plays an important role in transcribing the economic, social, cultural and political dynamics of the city into physical settings. The values and ideas of broad social structures such as patriarchy and capitalism are literally written into the urban landscape through architectural practice. The design of suburban landscapes and houses has reflected traditional assumptions about the productive and reproductive roles of women and men, while the recent proliferation of iconic buildings within cities designed by 'starchitects' has been seen as integral to the embedding of global capitalist processes into cities. Similarly, architecture is part of the desire of urban authorities to impose order on the city and cleanse the public realm of groups or behaviour deemed inappropriate or undesirable. Increasingly forms of fortress architecture are evident in cities, where people reside in secured access environments, exacerbating feelings of fragmentation and exclusion.

Follow-up activities

Essay title: 'Critically consider the ways in which power and cultural dominance can be seen to be symbolized in the urban landscape.'

Commentary on essay title

An effective answer would begin by exploring the idea of buildings and the urban landscape as 'produced' in that they can be seen to embody wider socio-cultural values and the values of those creating landscapes. Here the landscape can be seen as a text. It would also look to examine the issue of power and the idea that particular dominant cultural ideas and groups are more able than others to express their values in the urban landscape. Here case study examples could be drawn upon to illustrate the ways in which power, such as that exercised by the state, religious groups or large corporations, is expressed in specific urban forms and in the design and layout of urban landscapes. An excellent answer would then develop these themes and consider the arguments of Goss and Lees which highlight the problems of trying to uncover the meanings embodied in urban landscapes. It would consider the idea that the meanings of buildings are always multiple and contested and that meaning is not inherent but changes according to the social interaction of city dwellers with the urban environment.

Project idea

Undertake a 'reading' of an important public or commercial building in your city or a city that is familiar to you. Begin by developing a 'thick description' of the building, using the framework outlined by Domosh (1989) in her work on skyscrapers in New York. Here you will need to look for archive sources that give you information about when the building was developed, by whom and for what purpose. Good sources of information could be planning department records, local history sources and newspaper articles. When you have developed your 'thick description', develop this into a broader critical geography of architecture, drawing on the ideas of Goss (1993) and Lees (2001). One avenue for further work would be to undertake the first part of this project idea as a group, with each person producing his or her own 'thick description' and then comparing these to see the extent to which multiple interpretations of the building can exist and why this might be. Another avenue would be an examination of the users of the building and their reactions to it. This might either involve ethnographic research, as Lees undertook in the Vancouver library, or interviews with users, as in the case of Kaika's exploration of spectacular architecture in London.

Further reading

Books

- Coaffee, J. (2016) *Terrorism, Risk and the Global City: Towards Urban Resilience*, Abingdon: Routledge
  Focusing on London, the book extends the discussion of fortress landscapes to consider how terrorism is reshaping the contemporary city and the challenge of developing city-wide managerial measures and strategies.

- Ching, F.D.K., Jarzombek, M.M. and Prakash, V. (2006) *Global History of Architecture*, Chichester: John Wiley & Sons
  This book offers a comprehensive discussion of global architectural history, refreshingly not just from a western perspective.

- Fainstein, S.S. and Servon, L.J. (eds) (2005) *Gender and Planning: A Reader*, New Brunswick, NJ: Rutgers University Press
  An edited collection of key writings about gender and the design and planning of the city.

- Ingersoll, R. and Kostof, S. (2013) *World Architecture: A Cross-Cultural History*, Oxford: Oxford University Press
  This comprehensive student-friendly text extends discussion of architecture beyond the sole consideration of great monuments to find connections with ordinary dwellings, urbanism, and different cultures from around the world.

- King, A. (2004) *Spaces of Global Cultures: Architecture, Urbanism, Identity*, London: Routledge
  King's book examines how architectural and building cultures are affected by transnational processes and looks at the impacts of translating building types from one culture to another.

- Till, J. (2009) *Architecture Depends*, Cambridge, MA: MIT Press
  Written by an architect, this book is one of a number of recent texts offering a critical look at current architectural practice as out of touch with the 'ordinary' world and the users of buildings.

Journal articles

- Datta, A. (2006) 'From tenements to flats: gender, class and "modernization" in Bethnal Green Estate', *Social and Cultural Geography*, 7(5): 789–805
  Examines the constructions of spaces of domesticity through the interaction of residents and the buildings they inhabit.

- Jacobs, J.M., Cairns, S. and Strebel, I. (2012) 'Doing building work: methods at the interface of geography and architecture'. *Geographical Research*, 50(2): 126–140

Useful paper that explores the methodology for studying at the interface of geography and architecture and considering buildings as 'events', charting the diverse afterlives of two modernist housing schemes in Glasgow and Singapore.

- Jones, P. (2009) 'Putting architecture in its social place: a cultural political economy of architecture', *Urban Studies*, 46(12): 2519–2536
  Part of a special edition of the journal covering architecture, the paper explores the role of and practice of architecture as a profession.

- Lees, L. (2001) 'Towards a critical geography of architecture: the case of an ersatz colosseum', *Ecumene*, 8(1): 51–86
  Key paper underpinning the development of recent critical geographies of architecture.

- Moran, D., Turner, J. and Jewkes, Y. (2016) 'Becoming big things: building events and the architectural geographies of incarceration in England and Wales', *Transactions of the Institute of British Geographers*, 41(4): 416–428
  This extends geographies of architecture beyond 'signature' buildings to consider the production of the banal, carceral architecture of new prisons.

- Smigiel, C. (2013) 'The production of segregated urban landscapes: a critical analysis of gated communities in Sofia', *Cities*, 35: 125–135
  The article provides a case study of the emergence of gated residential suburbs in Central and Eastern European cities, and examines both their specific form and the political-economic processes underpinning their development.

Websites

- www.architecture.com/ – The website of the Royal Institute of British Architects (RIBA). A wealth of resources aimed at the architectural profession.

- www.american-architects.com/ – Some excellent visuals and profiles of American architects.

- www.worldarchitecturenews.com/ – Architectural news from around the world. Invaluable guide to the contemporary scene.

- www.women-in-architecture.com/index.php?id=35 – Women in Architecture. Resource exploring the issues faced by women in the architectural profession.

⑨ Images of the city

Introduction

All cities have images. These are simplified, generalized, often stereotypical, sometimes contrasting impressions that people hold about cities. It is impossible to know our environment in its entirety so our impressions of the world are inevitably partial and incomplete. Our knowledge of the world is heavily influenced by representations of places in such things as the news media, television, films, art, novels and poems or in conversations with friends and acquaintances.

Exercise

Our everyday lives are saturated with images and representations of a variety of places. Try to keep a diary of the ones you encounter for one day (think also about places that you don't encounter representations of and why this might be). Note the places whose representations you encounter, the media in which you encounter them, the nature of these representations and the ways in which they are partial and selective images of these places. In what ways do you think these representations shape your impressions of the world? How important are representations of cities, compared to those of other environments?

These representations typically exaggerate certain features, be they physical, social, cultural, economic, political or some combination of these, while reducing or even excluding others. It is easy to dismiss or ignore the significance of these representations. Indeed, it was rare, until recently, to find discussions of images and representation within urban geography textbooks. However, representations of cities are a vital aspect of our relationship to them and something that geographers and others have increasingly made the focus of their attention. Given that we are likely to encounter only a very small number of cities directly, compared to those that we encounter through representations, we could conclude that the majority of our knowledge and understanding of cities is derived not from direct experience but through representations – including those you will encounter in textbooks like this one. This chapter explores the nature and variety of these representations of the city and the images that they generate, and evaluates their significance.

Exercise

Think of a town or city that you know well, your home town or the place that you are studying in, for example. Does it have a clear, distinctive, dominant image? If so, how would you describe this? How has this image been formed? Has it changed over time do you know? Are there any alternatives to this image? Do residents share the same impressions as visitors? Who do you think shapes or determines this image? Have there been any recent attempts to change or update the image of the urban area you are studying? Reflect on your answers to these questions as you read the remainder of the chapter.

Urban images through time

Representations of the city have a very long history. We can trace positive and negative images of cities (often referred to as pro- and anti-urban myths) back to ancient Greek and Roman civilizations, find them in texts like the Bible and can see them as recurrent through every subsequent historical period (see Gold and Revill 2004; Short 2005; Hubbard 2006). It has been, and remains, common for images of the city to be constituted in opposition to those of the countryside and for these two environments to be seen to stand for very different sets of values. This is despite many arguing that this categorical distinction between city and country is difficult to sustain in actuality.

From this we can see that images of the city (and the countryside) are typically implicated in a series of wider moral debates. The cultural historian Raymond Williams has written extensively about the histories and wider cultural and moral significance of these images in his influential book *The City and the Country* (1973).

Typically, urban images are examined through the pro- and anti-urban framework adopted in table 9.1. While undoubtedly a very useful device for categorizing the variety of images from the plethora of representations of the city that have emerged through time, it does carry with it the danger of underplaying the complexity and ambiguity of many of these images. Not all urban images are categorically pro- or anti-urban. While most commentators are quick to recognize this, the deployment of the pro- and anti-urban lens through which to examine representations of the city has tended to underplay this ambiguity. Also, in focusing on this, accounts have tended to underplay other important dimensions of urban representations, such as the relationships between the formal qualities of representations and the experiences or qualities of the city they seek to depict. While not trying to demolish the pro- and anti-urban framework we want to use this section to draw attention to some

Table 9.1 *Pro- and anti-urban myths*

....

Pro-urban, the city as:	Anti-urban, the city as:
Civilized	Dangerous
Exciting	Ugly
Liberating	Corrupting
Romantic	Alienating
Modern	Polluted
Cultural	Commercial
Diverse	Fragmented
Planned	Sprawling
Managed	Chaotic

Source: Gold and Revill (2004); Short (2005); Hubbard (2006)

other dimensions of urban imagery. We will now consider this with reference to one example.

The city is a common theme explored in all sorts of popular music. There are many examples where songs have expressed seemingly unequivocally positive or negative views of cities. We can think, for example, of the edgy paranoia of a violent and bankrupt New York in Grandmaster Flash and the Furious Five's 'The Message' (1982) with its recurrent refrains of 'Don't push me 'cos I'm close to the edge' and 'it's like a jungle sometimes, it makes me wonder how I keep from going under'. An equivalent from the UK would be The Specials' 1981 hit 'Ghost Town' which paints a bleak picture of British cities at the time as fractured, violent, run-down places haunted by the ghosts of what they were. However, these are in sharp contrast to celebrations of the city in songs such as 'Paris' (2008) by Friendly Fires which sees the city as a place of romance, escape, hedonism, freedom and aspiration. In standing for these qualities, however, the city of Paris is seen as all that the British suburbs, from where it was written, are not (see chapter three; Silverstone 1997; Gold and Revill 2004; Short 2005). We should think, therefore, not only of the images of cities but also of which parts of the city these images refer to.

Country music, as the name would suggest, is a form of popular music that has been firmly associated with strong anti-urban sentiments. Cities in this genre are typically portrayed as lonely, alienating, corrupting places of loss and failure. Kris Kristofferson's (1969) song 'Sunday Morning Comin' Down' reflects this widely held view of the city:

On the Sunday morning sidewalk,
Wishing, Lord that I was stoned.
'Cos there's something in a Sunday,
Makes a body feel alone.

And there's nothin' short of dyin',
Half as lonesome as the sound,
On the sleepin' city sidewalks:
Sunday mornin' coming down.

One of the most innovative country music writers and performers was Gram Parsons (1946–1973). Ranked by *Rolling Stone* magazine in 2004 as one of the 100 most influential popular music artists of all time, he is widely credited with taking country music beyond its original constituencies and introducing it to a rock music audience. Parsons' music is typically interpreted as displaying a strongly anti-urban streak, like much other country music, which, drawing on long traditions within and beyond the genre of seeing cities as corrupting sites of moral decay, portrays them as places where dreams are wrecked over gambling tables, in run-down bars and on lonely streets. By contrast Parsons seems to idealize the countryside, associating it with the family, nostalgia for childhood, safety and belonging. For example, in his 1968 song 'Hickory Wind':

In South Carolina there are many tall pines
I remember the oak tree that we used to climb
But now when I'm lonesome, I always pretend
That I'm getting the feel of hickory wind.
…
It's hard to find out that trouble is real
In a far away city, with a far away feel
But it makes me feel better each time it begins
Callin' me home, hickory wind.

Confining analysis of Gram Parsons' music to the texts alone, the songs and their lyrics, would seem to confirm this anti-urban interpretation. However, if the sites of analysis are extended somewhat and the songs interpreted in the context of Parsons' personal biography, a potentially different and more ambiguous interpretation emerges. Parsons was certainly no backwoods hermit during his short life. From a wealthy southern family he spent much of his adult life in cities such as Boston, where he briefly attended Harvard, New York City and Los Angeles before spending time in France with members of The Rolling Stones. His passage through these predominantly major urban spaces was characterized by liberal, and ultimately fatal, indulgence in drink and drugs. Read in this context, Parsons' songs can be seen less as a simple expression of the anti-urban sentiment, more an attempt to both articulate and negotiate his own simultaneous attraction to and revulsion from the lures of the city. The implicit 'doubleness' of the urban imagery in Parsons' work can be seen in his evocation of the destructive, yet addictive, attractions of the city in his posthumously released song 'Ooh Las Vegas' (1974).

Ooh Las Vegas, ain't no place for a poor boy like me.
Every time I hit your crystal city
I know you're going to make a wreck out of me.

This example highlights the ways in which the meanings of urban imagery can shift with the contexts within which representations are examined. Traditionally though, analysis of urban imagery has tended to focus on the representations themselves at the neglect of other potential 'sites of meaning' (Rose 2006), such as their production or their consumption by audiences. Opening up representations to this wider analysis potentially results in both more nuanced and more complex interpretations.

Country music is not the only medium that demonstrates such ambiguity in its urban imagery. It can be seen across all media which take the city as their subject. For example, a number of researchers have noted 'polysemy', the tendency for a number of different meanings to co-exist within a single text, as characteristic of media coverage of the city. Jacqueline Burgess's (1985) analysis of the national press coverage of a number of street disturbances in British cities during the 1980s discerned notes of sympathy for inner-city populations, even in articles that elsewhere labelled these spaces through evocations of pathology and war.

It is important, therefore, when considering representations of all kinds not to uncritically accept overly one-dimensional interpretations. The case study that follows highlights ways in which the production process behind news texts contributes to the complexity of their meanings.

Case study: contesting the urban renaissance

The widespread regeneration of many post-industrial cities around the world has involved the reconstruction of their images around notions of an urban renaissance. This has been prevalent in representations such as promotional materials, the local media and speeches by politicians, developers and business leaders. However, by contrast, a rare public critique of this process emerged in sections of the UK national press in the mid-1990s. Criticisms highlighted here included the negative physical, social and economic impacts of projects of regeneration on local communities and the emphasis in these projects on image enhancement and physical and economic development rather than attention to issues of basic social welfare and community development. Typical was the following:

> It may have been the outbreak of scabies or it may have been the woman who was thrown through the toughened-glass screen of the housing benefit office in Birmingham. But in a city where a 'go ahead' Labour council was meant to have tackled economic decline by producing a

continued

'breathtaking civic renaissance' – complete with a new convention centre, sports arena, tourism office, orchestra hall, ballet company and opera company – the fact that thousands of citizens were facing eviction because of industrial action in a council benefits office was, to say the least, an image problem.

(Cohen 1993: 11)

It is rare to see such open criticism of urban regeneration in the British national media. It contrasts sharply with earlier press coverage of the process which was largely positive and highlighted the probable benefits to cities. There are a number of reasons for the emergence of this critique. Journalism is not a closed discourse, rather it is one that overlaps with a number of other discourses, in this case including academic analysis and debate about urban change. Academics, who were growing increasingly critical of the regeneration of many cities at the time, were often quoted in articles and cited as sources of expert opinion. Typically, they would introduce a very critical take on the process of regeneration. Second, a production process, often not acknowledged in analysis of their meanings, lies behind news texts. This process involves the production of copy by reporters, its editing by editors, the production of photographs by photographers and picture editors and the creation of headlines, captions and the design of layout by sub-editors. What emerges are complex texts that have passed through many hands and which are composed of many elements. While reporters might produce copy that adheres closely to notions of journalistic balance, albeit undercut sometimes by the placement of key sources at the start and ends of articles, the headlines and visual elements of articles typically project less balanced meanings. This was certainly the case in many of the articles that critically commented on urban regeneration in Britain in the 1990s. Examining news texts as a whole highlighted that their different elements projected subtly, and sometimes not so subtly, different meanings. For example, in John Arlidge's (1994) article on criticisms of the legacy of Glasgow's year as European Capital of Culture in 1990, while the article's main text largely balances opposing sources the other components of the article – the headline, headline paragraph, the picture and picture caption, which together make up the majority of the news text in terms of space and immediate visual impact – all emphasize the failure of the event to address deep-seated problems of poverty in the city. The picture shows a smiling Mr Happy figure (an icon from Glasgow's promotional campaign at the time) contrasted with a homeless man on waste ground under a crumbling Mr Happy mural. This message is reinforced by the main picture caption: 'No mean city? Since the original campaign poverty in Glasgow has worsened' and the headline 'Blob on the landscape'.

Source: Hall (2008)

It is common, then, to find a range of contrasting representations of places and associated processes and events in different media. It is not uncommon to find this even within a single medium, as the example above has demonstrated. Interpretations of these representations should acknowledge both the processes of production that lie behind them and the contexts within which they are presented. This is to say nothing of the interpretations of these texts by audiences which might produce a very different set of readings altogether. Recognizing these multiple sites of meaning will produce more rounded interpretations of these representations and their associated imagery.

Another dimension that it is important to acknowledge when analysing urban representations is the formal qualities of texts. We can see the significance of this if we consider the ways that some, particularly avant-garde, writers and artists have explored the relationships between the formal qualities of media and the nature of the urban experience. The late nineteenth and early twentieth centuries were times of huge change in the cities of Europe and North America. These changes included the restructuring of capitalism as it became increasingly industrial; the associated rise of the factory system and the consequent rapid growth of urban areas; the emergence of new building and architectural forms such as the first skyscrapers; technological changes in transport, telecommunications and media including the emergence of photography, film and advertising; and a greater heterogeneity to social forms such as the crowd and to urban social life more generally (Berman 1988). Modes of artistic representation that developed at this time sought to give expression to these experiences of capitalist modernity as it unfolded primarily through large cities. Modernist painters, photographers, film makers and writers shattered the formal limitations of the prevailing artistic mode of expression, realism, and introduced a series of bold, radical innovations intended to capture the exhilarating yet chaotic modern urban experience.

The imprint of capitalist modernity was felt most heavily in the great cities of Europe and North America such as London, Paris, Berlin, St Petersburg and New York. Much modernist art, literature and film was explicitly concerned with exploring and finding new ways of representing the development of an apparently distinctly modern urban consciousness in these cities. They were cities in which few of the settled certainties of the pre-modern world remained. The everyday experiences of these cities were at the same time exhilarating yet frightening, dynamic yet precarious, liberating yet disturbing (Berman 1988). This engendered a crisis of representation among writers and artists as the formal conventions of realism precluded its ability to give artistic expression to these experiences. Subsequently, a number of avant-garde artistic movements emerged that engaged directly with, and frequently took as their subject, the profound physical and social transformations of urban space. These included techniques such as Cubism, Futurism, Fauvism and Montage, which together radically altered the conventions and boundaries across a range of artistic media. Cubism's formal innovations included juxtaposition, whereby temporal and spatial order was disrupted and replaced by montages of temporal fragments and multiple perspectives within the same frame.

A number of painters employed cubist techniques in representations of urban subjects. Examples include Robert Delauney's *The City of Paris* (1912), Ludwig Meidner's *I and the City* (1913) and *Berlin* (1913), and George Grosz's *The City* (1916–17). One of the most potent symbols of modernity was new technology. Numerous artists took this as their subject, including Delauney whose *Eiffel Tower* (1910) viewed its subject from numerous perspectives simultaneously and Gino Severini's *Suburban Train Arriving in Paris* (1915) (figure 9.1) which depicted a train punching through the outskirts of the city scattering houses in its wake, a symbol of the disruptive impacts of new technologies on urban life.

Examples from other media include Paul Citroen's photo-collage *Metropolis* (1923), films such as Walter Ruttman's *Berlin, Symphony of a Great City* (1923), Dziga Vertov's *Man with a Movie Camera* (1929), poems such as T.S. Eliot's *The Waste Land* (1922) and a number of novels (see case study).

Case study: city of scrambled alphabets

John Dos Passos' (1925) novel *Manhattan Transfer* illustrates well the formal deconstruction that lay behind many modernist modes of representation. The novel is written using a montage technique and details the lives of a vast range of characters from a wide social spectrum in New York between 1892 and 1920. Short episodic scenes rapidly follow one another with little to indicate any transition. The novel covers three distinct time periods but is not presented in chronological order. Finally, the pace of the narration varies wildly from scene to scene. The following passage describes the experience of travelling through streets that are reminiscent of cubist paintings in that they are saturated with montages of wildly contrasting advertising imagery and their slogans:

> He walked north through the city
> of shiny windows, through the
> city of scrambled alphabets,
> through the city of gilt letter signs.

Spring rich in glutten … Chockfull of golden richness, delight in every bite, THE DADDY OF THEM ALL, spring rich in glutten. Nobody can buy better bread than PRINCE ALBERT. Wrought steel, monel, copper, nickel, wrought iron. *All the world loves natural beauty.* LOVE'S BARGAIN that suit in Gumpel's best value in town. Keep that schoolgirl complexion … JOE KISS, starting, lighting, ignition and generators.

(Dos Passos 1925: 351)

Formally *Manhattan Transfer* is more closely related to cinema than traditional forms of the novel. Using these techniques it conveys the intensity and diversity of city life in ways difficult to envisage within a more realistic narrative.

Source: Brosseau (1995)

Figure 9.1 *Gino Severini's* **Suburban Train Arriving in Paris** *(1915)*

Source: Tate Images, ADAGP, Paris and DACS, London.

The case study above highlights again the ambiguities inherent in much urban imagery. Indeed, it was the intention of a number of modernist artists to capture the contrasting nature of the urban experience in their work. Certainly it would be difficult to reduce many of these representations to simple pro- or anti-urban positions.

Representation, power and cultural politics

Representations do more than merely convey information and impressions about places. They are also implicated in struggles over notions of identity and the material development of places. For example, a series of negative media reports about an area may deter people from buying properties, visiting or investing there, contributing to a downward spiral of decline. They may also affect attitudes towards that place held by people such as local politicians, public officials and the police. The accuracy and validity of these negative images may be challenged by local residents and we may see the development of struggles around the ways in which places are represented. One such conflict developed around the representation of the South Bronx neighbourhood in New York City in the 1981 film *Fort Apache, the Bronx*. The film depicted the neighbourhood as violent and dangerous, peopled largely by prostitutes, criminals and drug addicts and barely controlled by a beleaguered police force. The film mirrored long-standing stereotypical attitudes towards cities, particularly their inner city areas. *Fort Apache, the Bronx*, while popular around the world, caused an outcry in the South Bronx where residents and community groups argued that it was a biased, inaccurate representation of their area and one that would prejudice attitudes towards it. The criticism of inaccuracy is commonly levelled against place representations of all kinds. However, all representations are inevitably partial and selective. They cannot hope to be completely comprehensive representations of even the smallest areas, even if this was their intention (Crang 1999; Holloway and Hubbard 2001).

Struggles over the ways in which places are represented are a form of cultural politics, conflicts, often involving unequal power relations, around issues such as the supposed image or identity of places and the rights to define these. It is important when thinking about representations to appreciate their implication in processes of cultural politics. The following questions offer ways in to exploring these issues:

● Who are the producers of representations?
● For what purposes are representations produced?
● What are the contents and nature of representations?
● Who are the audiences of representations?
● In what ways are representations interpreted or consumed by these audiences?

Those responsible for the production of representations of the city are many and varied. Some, such as global media corporations, are very powerful in that they control the channels through which representations are distributed. It is the images of cities produced by groups such as this that tend to be the best known and most widespread, the dominant representations of cities. However, as the case of *Fort Apache, the Bronx* demonstrates, such representations may be contested by groups positing alternative ways of seeing places.

Moving on to the second question above (for what purposes are representations produced?) we find that there are several reasons for this. Underlying all of the specific reasons, however, is a desire to control, or at least influence, how others see places. The reasons for this are numerous; they may include trying to attract visitors or investors, or in the cases of fictional or literary dramas, to evoke a particular atmosphere. It is important to remember, however, that this desire to influence how people view places is often implicated, both directly and indirectly, in the ongoing development of places. Even seemingly objective representational genre such as photographs have been commonly deployed to this end. In the nineteenth century, for example, it was common for the Medical Officers of Health in British cities to commission photographic surveys for the purpose of recording the conditions in different parts of the city. In 1890, the Medical Officer of Health in Leeds, a city in the North of England, commissioned such a survey in Quarry Hill, a district in the east of the city (Tagg 1993). In one sense this resulted in a comprehensive archive of the levels of poverty and ill health in the area. However, the photographs, drawing heavily on their reputation for accuracy of recording, were deployed as part of an effort to persuade a Parliamentary Select Committee to see Quarry Hill in a particular way, namely as a problem, and to try to convince the committee to sanction the area's demolition and replacement with new development. The photographic archive was presented to the Select Committee along with other evidence, such as official statistics and reports, that reinforced the impression of the problematic nature of the area at the exclusion of all other possible interpretations (Tagg 1993). Representations then, can be employed in determining the future paths of development of places. Indeed, they are central devices in many official discourses of urban development (chapter five).

Returning to the case of *Fort Apache, the Bronx*, its role in this regard, initially at least, is less clear. While it cannot be claimed that the film was directly embroiled in discourses shaping the future of the South Bronx, although it was frequently referenced in debates about the Bronx within New York City during the 1980s, the film did draw on and help reproduce deep-seated and long-standing cultural attitudes towards inner city areas, particularly those associated with black and ethnic minority groups. Such attitudes are in part fed and sustained by certain groups who do have an interest in the control and development of such neighbourhoods, groups such as politicians, police authorities, social reformers and developers for example (see Wilson 1996). So, although the producers of *Fort Apache, the Bronx* had no intention of intervening in the material development of the neighbourhood they depicted in the film, it did contribute to the perpetuation of stereotypical, partial and selective representations of inner city areas and the associated negative attitudes towards these areas that still tend to prevail. These attitudes ultimately shape the options facing neighbourhoods like the South Bronx.

Struggles around the definition and promotion of image at the city level have become central aspects of the economic development of urban areas and increasingly the physical reshaping of their landscapes. This often involves a cultural politics where competing visions of the city come into opposition. It is to the case of place promotion or city marketing and branding that we now turn.

Selling cities: image and urban change

As we noted in chapter four, as many cities around the world began to suffer the ill-effects of deindustrialization they began to turn their attention to the potential of sectors such as the service industry, sport, leisure, culture and tourism to generate new jobs and investment. This necessitated a much greater emphasis on the enhancement and promotion of positive urban images. It is clear that both the number of cities involved in this marketing, promotion or branding and the range of methods and media employed to this end have increased since the 1970s (Barke and Harrop 1994: 97). However, despite this, for many years cities remained 'undersold' (Holcomb 1994). Bailey (1989: 4), for example, cites the case of economic development advertising in the US media which totalled $46 million annually in the late 1980s, while at the same time around $100 million was being spent each year on the promotion of 'Miller Lite' beer. Similarly place promotion was, for many years at least, undertaken largely by members of local authorities with little specialist training. Consequently, Briavel Holcomb was able to argue in 1994 that 'the marketing of cities tends to be generic and repetitive' (p. 121). The increased importance of place promotion since then has seen both increasing funding and the greater professionalization of the process. Sophisticated city websites demonstrate that the Internet has emerged as a key platform upon which cities are now sold (Urban 2002), while city marketing has emerged as a recognized career trajectory among marketing professionals. Despite these developments, however, city marketing has not entirely escaped the criticisms of its earlier days. Commentators have noted that still cities tend to draw on a very narrow repertoire of images when promoting themselves. Chris Murray (2001: 6), for example, has argued that:

> the content of place marketing messages in promotional literature … reveals disturbing trends. Instead of a dynamic and challenging approach to local character, we are confronted with unrepresentative stereotypes and parodies of the past. Rather than an inclusive methodology that addresses local audiences, it is exclusively outward-looking, thereby ignoring whole sections of the population. Authenticity and reality are substituted for a burlesque caricature of place. The messages follow an insipid formula, which makes it difficult to distinguish one place from another.

Hudson et al. recognized many of the critical points above in their discussion of an alternative model of place branding which involved extensive community involvement in the development of place brands. They argue (2016: 1) that:

> Many places promote spectacular scenery, good quality of life, friendly people, and a sound business infrastructure. However, these factors are no longer differentiators, so places need a strong brand identity to stand out in order to attract people to live, work and play. But brand development is often driven by short-term, top-down approaches with limited community participation.

By contrast they discuss a model of brand development from the 'bottom-up' which builds the brand around the results of extensive workshops and in-depth interviews with local stakeholders, crucially the local community, who, they argue, are traditionally little included in tourism and brand development processes. Their approach, for the small town of Bluffton, South Carolina, offers a more democratic model of building a place brand. However, it is easy to see limitations here. For example, the authors are insufficiently critical of the mechanisms of community engagement they employ. While all community members were invited to participate in the brand development workshops, the 100–120 who participated are unlikely to be, in cross section, entirely representative of the town's 13,600 (2013) residents. Their discussion does not critically reflect on the potential absences of hard-to-reach groups within this process. Further, while this method of engaging the community in brand development might seem feasible and appropriate within a small-town setting with, perhaps, a relatively homogeneous community, it is less easy to imagine how it might be deployed in a much larger urban area with a range of more diverse neighbourhoods. Finally, the place attributes that emerged out of this process and which informed the development of Bluffton's new brand were not radically different to those that tend to inform more top-down place branding exercises.

Exercise

John Rennie Short (1996) has recognized four key sets of urban imagery that underpin a large proportion of city marketing campaigns. These are associated with the city as: multicultural, environmentally friendly, possessing a wealth of cultural attractions and an ideal location for new investment. The plethora of urban images that we encounter through marketing campaigns can be interpreted largely as versions and combinations of these basic sets of imagery. Look in detail at the promotional websites of two different cities. Can you classify the images they contain using the four categories outlined by Short? Can you find any images that do not fit these categories or could you devise an alternative set of your own categories? On the evidence of what you have observed on these two websites do you agree with Briavel Holcomb (1994: 121) that 'the marketing of cities tends to be generic and repetitive' and with the views of Chris Murray above?

Much criticism of city marketing by academics has sought to locate it within the cultural politics of the city. Many have commented on the apparent differences between the vibrant imagery of promotional campaigns and the persistence or worsening of social and economic conditions for large proportions of the populations of these cities (see Hambleton 1991; Watson 1991; Jones and Evans 2008). The enhancement of urban image, and the associated physical transformation of many city centre landscapes through projects of urban regeneration and development, have been labelled the 'carnival mask' of late capitalist urbanization (Harvey 1988: 35), the criticism being that while such city images create the impression of regeneration, change and vibrancy, they do little, if anything, to address the underlying social and economic problems that necessitated regeneration in the first place. Further, the images of cities generated through marketing campaigns may clash with those held by residents of cities, or at least some sections of them. Like all representations, marketing campaigns generate partial and selective impressions of cities. In being aimed at external audiences of potential investors, residents, tourists and companies looking to relocate, they are composed around elements likely to appeal to these groups. By contrast they tend to omit those elements of the city deemed more important to the everyday lives of residents, their senses of identity and collective memories of place. Aspects of cities that do not fit the profile projected to outsiders tend to be excluded from the dominant versions of the city. The cultural politics and conflict around Glasgow's image promotion since the 1980s has been extensively documented (see Boyle and Hughes 1991, 1995; MacLeod 2002; Hubbard 2006; Jones and Evans 2008). These are tales familiar in many other cities undergoing similar processes of regeneration and reimaging.

Exercise

Consider the following statement in the light of the promotional strategies for an urban area that you know well. 'The danger is clear: that city marketing strategies will deny any voice for, or celebration of, those elements of their indigenous culture that might impede the competitive search for wealth or jobs or a glossier image' (Barnett 1991: 168). Do you agree with the sentiment that city marketing is complicit in a process of exclusion? Do you feel that aspects of the city are excluded from the dominant images of this area? If so, what do you think are the implications of this?

City marketing has also been criticized for its apparent appropriation and sanitization of often complex and contested place and social histories and identities and their reconstruction within officially produced urban images. Where city marketing narratives do pay attention to 'alternative' or otherwise marginalized histories and identities it tends to do so by reducing them to a few stereotypical slogans or appealing images. The experiences of minority communities for example, tend to be reduced to that of a 'lively street life' or sanitized references to cosmopolitanism or diversity (see Dwyer 2005). Complex, contested identities, drawing on aspects of cities such as industry and work, memory, community and resistance tend to be stripped of any radical or confrontational associations and are represented through narratives such as skill, craft and tradition (Hall 1997). This is not to advocate an idealization of these alternative histories and identities, many of which, for example, may have been heavily gendered or exclusionary themselves, but rather to highlight their unproblematic (re)presentation through city marketing campaigns.

Finally, it is apparent that the reimaging of cities through marketing campaigns is increasingly driving the physical reshaping of cities through processes of regeneration and development. As image assumes ever greater importance in the post-industrial economy it is becoming clearer that the actual production of urban landscapes reflects the necessity for cities to present positive images of

Figure 9.2   Example, city marketing material
Source: New York City and Company.

themselves and that economic development is driven by programmes of promotion and marketing. It is important here to appreciate the distinction between 'selling' and 'marketing' or 'branding', for these are very different processes. Selling is a process whereby consumers are persuaded that they want what one has to sell. However, marketing or branding is a process whereby what one has to sell is shaped by some idea of what one thinks the consumer wants (Fretter 1993: 165; Holcomb 1994). The distinction between 'selling the city' and 'marketing or branding the city' is, therefore, crucial to understanding their relationship with urban development. 'Selling' the city is likely to impact most directly on the urban economy through increased visitor spending and outside investment. However, marketing or branding cities also impacts directly upon their landscapes and development. It would be true to say that prior to the 1980s cities were largely 'sold'. It is now truer to say that they are 'marketed' or 'branded'. City landscapes are increasingly shaped by the necessity to project positive urban images to the outside world (see chapter eight). Marketing or branding the city, therefore, is a process that is increasingly integral to the shaping of urban development, rather than incidental to it.

> 'Marketing' is starting to replace the concept of merely 'selling' … Place marketing has thus become much more than merely selling the area to attract mobile companies or tourists. It can now be viewed as a fundamental part of planning, a fundamental part of guiding the development of places in a desired fashion.
>
> (Fretter 1993: 165)

Case study: branding Montréal through design

City marketing initiatives have changed somewhat in recent years from those associated with major flagship projects of development to more diffuse initiatives involving a variety of actors. However, increasingly, these initiatives are tied to material changes in the urban landscape, leading to the creation of distinctive urban 'brands' that encompass both the images of cities and their urban spaces. One example is the annual competition for architecture and interior design launched in Montréal in Canada in 1995 (Commerce Design Montréal). This represents a new phase in the branding of cities in that the responsibility, and the associated costs and risks, are 'downloaded' to businesses and citizens. This approach also represents something of a shift in the nature of urban governance. While requiring relatively little public funding, the competition has generated a great deal of international media coverage and has been adopted by a number of other cities internationally. The aim of creative or design orientated approaches such as this is the differentiation of cities and the generation of economic, social and cultural value for cities or neighbourhoods. These initiatives extend the idea of the cultural destination from the art gallery or museum to include the commercial spaces of business.

The competition has undoubtedly benefited the winning businesses. Many have reported increased business and exposure as a result and designers within the city have reported similar upturns in business associated with the competition. Likewise, interest in design among the Montréal public has also grown as the numbers getting involved in public votes in the competition has increased greatly. Many winning businesses have been concentrated in particular neighbourhoods and the competition has helped in the creation of distinctive identities for these areas. Competition awards have not been a guarantee of commercial success, however, as a number of winners have subsequently closed. This has raised concerns that the competition places too much emphasis on the aesthetics of design rather than its functional and commercial aspects.

There is the potential though that initiatives such as this might generate negative external impacts. There is a danger that homogenization of design will result as a relatively small number of 'star' designers are employed by businesses who subscribe to

continued

international notions of 'coolness' rather than more diverse ideas of design. This can occur between cities as this model of city branding becomes exported internationally. Also, initiatives such as this can become implicated in processes of gentrification and the creation of socially exclusive urban spaces, indeed there is evidence of this in Montréal. Initiatives such as this have also been criticized for focusing on design as artefact and failing to incubate more substantial design linkages and cultures across the city. While the nature of city marketing has evolved and diffused across a broader range of actors within cities, it would seem that some of the earlier concerns expressed about its impacts and effectiveness remain.

Source: Rantisi and Leslie (2006)

Summary

Representations are a vital way in which we come to know cities. They are also, perhaps increasingly, bound up in the ways that cities change and develop in material ways. While history has been saturated with much unequivocally pro- or anti-urban imagery this is not to deny the significance of much ambiguous urban imagery and the values that have been associated with it. Almost inevitably representations of the city are implicated in processes of cultural politics. For this reason analysis of city images should go beyond merely noting bias, omissions and exaggerations. It should seek to uncover the ways in which representations are produced, the uses to which they are put and the ways in which they are interpreted by their audiences. Representations, then, are important. As Mike Crang (1999: 60) argues:

> We have to face the possibility that we may only be able to understand reality through words and images ... In this way we need to see images as actively creating the world rather than simply transmitting a prior reality.

Follow-up activities

Essay title: 'Cities are more concerned with enhancing their externally perceived image than they are with addressing basic issues of social and economic welfare.' Discuss.

Commentary on essay title

The question touches on the supposed current obsession among city governments with place promotion. An effective answer would note this but would suggest that the issue is not reducible to such a simple, bold statement that suggests an either/or situation. It might suggest, for example, that the relative emphasis varies between cities. The national government policy context might be important here or the nature and size of different cities. The emphasis in the answer would rest on the case studies chosen. A good answer might question the assumption that an emphasis in a city on image promotion necessarily means that basic social and economic welfare questions are less important. This might be the case but merely stating it would get little credit. More important would be the ability to convincingly demonstrate this link, drawing on evidence from case studies. Some of these issues are explored in the further reading outlined below.

Project idea

Select a town or city to study. Put together a portfolio of as many images of that place as you can find. These might come from the media, films, television, novels or poems, songs, advertising, place promotion campaigns or community or arts groups. Collect historical as well as contemporary images. Think of novel ways to present this portfolio.

Look at the nature and range of these images, noting similarities and differences between them. How does considering who produced these images, the purpose for which they were produced and for what audiences, help us understand the diversity in the nature of the images that you have collected?

Further reading

Books

- Gold, J.R. and Revill, G. (2004) *Representing the Environment*, London: Routledge (chapters seven and eight)
  Concise surveys of a range of urban representations. A great place to start.

- Hubbard, P. (2006) *City*, London: Routledge (chapter two)
  An excellent critical introduction.

- Pike, A. (ed.) (2011) *Brands and Branding Geographies*, Cheltenham: Edward Elgar (Part III)
  The take on brands in this edited collection is not limited to place branding. It looks at the geographical significance of brands to geography more generally. However, part III contains a number of chapters focusing on branding and space, most of them referencing urban spaces of various kinds.

- Rose, G. (2016) *Visual Methodologies: An Introduction to Researching with Visual Materials*, 4th edn, London: Sage
  The best guide currently available to a range of methods for deconstructing the meanings of representations and the use of visual methods within geography and the social sciences more generally.

- Ward, S.V. (1998) *Selling Places: The Marketing and Promotion of Towns and Cities 1850–2000*, London: E. & F.N. Spon
  The most comprehensive guide to place promotion /city marketing available. Lots of fascinating images of early place promotion materials. Great to see how the process has evolved over time.

- Williams, R. (1973) *The City and the Country*, London: Chatto & Windus
  A classic study, still very relevant today. Many subsequent editions available.

Journal articles

- Andersson, I. (2014) 'Placing place branding: an analysis of an emerging research field in human geography', *Geografisk Tidsskrift: Danish Journal of Geography*, 114(2): 143–155
  Most reviews of the place branding literature, such as the two cited below, are not written from a specifically geographical perspective. While this is not exclusively urban in its focus, its geographical take makes it particularly relevant here.

- Evans, G. (2003) 'Hard branding the cultural city: from Prado to Prada', *International Journal of Urban and Regional Research*, 27(2): 217–240
  Critique of culture-led regeneration and place marketing, drawing comparisons with commercial product branding. A widely cited piece of research well worth reading.

- Green, A., Grace, D. and Perkins, H. (2016) 'City branding, research and practice: an integrative review', *Journal of Brand Management*, 23(3): 252–272
  A broad overview of trends in both city branding practice and research and a discussion of the potentials to integrate these two, previously, largely disparate, endeavours. Provides a valuable review of key city branding literature.

- Jensen, O.B. (2007) 'Culture stories: understanding cultural urban branding', *Planning Theory*, 6(3): 211–236
  Outlines a narrative framework for examining the cultural branding of cities. Explores the competing stories around the redevelopment of the harbour area in Aalborg, Denmark.

- McCleery, A. (2004) 'So many Glasgows: from "personality of place" to "positionality in space and time"', *Scottish Geographical Journal*, 120(1/2): 3–18
  Considers the representation of Glasgow in regional novels.

- McFarlane, C. (2008) 'Postcolonial Bombay: decline of a cosmopolitan city?', *Environment and Planning D: Society and Space*, 26(3): 480–499
  Explores the cosmopolitanism of post-colonial Bombay through representations of the city in film.

- Oguztimur, S. and Akturan, U. (2016) 'Synthesis of city branding literature (1988–2014) as a research domain', *International Journal of Tourism Research*, 18(4): 357–372
  Another useful review of the now very extensive, city branding literature.

Websites

- www.seeglasgow.com; http://barcelonaturisme.com/; www.nyctourist.com – Examples of city promotional websites.

- www.imdb.com – The most comprehensive guide to films on the Internet.

- www.YouTube.com – Contains videos and performances from all of the artists mentioned in this chapter.

⑩ Experiencing the city

Introduction

This chapter is concerned with everyday life in cities. Inevitably an interest in the everyday permeates much of this book. For example, chapters six and eight discuss how the built form of the city, and the practice of planning more generally, can shape the lives of and exclude certain groups of urban dwellers, while chapter twelve discusses everyday physical movement around the city and the social dimensions of negotiating the city.

Rather than seeking to offer a comprehensive discussion of the many dimensions of everyday life in urban areas (for such discussions, from a variety of perspectives, see Jacobs 1961; Walmsley 1988; Jarvis et al. 2001; Highmore 2002a, 2002b; Hubbard 2006 (chapter three); Knox and Pinch 2010 (chapter seven)), it focuses on three aspects. First, it considers the values that people attach to urban landscapes. Second, it considers the roles that the senses, and bodies more generally, play in our perceptions, understandings of and being in the urban environment. In both cases these discussions focus on the distinctive approaches within human geography from where these concerns developed. Third, it will focus on the rhythms of everyday life in cities and the tensions manifest in the public realm between regularity, regulation and transgression. While focusing on the contemporary city it will attempt to ground some of these debates within more enduring theoretical perspectives on the urban. Underpinning all of this is the idea that the experience of the city is mediated through the social order. Thus, the city is experienced in radically different ways by different groups of people (Hubbard 2006).

Valued landscapes and everyday life

Everyday life, for the vast majority of people at least, is anchored around places that provide senses of safety, security, identity and belonging. The most obvious of these places are the home and the neighbourhood. Edward Relph (1976: 1), whose work we discuss below, has argued that 'to be human is to live in a world filled with significant places: to be human is to have and to know your place'. The emotional relationship between people and place, then, is an important aspect of everyday life but one to which human, and particularly urban, geographers have, at times, paid scant attention. The recognition of the importance of the emotions within people-place relationships did not emerge among human geographers until the 1960s and 1970s. One of the reactions against the dehumanizing effects of the quantitative revolution within human geography was the emergence of humanistic geography. Humanistic geography was a paradigm concerned with human experience and emotion and one that adopted qualitative rather than quantitative methodologies (for fuller discussions of humanistic geography, see Cloke et al. 1991; Crang 1998; Holloway and Hubbard 2001; Johnston and Sidaway 2004). Humanistic geographers were less concerned with the surface, measurable characteristics of place, that they argued had dominated human geography enquiry previously, but were more interested in uncovering the meanings and significance that people attached to places. Their concerns can be summed up as an interest in the places and landscapes that are valued by individuals and within societies and cultures. Given these concerns it is not surprising that the sources they turned to, such as the visual arts, literature, biography and poetry, were those more associated with the humanities than with the social sciences (Cosgrove 1989: 126–127).

It is possible to detect something of a contradictory tendency towards the city within humanistic geography. Edward Relph (1976, 1981, 1987), for example, pursued a sustained critique of the modern urban landscape. A major preoccupation of Relph's work was the supposed erosion of the deep emotional relationships between people and place that were said to be characteristic of earlier, pre-industrial societies. There are echoes here of earlier concerns with the urban voiced by sociologists, most notably Wirth (1938). Relph, for example, argued that the modern urban world has become increasingly characterized by inauthentic places and superficial relationships between people and place. He cited characteristics of the modern world, such as growing geographical mobility and travel, the centralization of planning, the increasing commercialization of urban landscapes and the mass production and serial reproduction of standard architectural designs, as all contributing towards the erosion of the deep emotional attachments people feel for places. He also argued that the encounters between people and place in the modern world have become predominantly shallow and transient, characterized by the spread of landscapes such as hotels, shopping malls and airports, or are centred increasingly around purposes such as entertainment or commerce. Some of Relph's arguments have their echoes in some recent anti-globalization discourses. He has said of modern urban landscapes:

> There is a widespread and familiar sentiment that the localism and variety
> of the places and landscapes that characterised pre-industrial societies and
> unselfconscious handicraft cultures are being diminished and perhaps
> eradicated. In their stead we are creating a 'flatscape'. Lacking intentional depth
> and providing possibilities only for commonplace and mediocre experiences.
>
> (Relph 1976: 79)

It would be wrong, however, to suggest that humanistic geographers were universally dismissive of modern urban landscapes. Donald Meinig's (1979) edited collection *The Interpretation of Ordinary Landscapes: Geographical Essays* offers a counterpoint to Relph's criticisms of the modern urban world. The book opens with an essay by Peirce Lewis in which he argues:

> It rarely occurs to most Americans to think of landscape as including everything from city skylines to farmers' silos, from golf courses to garbage dumps, from ski slopes to manure piles, from millionaires' mansions to the tract houses of Levittown, from famous historical landmarks to flashing electric signs that boast the creation of the 20 billionth hamburger, from mossy cemeteries to sleazy shops that sell pornography next door to big city bus stations ... Such common workaday landscape has very little to do with the skilled work of landscape architects, but it has a great deal to say about the United States as a country and Americans as people.
>
> (Lewis 1979: 11–12)

The book goes on to chart what the landscape can say about the people who create and use it and how this meaning can be interpreted.

Exercise

Think about your own relationship to ordinary, mundane, everyday places. Can you think of examples of such places that are important to you? What is the nature of these places? In what ways are they important to you? In what ways, and to what extent, do you feel they are part of your own personal geographies?

Do you feel that the planning, management and conservation of urban areas pays sufficient attention to the significance of these kinds of place? In what ways could these discourses be refigured to take account of the significance of these more ordinary places?

The arguments developed and put forward by humanistic geographers were important in the development of human geography during the 1970s and 1980s. They expanded considerably the theoretical and methodological terrain of the discipline and offered an alternative paradigm and set of practices to the Marxist and structuralist perspectives that were taking hold in human geography at roughly the same time.

Ultimately, the most enduring contribution of humanistic geography of relevance to the study of the city is a greater openness to the cultural value of ordinary landscapes. This undoubtedly owes something to the contributors to Meinig's collection and others working in a similar vein, not all of whom would identify themselves explicitly as humanistic geographers. In addition to

Meinig's work it is possible to identify a wider literature from a variety of disparate sources that has found richness and value in even the most mundane landscapes. The most influential of these is the American landscape historian J.B. Jackson, a contributor to Meinig's book, who was largely responsible for triggering a serious academic interest in the cultural value of the vernacular landscape, the landscape of shopping malls, commercial strips, roadways, trailer parks and suburban homes. This landscape was otherwise either dismissed as of no inherent value or interest or ignored within the prevailing academic discourse of the mid-twentieth century. Jackson founded the magazine *Landscape* in 1951 and remained editor until 1968. It was in *Landscape* that many of his most influential essays on the value of ordinary landscapes were published. He argued, in terms strongly redolent of Peirce Lewis above, that these landscapes offer a rich source of information about American culture and society:

> Over and over again I have said that the commonplace aspects of the contemporary landscape, the streets and houses and fields and places of work, could teach us a great deal not only about American history and American society but about ourselves and how we relate to the world.
>
> (Jackson 1984: ix–x)

Another influential study that wrote positively about a landscape otherwise largely dismissed among serious architectural critics was the study by Robert Venturi, Denise Scott Brown and Steven Izenour (1972) entitled *Learning from Las Vegas*. They argued that to dismiss Las Vegas' architectural landscape as merely gaudy, tacky and surreal was misguided and failed to appreciate that to many people it provided significant and enjoyable environments. More broadly, the authors were critical of architects and planners for failing to acknowledge ordinary people's experiences and desires and for seeking to speak for them from an apparently superior position (see chapter eight).

In Britain, it is possible to recognize a comparable, albeit somewhat later, literature that explored the cultural value of ordinary or commercial landscapes. For example, the historian Raphael Samuel (1994) has written in very positive terms about theme parks and heritage centres. Rather than being characterized by the gaudy and allowing only shallow engagements with place, he argued that they offer genuinely democratic routes to historical knowledge and experience. The cultural geographer David Crouch, in a widely cited study co-written with Colin Ward (1997), has written about allotments, landscapes composed of small plots of land cultivated by individuals which are usually provided by the local authority. Their study revealed the ways in which people form deeply rooted attachments to these most ordinary places.

We have seen a subsequent growth of community gardens, landshare schemes (www.landshare.net/about/) and smallholdings in urban areas in the UK and

Case study: ghostly everyday landscapes

Some recent work has tried to highlight the phantasmagoric or ghostly qualities of mundane, everyday spaces and landscapes. Drawing on the work of theorists of the everyday such as Michel de Certeau, this work has explored the ways in which everyday urban landscapes are replete with traces of the past and uses now gone, but which recent phases of development have only partially removed. Thus, the city can be thought of as a temporal collage in which traces of the past, and its obvious absences, 'haunt' the present-day city. These phantasmagoric landscapes are typically found away from the centres of cities and the sites of urban regeneration, where the pace of urban change is slower and its completeness less certain. Tim Edensor has used his daily commute to work in Manchester in the north of England as a vehicle to explore these qualities of the city. He considers the ghosts of the working-class, which these landscapes evoke, through the traces of this past that remain embedded in them and through the absences and gaps in the present-day landscape of the city where once vibrant activities took place. Edensor's work emphasizes the importance of the mundane spaces of the city and highlights an aspect of these spaces too frequently overlooked.

Figure 10.1 *Traces of the past: a former factory building, now a fashion and design school, East London*

Source: Author's photograph

continued

Edensor discusses a number of 'roadside hauntings' that he encounters on his commute. These include an old abandoned cinema, a railway line converted into a cycle path, an ex-council estate and the site of Manchester City Football Club's former ground Maine Road. The ghostly qualities of these spaces are many. Through traces of the past still visible, they evoke previous eras, communities and social, economic or political moments and activities now long gone; their shabbiness offers a counterpoint to both the gleaming modern city being reconstructed through regeneration and to their own former glory. These landscapes are often difficult to decipher, provoking uncertainties and questions about their former uses and significance. Traces of the past might look incongruous when set against more recent developments offering clues to the buildings, and communities and lifestyles associated with them, which lie under the foundations of the present cityscape. Alternatively, their current silence, in the case of the former railway line and the quiet streets around Maine Road, poignantly recall their contrasting noisier, more vibrant pasts. Edensor's work reiterates the significance of the most mundane of urban landscapes, but does so through a subjective, imaginative reading of urban space and one open to the uncertainties and poignancies that are written into its fabric.

Source: Edensor (2008)

examples of community organized agriculture reclaiming derelict land in cities such as Detroit in the US. This suggests that these relationships with the land, nurtured through cultivation, are a widespread phenomenon, perhaps manifest in different forms in different places. Informal urban agriculture is also an important feature of many cities of the Global South (Binns and Lynch 1999; Drakakis-Smith 2000; Lynch 2005).

Further, influential multi-disciplinary work in the 1990s saw something of a re-evaluation of perhaps the ultimate artificial landscape, the shopping mall (see Jackson and Holbrook 1995; Miller et al. 1998). Previous analyses of the shopping mall had often portrayed it as a manipulative, instrumental landscape that was both dedicated to the promotion of consumption and suffused with architectural fantasy and myths of escape (Gottdiener and Lagopoulos 1986; Hopkins 1990). Such analyses, while they offered very sophisticated deconstructions of the form and symbolism of shopping malls, did not tend to explore the meanings attached to them and produced in them by their many diverse groups of users. More recent work, by contrast, used methods such as focus group interviews and questionnaires to access and explore these meanings. This work revealed a hidden richness to the complex and multiple meanings that shoppers attached to malls and to the act of shopping. For example, it is argued in the abstract of one of the many papers produced from the project:

> Many studies of contemporary consumption have tended to reduce a complex and contested process to a momentary and isolated act of purchase. A similar

kind of reduction is common in many semiotic analyses of shopping malls and in studies of advertising which assume an audience's readings rather than investigating them empirically. Drawing on field research in north London, we provide evidence from focus group discussions of the social use of shopping centres and of the multiple meanings of such apparently mundane activities for the consumers themselves.

(Jackson and Holbrook 1995: 1913)

Finally, there is something of an echo of the concerns of this disparate literature in Nick Gallent's recent explorations of England's rural-urban fringe (Gallent 2006; Gallent et al. 2006; Gallent and Anderson 2007). This much-maligned landscape has long been seen as a problem in planning, management and aesthetic terms. One of the potential solutions that Gallent raises is not drawn from the usual planning or management driven approaches that tend to dominate discussions of this zone. Rather, he asks if the rural-urban fringe might be re-evaluated or reimagined through education that challenges our prevailing views of aesthetics and valued landscapes and encourages us to appreciate the functional elements of the landscape found in the rural-urban fringe (figure 10.2).

Figure 10.2 *Electricity substation, rural-urban fringe, UK*

Source: Author's photograph

In a way, Nick Gallent, and others who have worked with him, are raising similar questions about this most ordinary of landscapes to those raised by J.B. Jackson and Peirce Lewis above and are demonstrating that these often overlooked landscapes remain the subject of serious academic enquiry.

Exercise

One of the most mundane and undervalued landscapes in urban areas is the rural-urban fringe. It is a zone that encircles all urban areas and contains a number of distinctive land uses and functions. It is also the site of a number of particular issues that reflect the variety of competing land uses that come together here. The rural-urban fringe is generally seen as a problem in planning and management terms. To a large extent this stems from reactions to the aesthetics of the fringe which are overwhelmingly negative. A good guide to this zone is the work of Nick Gallent mentioned above.

Spend some time walking through the rural-urban fringe of an area that you know well. Try to take in as many of its different characteristics as possible. As you go around think about the questions below and collect evidence (perhaps in the form of photographs) to attempt to address them.

1 What is the nature of the rural-urban fringe? Is this zone:

- Rural?
- Urban?
- A transition zone moving from predominantly urban to predominantly rural functions?
- Something different, a zone that cannot be classified into the categories rural/urban?

What does the rural-urban fringe suggest about our understanding of the division between country and city? Does it challenge this division?

2 What is your reaction to the aesthetics of the rural-urban fringe?

- What aspects of this zone do you appreciate/value?
- What aspects of this zone do you dislike?
- Why do you think you hold these views about this zone?
- In what ways do non-visual senses play a part in your experience of the urban fringe?

What do you think your reactions tell us about the attitudes that exist in society towards ordinary landscapes? How does the rural-urban fringe fit into our views of valued landscapes?

3 What is to be done about the 'aesthetic problem' of the rural-urban fringe?

Should it:

- Be 'beautified' through physical change and planning (i.e. community forests, green infrastructure planning)?
- Be re-evaluated or reimagined, through education that challenges our prevailing views of aesthetics and valued landscapes and encourages us to appreciate the functional elements of the landscape found in the rural-urban fringe?

Compare your thoughts on these issues to those discussed in Gallent's work.

Senses, bodies and the city

It is a fact, all too easily overlooked, that the senses play a vital part in our experience and understanding of the environment. Urban geography though has rarely paid sufficient attention to the significance of the senses to our experience of the city. Work on the senses has tended to be 'ghettoized' somewhat within the sub-discipline of behavioural geography (Walmsley and Lewis 1993), an offshoot of **positivist** spatial science that has crept little onto the agendas of the other sub-disciplines of human geography. However, recent work from within cultural geography has refocused attention on the embodied nature of the urban experience, albeit more broadly than the behaviourists' interests in the senses (Pile 1996; Davidson 2000).

The senses pose challenges for geographers that they have not always been successful in rising to. In successive paradigms that shaped human geography over the twentieth century (see Johnston and Sidaway 2004) the roles of the senses have either tended to be ignored or else simplified through assumptions to the point of meaninglessness. For example, during the 1950s and 1960s, positivist spatial science was the dominant paradigm within human geography. A well-documented weakness of spatial science was that it made wildly inaccurate assumptions about the relationship between people and the environment, assuming simply that everyone had perfect knowledge of the environment (indeed its conception of the environment itself was deeply problematic). These assumptions conveniently sidestepped the fact that the senses are unable to take in all of the information they are bombarded with, resulting in the brain selectively filtering out a large amount of environmental information. They ignored also that the senses each have their own range. While many people may be able to see for many miles (albeit taking in progressively less detail the further away we look), our senses of hearing, smell, touch and taste are much more spatially restricted (Holloway and Hubbard 2001: 41–42). Finally, they ignored the massive variations in the sensory abilities between different people. These sensory differences are absolutely fundamental in shaping people's knowledge and experiences of the environment. It is vital, if geographical enquiry is to be meaningful, that these differences are acknowledged.

It was not until the emergence of behaviourism in the 1960s, in part reflecting frustration with the crude assumptions of spatial science, that the senses were taken seriously within human geography. Behavioural geography drew heavily upon techniques from environmental psychology, with which it had a fruitful cross-disciplinary dialogue. Initially it acknowledged and examined the partiality of people's environmental perception. However, later, more sophisticated, manifestations of behaviourism incorporated **cognition** and cognitive differences, recognizing also that people process environmental information in different ways (Walmsley and Lewis 1993; Holloway and Hubbard 2001; Knox and Pinch 2010). Despite behavioural geography's recognition of the importance of the senses, there has been a bias in geographical research towards the visual at the expense of the other senses, thus overlooking some important sources of environmental knowledge (Pocock 1993; Golledge and Stimson 1997). This has only been partially addressed through some imaginative research into smell- and soundscapes (Porteous 1985; Pocock 1989; Smith 2000).

Accessing the environmental information that we receive through the senses and subsequently process into images or impressions of the environment is not an easy task. Perhaps the most widely studied and well-known attempt to do this though, is through the process of mental, or cognitive, mapping. This is a technique pioneered and developed by the urban planner Kevin Lynch and published in his study *The Image of the City* (1960). Using this technique Lynch was able to analyse the nature of urban imagery that we possess and from this draw conclusions about the differential legibility of urban landscapes, findings that he argued could be applied in the design of cities. Lynch's technique was deliberately simple and consisted largely of collecting sketch maps that urban residents produced of their cities. These, Lynch argued, provided visual representations of the knowledge that residents possessed of their surroundings and were relatively stable, learnt images of the environment that are used to orientate and navigate (Holloway and Hubbard 2001). Mental maps tend to contain five elements – paths (such as streets and roads), edges (cliffs, walls), landmarks (such as prominent buildings or monuments), nodes (for example, busy public spaces or interchanges such as railway stations) and districts (such as neighbourhoods). What Lynch's work revealed, and which demolished the myth of complete environmental knowledge perpetuated by the assumptions of the models of spatial science, was that mental maps revealed that people's environmental knowledge was partial, simplified and often highly distorted.

Lynch found that while the mental maps produced by individuals are all unique he could recognize a series of broad similarities between them. He explored the factors that underpinned the similarities and differences that he observed between the mental maps of different people. He argued that people's perceptions of their environment tended to be affected by the differential 'legibility' of urban environments. The legibility of urban environments refers to the ease with which people are able to acquire and retain information about their environment and consequently to navigate these environments. It appears that some cities are easier to read than others, something he used to inform guidelines for legible urban design (Pacione 2009: 406–407). Further, Lynch recognized the influence of socio-economic background on the environmental knowledge of different groups of urban residents as expressed in the mental maps they produced. This was starkly revealed in a widely cited study of mental maps of residents of different areas of Los Angeles (Lynch 1960).

Lynch undertook research with residents of a number of different neighbourhoods in Los Angeles with varying socio-economic characteristics. From this he produced composite mental maps from each neighbourhood that revealed the vast differences in the perceptions and knowledge of Los Angeles held by residents of different areas. In Westwood, for example, a wealthy, predominantly white neighbourhood located near Santa Monica and Beverley Hills, Lynch found that the mental maps produced by its residents were detailed and extensive, covering much of the Los Angeles urban area. He did find though that the levels of detail attached to different areas varied a great deal, reflecting the fact that Westwood's residents knew some areas only vaguely, perhaps being aware of their location but being able to provide little, if any, internal detail. By contrast, residents of the poorer, Spanish-speaking neighbourhood of Boyle Heights, located near downtown Los Angeles, produced maps that were less extensive relative to those of Westwood residents. These maps still contained a great deal of detail but it was restricted predominately to a very small area around their immediate neighbourhood. This reflected socio-economic differences between the residents of the two neighbourhoods, specifically their differential levels of mobility throughout the city. While Westwood residents enjoyed car-based mobility to spatially dispersed sites of employment, education, leisure and retail across the city, the residents of Boyle Heights largely relied on the city's poor public transport system to access more restricted opportunities that tended to be located in or near their own neighbourhood.

Lynch's work was pioneering in the techniques it employed and also in the insights it revealed. It has been subject to subsequent criticism, however. First, it reflects the bias towards the visual sense noted above. It is debatable whether the simple, two-dimensional sketch map that Lynch used as the basis of his research and as a representation of people's environmental knowledge is really able to capture the complexity of their relationships and interactions with their surroundings. It is certainly unable to capture the contribution of any of the senses other than the visual. It can offer nothing, for example, for those interested in the environmental perceptions of people with severe visual impairments. Second, while reflecting the different levels of environmental knowledge that people possess, mental maps also reflect their differing abilities to draw. Thus, the differences revealed in mental maps might be as much a reflection of the varying artistic and cartographic abilities of research participants as they are of varying levels of environmental knowledge. Finally, mental maps themselves tell us little, if anything, about how environmental knowledge is acquired and processed, although some studies have employed mental mapping techniques with children of different ages and have revealed how spatial awareness develops over time (Holloway and Hubbard 2001: 52–55).

Interest among geographers in the body has moved beyond the purely sensory in recent years (Rodaway 1994; Pile 1996; Nast and Pile 1998; Longhurst 2000; Valentine 2001). While this work has explored the ways that bodily abilities differ between individuals, echoing some of the concerns of behavioural geographers noted above, it has gone beyond the purely biological to also explore the social worlds in which the body is embedded. This work is significant here because it views the body as providing a bridge between the biological and the social, the private and the public. It has considered the ways in which the body is presented in different social contexts but also the ways in which the social acts upon the body, for example, through the exclusion of certain types of body from some spaces or the pressure on bodies to conform to a narrow range of 'ideal' types and to be presented 'appropriately' within certain social settings. This work has frequently noted how urban spaces are suffused with images of such ideal bodily types, for example, through the prevalence of advertising imagery (figure 10.3), or designed to include and accommodate only certain types of body. One effect of this has been the social unease and sense of exclusion among those whose bodies do not fit these ideal types or who find it difficult to access certain spaces, for example, because of restricted mobility.

The contributions of this work are well illustrated through the considerable and significant literature on the geographies of disability which has noted

Figure 10.3   *Ideal bodies: advertising imagery*

Source: Author's photograph

both the ways that disabled bodies are physically excluded from some urban spaces and also the social pressures that make disabled bodies feel 'out of place' in some locations, for example, those of leisure, consumption and even education (Gleeson 1996; Imrie 1996; Butler and Bowlby 1997; Butler and Parr 1999; Kitchin 2000; Hall et al. 2002; Fuller et al. 2004; Imrie and Edwards 2007; Chouinard et al. 2010). Work in this vein has also explored other bodily dimensions and their relationships to the social world, including the social geographies of age, illness, gender, pregnancy and sexuality (for an overview of this work see Hubbard et al. 2002, chapter 4).

The urban public realm: regulation, rhythm and transgression

The chapter now moves from the private, biological and psychological worlds of the city discussed above to consider the social worlds of the city's public realm. Here we conceive of the public realm as the space where the body comes into contact and interacts with a range of social worlds and the city's **materiality**. The public realm has long been a contested terrain within the city. We might interpret the everyday public realm of the city with reference to three presences that work, sometimes together, sometimes against each other. First, it is an arena for the inscription of authority and regulation. Second, it is the site of multiple rhythms of daily life through which wider structures are maintained and reproduced (Hubbard 2006: 100). Third though, it is the site of resistance and transgression, opposition to the authorities inscribed onto the public realm and the disruption of the banal rhythms of everyday life. We briefly consider each of these in turn. The public realm of cities is a terrain onto which order and authority are inscribed. The landscapes of the city can be read as expressions of the attempts of planners, architects and urban designers to control bodies and bodies of people (see also chapters six and eight). Often these attempts are explicit, taking the form of signs inserted into the landscape that outline permitted and restricted behaviours (figure 10.4).

However, in other cases they can take a more subtle material form. For example, the design of seating in urban areas is careful to produce forms that do not allow homeless people to use them as places to sleep (figure 10.5).

The geographer David Sibley (1995, 1999) has written extensively about the ways in which the spaces of the cities of the Global North are purified, cleansed of forms of behaviour and social groups not deemed appropriate or desirable, and has highlighted the importance of the design of spaces in this regard. The urban landscape, then, is suffused with strategies of ordering and exclusion by those responsible for its production and maintenance. These strategies are underpinned by the practices of the management, surveillance and policing of these spaces.

Figure 10.4 *Control of behaviour, Cabot Circus, Bristol, UK*

Source: Author's photograph

The ordering and purification of urban space, though, goes beyond its design and regulation and can be seen to permeate it entirely. The use of cultural regeneration to refashion the centres of major cities since the 1980s illustrates this point well. Since the widespread processes of deindustrialization during the 1980s the centres of cities have been extensively redesigned to make them attractive environments and assets in the attraction of external capital, in the form of tourists, business tourists and companies. As well as being physically redesigned, this has altered the social geographies of these spaces, making them attractive for certain social groups while leading to others, who do not fit the profile for whom these spaces have been designed, feeling and being excluded.

So, the experience of the city is made, in part, through the inscription of authority, power and regulation from above. However, it is also made (negotiated) at ground level through the daily rhythms of movement across

Figure 10.5 *Seating, Stuyvesant Square, New York City*
Source: Author's photograph

and through the city. Without this animation of the urban by the world's billions of urban citizens the city would merely resemble a sterile film set after the cameras have stopped rolling. To only consider the city through plans, architects' drawings or the operations of the powerful, then, is to miss the ways in which the city is made through the actions of its inhabitants. The world of everyday experience that emerges at ground level is a very different city to that which can be discerned through reading the texts of the powerful and the inscription of authority across urban space. These representations of the city tend to be too clean and to deny the agency of people, sometimes denying their presence altogether (Miles 2002). Rather, the experience of the city that emerges on the street is more messy and complex than the producers and regulators of the city necessarily imagined. The everyday public realm of the city then is 'a collection of repetitive and banal actions which reproduce the status quo, yet a site of resistance, revolution and ceaseless transformation' (Hubbard 2006: 100).

There is a rich tradition of theoretical exploration of the everyday, especially from a broadly sociological perspective. This is a tradition, though, that has

caught the attention of urban geographers, who have been particularly drawn
to the role of space at the heart of much of this theory (Hubbard 2006).
Among the most influential of these theorists have been Georg Simmel, Walter
Benjamin, Henri Lefebvre and Michel de Certeau (for fuller discussions of
their work, see Highmore 2002a; Hubbard 2006; and for examples of their
work, Highmore 2002b). Much of their work sought, albeit through a different
lens and apparatus, to understand the 'ambivalence' of everyday life in the
city (the coexistence of the banal and the transformative) (Hubbard 2006: 100).
These authors explored the tensions between the capitalist mode of production
– which sought constantly, through the production of space, to repress and order
the potentials of the body – and both the body's attraction to the rhythms of the
city under capitalist modernity and its potential to destabilize these very
rhythms through acts of resistance.

The most methodologically sophisticated exploration of the temporal and
spatial geographies of everyday life are those developed within time-geography.
This approach originated among Swedish geographers in the 1960s and became
most developed in the work of Torsten Hägerstrand (1982) (Holloway and
Hubbard 2001: 29). The basis of time-geography is the simple premise that
doing things ('projects' to use the terminology) both takes time and involves
moving through space (time-geographers used the term 'resources' to refer to
time and space). A reflection on your own daily activities highlights this. To get
to lectures you need to move through space (unless of course you are studying
geography online by distance learning, although we will come to the impacts
of technology later). Members of the class will move through space covering
different distances at different rates, some will walk, some cycle, some drive
and some travel by public transport. For some, their negotiation of space-time
(time-geographers emphasized the interdependence of the two) will be affected
by their physical condition or perhaps a disability. For many the determining
factor will be economic, a lack of money to buy a car necessitating travel to
university by bus or bicycle, for example (see chapter twelve for discussions
of the connections between social exclusion, social difference and mobility).
Time-geographers were interested in mapping these daily paths of individuals
and represented them through time-space diagrams with space as the horizontal
plane and time the vertical (see figure 10.6 for an example of a recent adaption
of this technique).

This mapping work was valuable in that it revealed that the daily paths of many,
particularly women, were more complex than others, involving multiple journeys
to different stations (for example, shops, schools, places of employment, friends
and relatives and childcare facilities). Often this coincided with a lack of access
to cars for personal mobility, increasing the time taken to negotiate these tasks.
Also, time-geography articulated a number of constraints that are inherent to the
daily paths of individuals. Three such constraints were identified: capability

constraints (the limits on how far we are able to travel in a given time), coupling constraints (the necessity for people to be spatially co-present to achieve certain tasks) and authority constraints (legal or quasi-legal restrictions on access to certain spaces such as bar opening hours or restricted areas in, for example, institutions, shops or banks) (Holloway and Hubbard 2001: 31).

Recent advances in technology have opened up a debate around the ways in which people's daily paths, and the constraints inherent to these, might be affected as communication technologies potentially reduce the necessity for face-to-face contact between people. It is possible to envisage ways in which coupling constraints might be loosened, for example, by the impacts of successive waves of communications technology. We could think, for example, of the impacts of the telephone and more recently the mobile telephone, the internet and videoconferencing and the associated coupling practices that have evolved alongside these technologies that have removed the universal need for spatial co-presence for instantaneous communication. However, my own (Tim's) experiences as I write this passage of text highlight the complexities of the relationships between technology and face-to-face communication. The development of mobile computing technology allows me the freedom to work away from my office, at home or in a variety of remote locations such as libraries, archives or fieldwork sites. I do not need to be at work to do work or to communicate with my colleagues. Technology has clearly allowed the loosening of a coupling constraint here. At the moment, I am writing this as I speed towards London for an editorial board meeting of a journal I am involved in. It would have been difficult to work this way when I wrote the first edition of this book in 1995 and 1996. Then, computers were too bulky or expensive to practically work from while on the move, their batteries were also notoriously short-lived. Writing by hand is also next to impossible on a busy, bumpy speeding train, ruling out that option for drafting text. Now all I have to worry about is spilling coffee on my iPad.

The meeting I am going to, while mainly involving people based in the UK, also has attendees contributing from Australia, Singapore and the US. It is technically possible for the meeting to take place via videoconference, although the affordability of effective technology might be an issue even for higher education institutions and co-ordination between different time zones might be difficult for a three-hour meeting. What videoconferencing cannot replicate, however, is the informal sociability and bonhomie, not to mention the excellent food, of the post-meeting meal in the Bombay Brasserie Indian restaurant. This is justified to the publishers who pick up the tab, on the grounds that it is crucial to the building and maintenance of social bonds between editorial board members. The apparently sterile, dehumanized world of time-geographies and their representations then, is actually suffused with a social richness that undercuts the potentials of technology to loosen the functional imperatives of coupling constraints. Until these social dimensions of encounter can be convincingly reproduced without the necessity for spatial co-presence, face-to-face meetings will, it seems, remain resilient in the face of technology.

The incorporation of the social within time-geographies is something that Alan Latham (2004) explores in the case study below. Here he tries to couple a technique rooted in the quantitative revolution and spatial science approaches, with all of the criticisms that have been made of them, with more contemporary cultural geographies. He attempts to produce a more humanized time-geography, echoing to an extent some earlier work by Alan Pred (1990).

While time-geography's focus is on the 'individual, short-term and small-scale' (Holloway and Hubbard 2001: 33) we should not lose sight of the fact that these individual daily paths are actually part of the aggregate rhythms of the city. Further, we should remember that these rhythms, the pulse of the city, to shift the metaphor a little, are reflective of the processes and the modes of production that gave rise to them. We can read the city then through its daily rhythms. The rhythms of a former socialist city or a colonial capital, and the thousands, perhaps millions of individual daily paths that make up these rhythms, will be different to those of a city that developed under industrial capitalism during the nineteenth and twentieth centuries or one that is largely the product of the current era of disorganized global capitalism or one that is striving to become more sustainable (see chapter 13). We should also remember though not to be too deterministic. As we shall see the city is also the site of resistance, transgression and disruption.

Case study: representing the subjective dimensions of time-geographies

There has been a renewed interest recently among social scientists in sociality. This refers to the routine, often banal, but frequently playful encounters and interactions that make up much of daily life in the city. It has been argued that this has taken on a heightened importance in contemporary society, perhaps replacing traditional social divisions around class or kinship. Alan Latham (2004: 118), in reflecting on the contemporary significance of sociality, argues that social science has undergone something of a 'theoretical refocusing that is organized around a heightened appreciation of the significance of everyday interactions in the production and reproduction of wider social structures'. As well as conceptually upgrading the importance of the everyday this also poses methodological challenges to social scientists. Namely, how do we capture the fine-grained, elusive, trivial details of everyday routines and interpret their significance? In approaching this challenge, Latham draws on the potentials of time-geography. He recognizes its limitations, most notably that it is primarily concerned with the collection of quantitative data about people's movements through time and space. Given this, time-geography has said

continued

very little about the subjective, sensuous, qualitative dimensions of these movements and the encounters that take place within various stations. He proposes a mixture of techniques to address this including interviews, researcher observations, detailed diaries kept by research participants and photographs taken by participants as ways of layering qualitative detail across the skeletal framework of time geography (see figure 10.6). This combination of techniques he refers to as the diary-photo diary-interview method. This method, he argues 'offer[s] them [research participants] the resources so that they can tell a narrative about themselves that retains a strong sense of social and personal context' (p. 126). It draws, for example, on the evocative qualities of photographs and reflective material from diary entries and interview transcripts to 'convey a much greater sense of the sensual and affective elements' (p. 127) of individuals' movements through the city. Latham's proposal potentially offers a method of representing the rhythms of individuals' daily movements through the city in ways that capture at least something of the rich banality of everyday life upon which we depend, typically without reflecting upon it, and that plays such an important role in the wider structuring of society.

Source: Latham (2004)

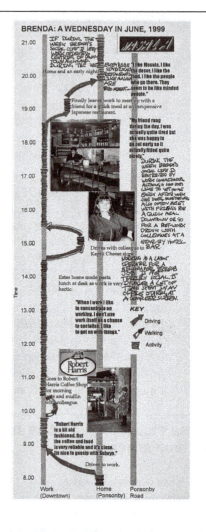

Figure 10.6 *An attempt to represent the subjective dimensions of daily paths using photographs, diary entries, interviews and researcher observations*

Source: Latham (2004: 128)

There is a distant echo in Latham's work of the concerns of humanistic geographers outlined at the start of the chapter. Although Latham's concern is more with the relationship between sociality, the materiality of the city and the reproduction of wider structures, what emerges here again is the subjective significance of the everyday and the mundane landscapes of the city.

Time-geography made a valuable contribution to the excavation of the complex rhythms of the city. These had previously remained largely unacknowledged in human geography or had been reduced to something of a robotic impulse in, for example, the models of spatial science. What time-geography could say little about though, although the work of people such as Pred (1990) and Latham (2004) got closer, was what the intentions behind these movements and the rhythms they engendered were. Time-geography was particularly effective in revealing the rhythmic reproduction of the status quo (Holloway and Hubbard 2001: 31). However, a growing number of geographers have recognized the transgressive potentials of the body in recent years and have explored the city as a key site in which these potentials are both revealed and repressed.

The social categorization of space and the rhythms of everyday life discussed above are not necessarily accepted by all groups. Deliberate acts of resistance are common as are acts of transgression, the unknowing crossing of social boundaries or breaches of accepted norms of behaviour. Some examples of resistance, such as riots or protests, often include an explicit spatial aspect. Such acts often target spaces associated with dominant groups or the issue against which the act of resistance is directed. Occupation of spaces associated with dominant groups is a powerful statement of resistance to the social rules and norms of this group (Watson 1999). However, many acts of transgression are far subtler than this and occur through everyday actions, such as cycling in pedestrian areas, for example.

Skateboarding offers a widely cited example of the transgressive use of urban space (Borden 2001, 2004; Woolley and Johns 2001; Hubbard 2006). Skateboarding on and near war memorials, for example, may be seen as a blatant act of disrespect and resistance by some. However, to skateboarders themselves their actions may have nothing to do with conscious acts of resistance. The memorial may simply provide a convenient and challenging surface for the boarder to practise and display his or her skills. Such acts of resistance and transgression typically prompt responses from other social groups designed to restrict or control what they see as inappropriate behaviour. Internationally, skateboarders have been subject to censure in the media, public hostility, the erection of physical barriers to their activity and local police action, for example (Borden 2001). The reason for this is that their activities in the centre of town have been deemed 'out of place' and in conflict with the uses for which the spaces were designed. Skateboarding, it seems, is an activity deemed more appropriate to a skate park located in some distant corner of the city where it does not clash with the commercial and consumption based leisure activities that dominate central urban areas. As the city becomes increasingly an arena shaped by the inscription of authority it also becomes one of transgression and resistance. This is a tension that has caught the attention of a growing number of geographers (Cresswell 1996; Daly 1998; Young 2003; Hubbard 2006: 111).

Summary

The experience of the city encompasses many diverse aspects. Most broadly this necessitates an understanding of the city's private and public worlds and the many ways in which they come together. This chapter has moved across the private, emotional and biological dimensions of everyday life in the city to its more public dimensions. The example of skateboarding, discussed at the end of this chapter, reminds us of the constant tension enacted in the public realm between the desire to impose order through design, management and maintenance and the desire to resist and transgress these regulatory strategies. Many mundane, everyday, bodily acts that take place in the public realm, such as walking, cycling, driving, eating, drinking and smoking (Fischer and Pollard 1998), are framed by the negotiation of this tension.

Follow-up activities

Essay title: 'Everyday life in the city is shaped by the body's mediation of the biological and the social.' Discuss.

Commentary on essay title

A good answer would draw heavily upon recent work on the geographies of the body which has, in many different ways, focused on the ways in which the body acts on the social but also the ways in which the social acts on the body. It would explore the work of both key theorists such as Pile and Longhurst and also the rich empirical work in this area. It might, for example, draw on the extensive and diverse literatures of the geographies of disability that are cited above. It is crucial though that an effective essay maintains its specific focus on the city within all of this. It might do this in two ways. First it might include a range of examples from urban areas that would demonstrate how the city is an environment that is shaped largely around the assumption of, for example, the able-bodied citizen and the ways that this excludes disabled bodies. It would also look at the issue of the social geographies of urban spaces and the ways in which the images that saturate these spaces and the activities that go on within them can be equally exclusionary. Second, it might look at theorists who have argued that the city is a key agent in either maintaining or collapsing the distinctions between private and public, social and biological. One such theorist is Richard Sennett, whose work is cited below. Finally, and only if there is sufficient space, the essay might cast an eye back to the history of human geography and contrast the sophistication of recent work on the body with cruder earlier work in this area. It should not seek to be overly critical of this work though, but might wish to discuss the ways in which it was significant in opening up new research pathways within the subject.

Project idea

Alan Latham's (2004), diary-photo, diary-interview method discussed earlier in the chapter demonstrates the ways in which qualitative detail can be layered across the quantitative framework developed within time-geography. However, the method outlined by Latham is overwhelmingly visual in its orientation. Does it fail to capture the other sensory elements of everyday life? Think about the role that sound plays in your daily experiences of the city. Can you think of ways, perhaps using digital technology, that Latham's method could be extended to capture the daily sounds of the city that are important to your experiences of it? Record some audio files of sounds that you think are important to your daily experiences of the city. Why and how are these sounds important? How do they work in relation to other sensory elements? Try to construct a text which incorporates these sounds into a mapping of your daily path. What does this text reveal about the role of the non-visual senses to your experiences of the city?

Further reading

Books

- Bates, C., Imrie, R. and Kulman, K. (eds) (2017) *Care and Design: Bodies, Buildings, Cities*, Chichester: Wiley-Blackwell
  A collection of essays that consider the interface between bodies, minds and design, at a variety of scales within urban environments, and explores design through the lens of various forms of care.

- Davidson, J., Bondi, L. and Smith, M. (eds) (2016) *Emotional Geographies*, Abingdon: Routledge
  Some key authors contribute chapters to this edited collection which charts the ongoing fascination of geographers with the significance of the emotions to our relationships with place and landscape. Not specifically urban but there is much here that is applicable to the study of the emotions in city life.

- Highmore, B. (2010) *Ordinary Lives: Studies in the Everyday*, Abingdon: Routledge
  While not specifically urban or geographical, it, nonetheless, brings together empirical investigation and theoretical reflection grounded around the study of everyday objects and routines.

● Hubbard, P. (2006) *City*, Abingdon: Routledge (chapter three, 'The everyday city')
A wide-ranging discussion drawing heavily on cultural theories of everyday life in the city.

● Meinig, D.W. (1979) *The Interpretation of Ordinary Landscapes*: *Geographical Essays*, New York: Oxford University Press
Now quite dated but a key collection responsible for maintaining serious academic interest in the value of ordinary landscapes.

● Sennett, R. (1977) *The Fall of Public Man*, New York: Alfred A. Knopf
A key theorist. Discusses the development of the city and its key role in the erosion of the boundaries between the private and the public. Many subsequent editions available.

● Valentine, G. (2001) *Social Geographies: Space and Society*, Harlow: Pearson (chapter two 'The Body')
Although not specifically urban, an excellent, wide-ranging introduction to geographical perspectives on the body.

Journal articles

● Baum, H. (2015) 'Planning with half a mind: why planners resist emotion', *Planning Theory and Practice*, 16(4): 498–516
Planning, a practice crucial to the shaping of the urban environment (see chapter 6), has tended to imagine that people act and think rationally. Traditionally planning has made little space for the emotions of those living in and using the city. This article is a call for planning to make room for the emotions.

● Edensor, T. (2005) 'The ghosts of industrial ruins: ordering and disordering memory in excessive space', *Environment and Planning D: Society and Space*, 23(6): 829–849
Another example of Edensor's imaginative engagement with the mundane and forgotten landscapes of the city utilizing the evocative qualities of photography as a research method.

● Gallent, N. and Anderson, J. (2007) 'Representing England's rural-urban fringe', *Landscape Research*, 32(1): 1–21
An interesting attempt to critically engage with attitudes towards this much-maligned aspect of the landscape of the contemporary city. Gallent's work is discussed briefly above.

● Hopkins, P. (2008) 'Critical geographies of body size', *Geography Compass*, 2(6): 2111–2126
Explores the role of body size in shaping social identities through the negotiation of everyday space.

● Longhurst, R. (2000) 'Corporeographies of pregnancy: "bikini babes"', *Environment and Planning D: Society and Space*, 18(4): 453–472
Discusses the exposure of the pregnant body in public space and the various ways in which it is socially regulated through assumptions about its 'appropriate' presentation in such spaces.

● Pinder, D. (2005) 'Arts of urban exploration', *Cultural Geographies*, 12(4): 383–411
Considers artistic interventions in the city and what they reveal about the urban everyday.

11 Housing and residential segregation

Introduction

One of the most obvious points to note about cities is that they are home to lots of people! Indeed, housing forms the most substantive land use within contemporary cities. Shelter is a basic human need and access to adequate housing is an important human right. Yet a home is much more than merely a physical dwelling within the city, it is also an important commodity to be traded and in addition represents a key factor in shaping the identity and place of people and households within the city. Housing is an important determinant of personal security, comfort, wealth and status, and ownership of housing can also be important in structuring access to other scarce resources within the city such as employment opportunities, education and healthcare facilities. However, it is clear that the ability of people to secure a home in the city is highly unequal, with the problems of lack of access to adequate and affordable housing for many urban dwellers most starkly illustrated in the world's biggest cities in the form of homeless people on the streets. As more people are drawn into cities, as a result of the economic and other opportunities perceived to be available within them, the ability of cities to meet the housing needs of growing and changing urban populations has become increasingly problematic. No city is free of housing problems, yet their nature and scale is highly varied around the world, from increasing problems of housing affordability in advanced industrial nations to how to deal with the growth of large informal shanty settlements in many large cities of the Global South.

Where people live within cities, and why, have been long-standing concerns of urban geography, stretching back to the early work of the Chicago School (see chapters two and three). As we have noted, the modern city is a highly complex, disorganized and heterogeneous place, yet these early researchers discovered that rather than resulting in a jumbled mass of people, the modern city displayed a distinct mosaic, with certain household types occupying particular niches in the city. This early work has stimulated a considerable body of urban research considering residential patterns within cities, the residential mobility of urban households and the operation of housing markets. This work has been underpinned by some key questions:

- Why do city populations get sifted out according to a range of social criteria, such as class, race, stage in the lifecycle, to produce distinctive neighbourhoods?
- What are the processes responsible for this sifting?
- How do people choose where to live, and what are the constraints on their choices?
- Which groups are able to manipulate the geography of the city and to whose advantage?

The remainder of this chapter examines why people live where they do in cities and the role that housing plays in shaping the form and life of cities. First, it provides an overview of some of the work examining residential patterns within cities. The chapter will then go on to consider the processes underpinning these patterns, focusing on ideas of choice and constraint through a consideration of residential mobility and decision making and the operation of housing markets. Finally, the chapter considers the key issues of access to, and affordability of, housing, highlighting a range of contemporary urban issues, such as homelessness and squatting.

Residential patterns in cities

Within his chapter on Boston in the book *The Soft City*, Jonathan Raban stops to ask:

> Why should the Italians all cram themselves behind the expressway in the North End? Why should [African-Americans] live in Roxbury and Jews in Chelsea? By what law do Boston suburbs turn into rigidly circumscribed ghettos when they look so much alike, so quaintly attractive, so prim, so dull? For it is as if someone had taken a map of the city and, resolutely blind to its topography, had coloured in irregularly-shaped lumps labelled 'Blacks', 'Jews', 'Irish', 'Academics', 'Gentry', 'Italians', 'Chinese', 'Assorted Others'.
>
> (Raban 1975: 216)

A similar patterning or clustering of particular groups had been highlighted by the Chicago School in their study of the city in the 1920s. Their detailed work identified a mosaic of community clusters which they termed natural areas, where they argued that competition linked to people's ability to pay for housing determined the location of particular groups within the city. In this context, the key contributor to patterns of difference and inequality was household income, although it was not the only factor in determining differentiation, with family structure and ethnic background also seen to be influential. This study formed the basis for a range of empirical research seeking to test the general applicability of these ideas and whether patterns evident in Chicago were present in other cities, known as social area analysis or factorial ecology (see also chapter two).

The key points to emerge from the huge volume of empirical research undertaken were the affirmation of socio-economic status, family status and ethnic status as key dimensions underpinning residential differentiation and a general consistency in the spatial expression of these dimensions in the great majority of cities in the developed, industrial world (Knox and Pinch 2010). Socio-economic status displayed an essentially sectoral pattern, family status a zonal gradient and ethnicity a clustered pattern. These factors were incorporated into an idealized model of urban ecological structure by Murdie (1969), although his model also acknowledged that in reality the city's ecological structure was the result of detailed interaction with the city's morphology and other local conditions (figure 11.1). Factorial ecology study peaked in urban geography in the early 1970s, declining in significance after this, although the ideas and techniques employed are still utilized commercially in identifying particular residential areas for the purposes of target marketing (Short 1996; Pacione 2009).

Three key reasons can be cited for the shift away from ecological research. First, the development of more sophisticated analyses were limited by methodological concerns, chiefly the availability, currency and detail of census data and the territorial units used for its collection. Concern was expressed that similarities in patterns were the product of the similarities in data variables and territorial units employed within censuses, which masked other patterns of differentiation. Equally, problems of the ecological fallacy were highlighted, where assumptions about the homogeneity of socio-economic characteristics within territorial areas masked the reality of greater heterogeneity at the micro scale. Linked to this, research in cities beyond North America revealed quite different residential patterns, suggesting the importance of other factors in determining residential structure. Second, as we have previously noted (see chapter two), the theoretical terrain of urban geography began to alter in the 1970s. Researchers looked to move beyond the quantitative analysis of patterns to examine the processes underpinning these patterns, opening up new ways of approaching the study of residential differentiation. In particular, research into residential differentiation began to focus on issues of choice and constraint in access to housing, considering household mobility and decision making and the operation of housing markets (see below). Finally, as we have noted throughout this book, the city itself changed in the late twentieth century as a result of a range of global economic, technological, political and cultural changes, challenging the validity of traditional models and theories of the city and opening up new issues and questions for researchers to address. These new directions in the consideration of residential patterns are clearly exemplified by the development of research into gentrification, a process that was seen to be

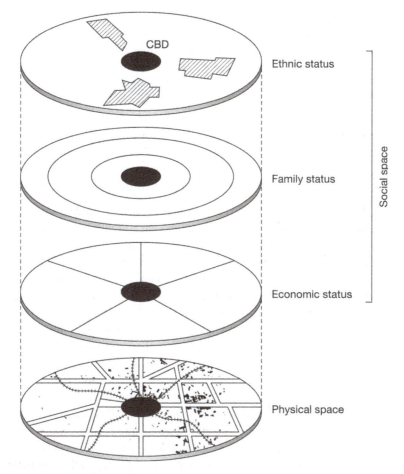

Figure 11.1 *Murdie's idealized model of urban ecological structure*
Source: Knox and Pinch (2010: 72)

reshaping classical models of urban structure and creating new tensions within cities. Examination of the processes fuelling gentrification became a key focus for debates surrounding the relative importance of economic and cultural influences and structure and agency in determining residential patterns in cities.

Gentrification

Gentrification has certainly been a hot topic in urban geography for some time and the volume of work that has been produced could fill a textbook on its own (indeed Lees et al. (2008) have produced such a book). Broadly, gentrification can be defined as 'the transformation of a working-class or vacant area of the central city into middle-class residential and/or commercial use' (Lees et al. 2008: xv). However, the key issue for researchers has been the issue of the displacement of these lower income groups, which has frequently been viewed negatively and has generated considerable reaction and protest by those displaced (Short 1989; Smith 1996). Initially, research interest in gentrification was prompted by the challenge that it posed to accepted ideas about residential structure, with more affluent households moving back into the city rather than moving outwards to the suburbs, the inverse of traditional models. However, research into gentrification has developed to encompass a number of key theoretical debates within urban geography generally and also within research into residential differentiation more specifically.

At the heart of the early debates into the processes underpinning gentrification has been a division between those who have highlighted the importance of economic processes and those who have highlighted the role of human agency and consumer preferences. A key figure in the development of economic explanations of gentrification has been Neil Smith. Drawing on Marxist understandings of the city, Smith (1996) has argued that gentrification occurs where a 'rent gap' exists, which is the disparity between the potential rents that could be commanded by dilapidated inner-city properties or vacant land when redeveloped and the actual rents being obtained. Where a significant gap opens up as a result of inner-city decline, it becomes profitable for developers to buy up properties cheaply, refurbish them and sell them on for a significant profit. However, critics of this perspective note that while rent gaps exist in many cities and city areas, not all become gentrified, so suggesting that other factors are important in these processes. In contrast to Smith, researchers such as David Ley (1996) have highlighted the role of consumer preference and human agency in the production of gentrification. Ley points to the importance of the emergence of a new middle-class, or creative class (Florida 2004) in many cities, linked to the growth of producer service jobs and cultural industries within these cities (see chapter four). These groups seek lifestyles different to the conformity of the suburbs and are drawn to the diverse and vibrant cultural opportunities available in urban centres, seeking to live close to these opportunities and therefore leading to the gentrification of inner-city neighbourhoods. Indeed, it is argued that processes of gentrification are often started as a result of the location decisions of pioneer gentrifiers, such as artists and other cultural innovators. For example, the gentrification of areas such as Hoxton in London or SoHo in New York was seen to have been precipitated by their notoriety as the location of vibrant arts scenes (Zukin 1989; Hubbard 2006). Other important pioneer groups have been identified, including single professional women (Rose 1984), gay people (Knopp 1990) and students (apprentice gentrifiers) (Smith and Holt 2007).

Currently, writers on gentrification generally acknowledge that both economic and cultural processes are important, with the key questions lying in the relative influence of these forces in understanding gentrification in particular urban contexts (Lees et al. 2008). Equally other key themes have emerged, in particular the facilitation of gentrification through the actions of local urban managers and policies, where gentrification is championed as a key component of the urban renaissance ambitions of those cities seeking to revive their fortunes (Lees 2003; Wyly and Hammel 2005). Indeed, Smith (2002) notes that so popular is this discourse that it has become a key global urban strategy. The popularity of gentrification with urban policy makers is in sharp contrast to the overwhelmingly negative perception of the process and its impacts on urban populations from within academic and popular writing on the issue (Lees et al. 2008). Despite some earlier declarations in the 1980s that the process would be short-lived, gentrification continues apace and has been identified in cities around the world and at lower levels of the urban hierarchy within provincial cities, beyond the large global cities, such as London and New York, where it was initially evident (see for example Lees et al. 2015; Lees et al. 2016). In many central city areas it has contributed considerably to problems of affordability as house prices have been pushed up, frequently beyond the reach of all but the most well paid professional workers. At the same time gentrification is mutating into new forms, with some areas of major cities that were the first to gentrify undergoing new waves of gentrification (see case study).

Case study: super-gentrification in Barnsbury, London

Barnsbury, a residential neighbourhood in Islington, North London about two miles from the City of London, provided one of the early case studies of classic or 'first wave' gentrification. Composed of large, early nineteenth-century houses, this inner-city area had declined in the 1950s and 1960s, with most of the housing being privately rented, in multiple-occupancy, overcrowded and in poor condition. Pioneer gentrifiers, such as architects and academics, began moving into the area in the 1960s and 1970s, generating much local opposition and conflict and eventually leading to a marked change in the socio-economic character of the area and rising property values by the 1980s. In the 1990s the area underwent a

'second wave' of gentrification, driven by London's role as a global financial centre. In this wave of 'corporate' gentrification, the gentrifiers were increasingly those holding jobs in London's rapidly expanding financial services sector, fuelling rapid increases in house prices. Further gentrification was also supported by increasing private sector investment and urban policies.

Butler and Lees argue that more recently a further (third) level of intensified gentrification has been superimposed on this already gentrified neighbourhood. Terming this super-gentrification, or financification, they note that this process requires higher financial investment than previous rounds of gentrification, and has been driven by

continued

globally connected workers, drawn from Britain's social elite, working in London's financial sector who enjoy large salaries and might be seen as super-rich. These super-gentrifiers connect global capital flows to the neighbourhood level, with their global identity being projected on to the local scene. Through their analysis of census data, Butler and Lees note that a tripartite socio-economic division now exists between these super-wealthy professionals, middle-class professionals and the working-class and economically inactive in the area, creating increasing cultural divisions and tensions.

The trends evident in Barnsbury are similar to those identified in Brooklyn Heights in New York, linked to its key position as a global financial centre (Lees 2003).

The mutation of gentrification into this new form of super-gentrification has also questioned the early stage models of gentrification that assumed an endpoint of mature gentrification in neighbourhoods. Equally, it questions ideas of a rent gap, as there is no decline in value and an already prosperous neighbourhood becomes even more exclusive and expensive.

Source: Butler and Lees (2006)

Therefore, the study of gentrification as a phenomenon provides a useful case study highlighting the broader issues underpinning the question of why people live where they do within cities. At the heart of this question are long-standing debates as to whether this is the result of the action of people and the housing choices they make or a product of the operation of economics and housing markets. In reality, most researchers would concede that it is a combination of these factors, and that both of these areas therefore form important topics of enquiry when considering housing and residential segregation within cities. Equally, the different methodologies employed to examine gentrification highlight the issue of the scale at which research is undertaken, with different issues and insights revealed by macro scale and micro scale perspectives (Lees et al. 2008).

Choosing where to live: residential mobility

Broad patterns of residential segregation within cities can appear relatively fixed and enduring. However, looking below this at the micro-scale the residential landscape is far from fixed and is rather in a constant state of change, where these broader patterns are the result of the combination of a large number of individual household decisions about where to live in response to the housing opportunities available. Research into residential mobility has revealed that the amount of movement by individual households is quite significant, with between five and ten per cent of all urban households moving in one year in European cities and between fifteen and twenty per cent in cities in North America, Australia and New Zealand (Knox and Pinch 2010: 254). However, within this, rates for individual cities vary considerably, linked to the wider economic, social and political forces within which the city is positioned and which influence the local housing markets within which decisions to move are made. In addition, not all households are equally mobile, with some more likely to move than others; such as younger rather than older households, private renters rather than owner occupiers. Equally, there is also an effect resulting from the duration of residence, whereby the longer a household remains in a dwelling the less likely they are to move, developing an attachment to the dwelling and the neighbourhood. Spatially, empirical studies reveal a three-fold division of the city in terms of rates of mobility: an inner-city zone with high rates of movement fuelled by new migrants to the city and the establishment of new households with a high proportion of renters; an outer zone, also with a high level of mobility, with households taking up new suburban housing opportunities as a result of changing aspirations linked to increasing income and changing family status; and an older more established zone of relative stability in between where household's needs seem to have been satisfied (Knox and Pinch 2010: 256). Generally, a circular and reinforcing relationship exists between urban residential structure and household mobility. Here, household mobility is the outcome of a set of decision making processes relating to the housing needs and expectations of households, influenced by income, lifestyle, family status and knowledge and perceptions of different neighbourhoods and the housing opportunities available in the form of new or vacant dwellings. These individual household decisions and moves then shape and reshape the social ecology of the city which forms the framework of housing opportunities and neighbourhood perceptions that influence subsequent rounds of household decision making.

Exploration of residential decision-making processes developed in the 1970s, as an area of enquiry drawing on behavioural approaches in urban geography (see also chapter ten). The decision to move to a new residence can be either voluntary or involuntary. Involuntary moves result mainly from property demolition or eviction, and formed about a quarter of all moves in a classic study of migration in Philadelphia by Rossi (1955). Involuntary moves can also be forced by changes in family status or lifestyle (e.g. marriage/divorce, ill health, bereavement) or changes in economic circumstances (e.g. long-distance job change or retirement). However, the majority of moves are voluntary (Pacione 2009).

The decision to move is underpinned by two key aspects of household behaviour: the decision to seek a new residence and the search for and selection of a new residence. The decision to seek a new residence is triggered by housing stress where the current accommodation no longer meets the needs or expectations of the household. In reviewing the literature on mobility, Clark et al. (2006) argue that the overall trigger for mobility is a difference between actual and desired housing characteristics, either as a result of disequilibrium between current and potential housing consumption (equilibrium approach) or as a result of dissatisfaction with the characteristics of home and neighbourhood (dissatisfaction approach). Important triggers can be environmental, where problems exist with the current dwelling or neighbourhood, such as the house being too small or in poor condition; a deteriorating urban environment or loss of amenities; or linked to changes in household circumstances, such as income changes (gaining/losing employment), changes in family status/life course (changing household size or composition) or lifestyle changes (e.g. focus on career, family, consumption, etc.) (Levy et al. 2008).

Exercise

Reflect on your own household's recent move(s) and the triggers that underpinned this/these. Was your move involuntary or voluntary? What was the main prompt for the move: house, neighbourhood, lifecycle, job or something else? In your search for a new residence, what were you looking for and what search criteria did you use (in a particular area, a particular type of home, proximity to employment/facilities, etc.)?

You could research this further by asking fellow students, friends, family members about their household moves. Does this reveal any broader patterns about migration in your city?

What are the main problems/challenges in collecting data on household migration decisions (consider who decides, how decisions are made, when a move took place, etc.)?

In response to housing stress, households can choose to stay put, seeking to improve their environment, such as through up-grading their accommodation or seeking improvements to their neighbourhood, redefine their aspirations or make the decision to seek an alternative residence. Where an available vacancy can be matched to the household's expectations then a move to a new home can occur. Generally, households define the criteria and area within which they will search for accommodation based on their personal needs and knowledge of local housing markets, although it is only relatively recently that there has been more detailed research into the behaviour of households in looking for a new home (see case study below). Choice can be constrained by the time that households have to search and householders' knowledge of the locally available housing opportunities, either personal knowledge of local markets or knowledge obtained through various information sources, such as estate/letting agents (Pacione 2009). Indeed, far from being a rational process, as many earlier behavioural studies assumed, the search for a new residence is 'messy' and underpinned by strong emotions and non-rational behaviour (see Smith et al. 2006 and case study following).

Case study: family decision making processes when buying a house

Levy et al. (2008) explore the internal family decision making processes involved in buying a house, which constitutes an infrequent but significant purchase for households. They argue that this has been a relatively neglected area of research in residential mobility studies where attention has focused more on the external socio-economic influences on household decision making. Their research uses in-depth interviews with recent adult house purchasers and estate agents to explore their understandings and experiences of buying a house in Auckland, New Zealand.

Their research highlights the messy nature of the decision-making process. While house purchase is an economic activity, it is not a rational economic process, but rather is a social activity involving emotion, discussion and negotiation of the family's needs, interaction with housing exchange professionals and the interpretation of the wider housing market. In this sense housing transactions are performed in power-filled negotiations between adult family members, buyers, sellers and housing market professionals.

These family decision-making activities are shaped by family structures, gender roles, ethnicity and socio-economic status. In addition to adults in the household, extended family members and friends can participate in the decision-making process, sometimes influencing decisions to take an expected course. However, the extent to which wider social collectives influence decision-making varies by the ethnic background of the purchasers, with Pacific Island and Asian families seen as more open to these influences. In the context of the specific structure of the New Zealand housing market, estate agents can also significantly influence the decision-making process. Here, agents are not simply information intermediaries but are active market makers, potentially leading to confusion and conflict in their dual roles acting for sellers and buyers.

With regard to the internal family negotiations and power relations involved in the decision-making process, both estate agents and adult family members identified gendered dimensions to the process. Generally, men were concerned with financial and status aspects of home ownership while women were more concerned with family needs, although in reality the links were more complex and fluid depending on the roles adopted by men and women in the family structure. Links between gender and the decision-making process were also affected by other dimensions such as ethnicity, personality and experience and in addition varied according to the stage within the search and purchase process. Both estate agents and families also highlighted the important role of emotions and feelings in the final decisions to purchase. While initial house hunting might be more structured and systematic, final decisions were often based on a general feeling which families found difficult to explain and which could also be influenced by wider collectives of extended family, friends and professionals. The study therefore serves to highlight the messy complexity of residential decision-making processes and the importance of both individual and collective feelings and characteristics and broader structural housing market conditions in influencing them.

Source: Levy et al. (2008)

Constraints: housing markets

While behavioural approaches provided important insights into residential mobility and spatial patterns, their emphasis on individual or household decision making has often tended to downplay the constraints within which these decisions are taken. For many households in the city, particularly the most disadvantaged, their residential location and decision making is highly constrained, both by their personal circumstances and also by wider structural factors which shape their lives and the operation of local housing markets. The availability of housing, as with other elements of the built environment, is not simply a reflection of the economics of supply and demand but is affected by institutional constraints and the interactions of various actors (figure 11.2). Housing is therefore socially produced and the dynamics of housing supply in the city are constructed through distinctive housing sub-markets where 'different socioeconomic groups are matched to particular types of housing through a series of different market arrangements' (Knox and Pinch 2010: 116).

These housing sub-markets have been traditionally defined by housing attributes (dwelling type/tenancy/price), household type (family status/ economic status/ethnicity) or location within the city (inner-city/suburban) (Bourne 1981: 89), although in reality urban housing markets have rarely been found to fit neatly into these categories. Within western cities, key trends throughout the twentieth century have been the growth of the owner-occupied market, an increase in the letting of dwellings by public authorities, mainly in

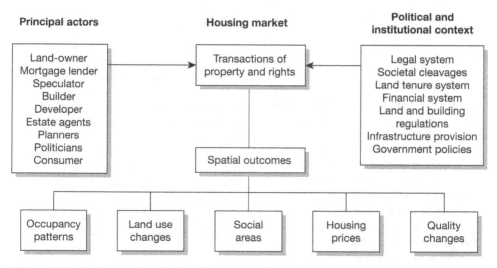

Figure 11.2 *Actors and institutions in the housing market (Bourne 1981, Figure 4.8, p. 85)*

Source: Adapted from Knox and Pinch (2010: 117)

Europe and particularly in the UK, and a decrease in the availability of cheaper privately rented dwellings. However, more recently public authority involvement has declined in many western cities, driven by neo-liberal political agendas to 'roll back the state', which has led to contractions in the public letting market (Pacione 2009). Within many cities of the Global South the trend has been the growth of the informal housing market beyond smaller private owner occupied and public housing markets. Only within a limited number of newly industrializing nations, such as Singapore, has there been substantial public intervention and investment in housing with governments in many other developing nations either unable or unwilling to intervene (Drakakis-Smith 2000).

A key focus of research into the operation of housing sub-markets has been the analysis of the role of the various agents who shape the opportunities available within these markets, known as the managerialist perspective. This work has drawn on the ideas developed by the sociologist Max Weber, who sought to explain social systems by considering the actions of the people who make and sustain them, and which were utilized in the context of urban research by Ray Pahl (1969). Pahl argued that the key to examining social constraints lay in the exploration of the ideologies, policies and actions of those managing the urban system, the gatekeepers within particular institutions. These gatekeepers act as mediators between a client population and the available resources, and research has highlighted their influence in shaping people's sense of possibilities through the decision making or eligibility criteria which they apply in the allocation of resources (Knox and Pinch 2010). These criteria can be explicit in the form of policy documents or can be implicit or tacit knowledge used within the organization. However, while managerialist research has demonstrated the important influence exerted by urban managers, researchers working from a political-economy perspective have questioned the relative power of gatekeepers, pointing out that the decisions of managers are themselves subject to wider structural constraints and that residential patterns are shaped by economic, political and cultural forces beyond the control of urban managers (see also chapter two).

Exercise

Undertake a survey of the property pages of your local newspaper or the properties advertised by various estate agents in your local area. Consider the location, type and price of property available and whether properties appear to be targeted at particular groups or lifestyles. Can you identify particular housing sub-markets within your city and do particular parts of the city appear to have more dynamic markets than others? Do some estate agents focus on particular markets, in terms of the location, type or price of homes available?

Within the context of the operation of housing sub-markets, key managers include important personnel from financial institutions such as banks and building societies, exchange professionals such as estate agents (see case study above) and lawyers, agents of the local state such as housing managers and planners, managers of voluntary and charitable organizations such as housing associations or homeless charities, and property owners and developers such as landowners, landlords and builders. Examples of the influence of landowners, developers and builders in housing supply have been highlighted by recent urban morphological research into the development of suburban landscapes, such as the role of these agents in shaping the phasing and style of Britain's twentieth century suburbs (Whitehand and Carr 2001; see also chapter three). However, much of the work on housing market gatekeepers has focused on the biases associated with the practices of mortgage finance institutions, estate agents and public housing authorities. This work has highlighted biases against both people and property. Within private property markets, biases against people occur where lenders and agents either do not deal with particular groups, usually those who are poorer or marginalized, or only highlight specific loan or housing opportunities, directing people to particular areas, known as 'steering' (Gottdiener and Budd 2005). While many countries have introduced anti-discrimination legislation to outlaw overt biases, financial and social exclusion still occurs with agents operating tacit or hidden practices, or in the case of lending institutions employing strict lending or credit rating criteria which, while being a legitimate business practice, will exclude many groups (Leyshon and Thrift 1995). Biases against property include the practice of redlining, where funds are not loaned on properties in neighbourhoods identified as a bad risk, particularly those in poorer inner-city districts (Gottdiener and Budd 2005 and see also chapter seven). This institutional discrimination within housing markets has been seen as an important factor underpinning the persistence of patterns of poverty and racial segregation in cities, particularly the segregation of black and white populations in American cities (Wilson 2007; Hernandez 2009).

Within countries with large public housing sectors, such as Britain, research has examined the impacts of the allocation policies of local housing authorities. Most authorities operate eligibility rules and priority systems which filter access to public housing; for example young single people without dependents, newcomers to an area or former owner occupiers can find themselves with least access to public housing in British cities, fuelling pressure on private rental markets and leading to problems of homelessness (see below). Equally, eligibility criteria can influence the location and type of public housing offered, with good tenants offered the best housing and 'problem households' being directed to 'dump estates' (Pacione 2009). The direction of certain households to dump estates can lead to the stigmatization of the area and its population, reinforcing negative attitudes and making housing difficult to let. Once established, negative images of estates are persistent and can remain even following regeneration intervention by agencies, where activities reinforce rather than challenge neighbourhood images so perpetuating the stigmatization of the area and its residents (Hastings 2004). To this end Hastings has argued in her research into stigmatized housing estates in the UK, that the issue of image needs to be managed more explicitly within housing and regeneration policies and activities.

Housing crises

As the discussion above should have hopefully revealed, there are many factors influencing where people live and their access to housing options. What should also be clear is that these options are quite variable and that inequalities exist in access to housing. A key issue facing cities is that not all the people who want a home have access to one that adequately meets their needs. Globally, the scale of the problem is enormous with the United Nations estimating that over 100 million people lack any home, while over one billion reside in sub-standard and insecure accommodation with no services (UN-HABITAT 2003). This has been an enduring problem for many cities around the world; indeed, as we noted in chapter six the development of systematic attempts to plan the growing industrial cities of the late nineteenth and early twentieth centuries stemmed from concerns about the provision of adequate housing for their growing populations. Then, as now, the lack of good quality, low-cost social housing can be linked to enduring problems of a mismatch between the demand for, and supply of, this type of housing.

Key factors underpinning this continuing mismatch between supply and demand are the issue of affordability and the changing role of the state in housing supply. Affordability is basically the proportion of household income spent on obtaining housing. For many in the Global South household incomes are nowhere near adequate to secure even the most basic formal housing. In Europe and North America recent rises in rents and house prices have led to more households experiencing problems in affording housing, classed as those households spending over thirty per cent of their income on housing in the US and twenty per cent in the UK (Pacione 2009). However, affordability is a problem of places as well as people, where in over-heated housing markets, such as in global cities, even people with average wages can experience affordability problems, such as key public sector workers including health workers, teachers and police officers. These affordability problems were exposed by the global financial crisis of 2007, when over-heated property markets collapsed and house repossessions rose (Knox and Pinch 2010; García Lamarca and Kaika 2016; Waldron 2016).

The affordability problem has been underpinned by a diminishing supply of low-cost social housing below market rents. Here the role of the state in housing provision is critical. Broadly speaking, the adoption of neo-liberal governance strategies around the world in recent years has led to a reduced role for the state in the provision of low-cost social housing with the encouragement of provision through the market or via third-sector charitable organizations, such as housing associations (Pacione 2009). However, the ability and willingness of private and charitable organizations to meet this low-cost social housing need has been highly variable and often insufficient to meet rising needs and the retreat of state provision. While the issues arising from the mismatch in the supply of and demand for affordable housing are wide-ranging, there are perhaps two contemporary issues that most clearly illustrate the housing problems facing cities in the first half of the twenty-first century: the growth in informal housing areas, often referred to as squatter settlements, in cities of the Global South and the growth in homelessness in cities around the world.

The presence of large informal housing areas has been a long-standing feature of rapidly urbanizing areas, particularly the megacities of the Global South (Gilbert 2000; UN-HABITAT 2003). Within these cities, considerable unmet low-cost housing demand exists as a result of the rapid population increases these cities experience. However, the limited and precarious income earning opportunities that exist in these cities (see chapter four) mean that many poorer households are unable to obtain a high enough income to afford to access the restricted formal housing opportunities available. Consequently, for many, the only option is the informal or popular housing sector (see Drakakis-Smith 2000 for a good overview of the issues).

The attitude of governments and development organizations to the presence of squatter settlements has varied considerably, ranging from hostility, to suppression, to toleration and occasional support (Pacione 2009). Stokes (1962) characterized these areas as either 'slums of despair' or 'slums of hope' reflecting opposing views as to whether squatter housing is a burden or a benefit to the urban poor. Negatively, the continuance of squatter settlements as home to many urban dwellers highlights the chronic economic problems faced by poor urban populations, marginalizing these groups within the city and trapping them in poverty. However, more recently there has been some support for these settlements, offering as they do some support for people by providing a place in the city and economic opportunities with little government effort or investment. In some instances, settlements have been legalized and offered improvements in services, such as in some of Rio de Janeiro's favelas (Riley et al. 2001). However, while many view support of these self-build settlements as a positive development, others argue that it represents a neo-liberal development agenda which looks to absolve the state of responsibility for adequate housing provision and increasingly financialize housing in cities of the Global South (Gruffydd Jones 2012).

The presence of homeless people on the streets of a city is a highly visible manifestation of housing affordability problems. Homelessness is an issue for cities of both the Global North and South; within cities of the Global South the presence of large numbers of homeless people on the streets has been an enduring concern, while in the Global North although numbers are lower they have increased significantly in recent years (Daly 2008; Pacione 2009). However, comparative measures of the extent of homelessness within cities are difficult to achieve as definitions of homelessness vary between countries and homeless populations are highly mobile and fluid, moving around cities, into and out of the gaze of urban authorities and in and out of being in a state of homelessness. Generally, figures quoted are seen as an underestimate as many homeless either do not count under official definitions or have temporary, yet unsecure, accommodation in hostels or with family or friends, rendering them 'invisible' (Langegger and Koester 2016). Broadly speaking the homeless are those who cannot afford shelter by themselves and most often consist of marginalized groups such as the unemployed, recent migrants, substance abusers, mentally ill people, ethnic minorities, battered women, runaway youths and street children (Gottdiener and Budd 2005).

The causes of homelessness are as diverse as those people who are homeless, and are the result of a combination of personal and structural factors. Key structural factors which have resulted in increased homelessness are global economic changes, leading to increased migration and unemployment, decline in welfare help as a result of rolling back the state and deinstitutionalization of groups such as the mentally ill, linked to welfare changes (Dear and Wolch 1987; Daly 2008). Attitudes to the presence of homeless populations is highly variable between city authorities, ranging from hostility and forced removal to more compassionate strategies of providing shelter and support. However, recent research into the nature of homelessness and homeless populations has stressed that the problem needs to be viewed through the eyes of the homeless, otherwise there is a concern that strategies in place will fail to meet the needs of homeless groups (Daly 2008; DeVerteuil et al. 2009) (see also case study below). In reviewing recent geographical work on homelessness, DeVerteuil et al. (2009) emphasize the need for research to recognize the diversity of homeless populations, particularly differences due to gender and ethnicity, and the need to give a voice to the homeless in research, utilizing methods such as ethnography, participant observation, life histories and autophotography in tracing homeless peoples' experiences in, and geographies of, the city. Additionally, they argue that research into the homeless has been too focused on viewing policies as punitive towards these populations in the city and that there is a need to recognize that policies towards the homeless are in reality multi-faceted and ambiguous, requiring more critical and nuanced examination (see case study below).

Case study: strategies for managing the homeless in San Francisco

Murphy (2009) examines the ambiguities and contradictions in recent policies enacted by the city of San Francisco to address its homelessness problem. The city was seen to have a high rate of homelessness and was also viewed as one of the most punitive in terms of its anti-homeless measures. However, in 2004 the newly elected Mayor, Gavin Newsom, began to usher in a 'kinder and gentler' policy regime that aimed to provide legitimate and compassionate assistance for the homeless. The 'Care not Cash' policy offered referrals to housing and services for homeless adults rather than cash handouts. While this was seen as a controversial move by some, it was also viewed as a successful strategy by many in the city and beyond to reduce the numbers of the city's homeless. However, based on her research Murphy highlights the deeper complexity of this initiative and argues that while being less punitive than their predecessors, the policies have introduced new exclusions in service delivery, new definitions of the deserving poor and new degrees of homeless marginalization, which are obscured by the language of compassion. Broadly, these softer strategies are reserved for those willing to comply with policy programme requirements, while those homeless seen as non-compliant continue to be targeted with more punitive tactics.

Murphy highlights the ways in which these changes are played out in the city's spaces of service provision, identifying three key homeless geographies in the city: institutional geographies of housing and hotels, protracted geographies of referral and waiting and spectral geographies of the hidden and mobile homeless. The aim of the programme to refer homeless clients to housing units has served to concentrate homeless people registered on the programme in city-funded, single room occupancy hotels (SROs) in poor central areas of the city, as a result of the lack of other available housing options. While for some this accommodation represents an improvement over living on the streets, others have noted that much of this accommodation remains sub-standard and imposes restrictions on those in the units, viewing them as clients to be monitored rather than as tenants. In addition, requirements to look for work while being a resident represent a tradeoff that many homeless people will not accept and therefore serves to remove them from programme lists.

Despite these problems with the SROs there is a waiting list to be placed in a hotel and many homeless men and women remain waiting in the city's emergency shelter system. The problems faced by many homeless people in accessing or staying in these emergency shelters mean that further people disenroll from the programme. Many homeless people view the shelter system as stressful, fragmented, frightening and difficult to navigate, characterized by transience and flux with shifting rules and regulations and centres frequently closing down and opening up in different areas. This constant waiting and transience means that many homeless people stop seeking support in the shelters and remain on the streets. This can be wrongly perceived as 'service-resistance' on the part of the authorities. It is these groups who continue to be the

continued

target for punitive measures by the police as their lives are visible on the street and therefore they inhabit spectral geographies of constant hiding and movement in order to avoid harassment and to feel safe. Therefore, by examining these geographies of homelessness in San Francisco, and how the city's homeless navigate their way through them, it is possible to observe the ways in which the city's homeless are sifted and sorted into the deserving and undeserving poor and how different soft or hard policies are applied.

Source: Murphy (2009)

Summary

In this chapter we have considered the issue of why people live where they do within cities. The city can be viewed as a residential mosaic, with different household types occupying particular housing niches within it. Housing is an important component of the urban landscape and access to adequate accommodation is an important aspect of people's life and place within the city. However, access to housing and the operation of housing markets in cities is highly diversified and also unequal, underpinned by a complex mix of individual household choices and decisions operating within wider institutional and structural constraints. Housing problems exist within all cities, with growing problems of housing affordability and homelessness in many cities around the world. In many cities of the Global North, rising property prices, processes such as gentrification, and a decline in the availability of social housing and private rented accommodation, have fuelled problems of housing affordability for lower, and some middle, income groups and has also led to increasing homelessness. In many megacities of the Global South, the growth of informal squatter settlements and the visible homeless starkly illustrate the widening gulf between the demand for, and provision of, good standard affordable housing. Consequently, housing provision remains a significant challenge for urban managers in the twenty-first century.

Follow-up activities

Essay title: 'To what extent are economic factors responsible for residential segregation within cities?'

Commentary on essay title

An effective answer would outline the ways in which residential areas of cities can be segregated, drawing on classical urban models of residential structure. It would then review the variety of factors that are seen to underpin segregation (economic,

*political, cultural) and consider the issue of choice and constraint in the residential
decision making undertaken by households, evaluating the relative significance of
these factors in producing patterns of segregation. An excellent answer would
further examine the complexities of the residential decision-making process and
the dynamic relationships between broader structures and the actions of various
actors in the process. The case study by Levy et al. (2008) would be relevant here
in examining these complexities from the micro scale perspective.*

Project idea

Consider whether there is evidence of a 'housing crisis' in your city or a city that is familiar to you. You can examine this issue in a number of ways, drawing on some of the themes introduced in this chapter, for example:

● Using data from local property agents or the local press consider the prices for housing in the city compared to average wages (use local census data or again use the local press to look at wage rates for jobs in the city). How big is the affordability gap (rough definitions are provided in this chapter)?

● Does the city have a high number of homeless people? You might talk to the local housing authority to see if they have information on numbers of people waiting for homes and the types of policy adopted to tackle homelessness problems. Consider whether the local authority provides housing or not. You might also find out whether homeless charities or shelters operate in your city and might interview those running these organizations to gain their views on the housing issues facing your city.

Further reading

Books

● Knox, P. and Pinch, S. (2010) *Urban Social Geography: An Introduction*, 6th edn, Harlow: Prentice Hall
One of the definitive textbooks on urban social geography. Provides an excellent overview of research undertaken into residential patterning, residential mobility and housing markets, with many useful examples.

● Madden, D. and Marcuse, P. (2016) *In Defense of Housing*, New York: Verso
Offers a detailed critical analysis of the current housing crisis in cities resulting from the increasing commodification and financialization of housing.

● Maloutas, T. and Fujita, K. (eds) (2016) *Residential Segregation in Comparative Perspective: Making Sense of Contextual Diversity*, Abingdon: Routledge

A book that offers a welcome comparative perspective on residential segregation, with case studies from a number of cities from around the world.

- Lees, L., Shin, H.B. and López-Morales, E. (2016) *Planetary Gentrification*, Cambridge: Polity Press
  Acknowledging the increasing global extent of gentrification, the book offers a trans-urban analysis of gentrification, extending analysis beyond the major cities of the Global North.

- von Mahs, J. (2015) *Down and Out in Los Angeles and Berlin: The Sociospatial Exclusion of Homeless People*, Philadelphia: Temple University Press
  Drawing on ethnographic accounts the book offers an illuminating comparative analysis of the impact of social welfare policy on homelessness in these two cities with high homeless populations.

- Woldoff, R.A. (2011) *White Flight / Black Flight: The Dynamics of Racial Change in an American Neighborhood*, Ithaca, NY: Cornell University Press
  An in-depth, longitudinal study of racial change in a specific urban neighbourhood focusing on the perspectives of both newly arriving black residents and white residents who have stayed in the neighbourhood.

Journal articles

- Choi, N. (2016) 'Metro Manila through the gentrification lens: disparities in urban planning and displacement risks', *Urban Studies*, 53(3): 577–592
  Extends discussion of gentrification by looking to apply theories developed in the Global North to a city in the 'Global East'.

- Crowder, K. and Krysan, M. (2016) 'Moving beyond the big three: a call for new approaches to studying racial residential segregation', *City & Community*, 15(1):18–22
  A theoretical essay which calls for explanations of segregation of Black and White residents to move beyond traditional explanations, based on income, preference and prejudice.

- García Lamarca, M. and Kaika, M. (2016) '"Mortgaged lives": the biopolitics of debt and housing financialisation', *Transactions of the Institute of British Geographers* 41(3): 313–327
  A different 'take' on the issue of housing financialization looking at the materiality of mortgage contract and financialization as an embodied practice, focusing on the mortgage defaults and eviction crisis in Spain.

- Ghertner, D.A. (2015) 'Why gentrification theory fails in "much of the world"', *City*, 19(4): 552–563
  Questions the applicability of theories of gentrification as something that can explain urban change around the world, particularly beyond the Global North.

- Ley, D. (2017) 'Global China and the making of Vancouver's residential property market', *International Journal of Housing Policy*, 17(1): 15–34
  Examines the role of international capital and property development in fueling a crisis of affordability in the city.

- May, J. (2015) 'Racial vibrations, masculine performances: experiences of homelessness among young men of colour in the Greater Toronto Area', *Gender, Place & Culture,* 22(3): 405–421
  Ethnographic study which importantly draws out the intersections between gender and race in terms of the lived experiences of homelessness.

- May, J. and Cloke, P. (2014) 'Modes of attentiveness: reading for difference in geographies of homelessness', *Antipode*, 46(4): 894–920
  Explores the British homeless services system, but draws a comparison with Murphy's (2009) study featured as a case study in this chapter.

- Musterd, S., van Gent, W.P., Das, M. and Latten, J. (2016) 'Adaptive behaviour in urban space: residential mobility in response to social distance', *Urban Studies*, 53(2): 227–246
  Looks at the social distance between an individual and the social position of their residential neighbourhood as a trigger for residential movement, focusing on four Dutch cities.

Websites

- www.unhabitat.org/categories.asp?catid=282 – UN Housing Rights website.

- https://portal.hud.gov/hudportal/HUD – Website for US Department of Housing and Urban Development.

- www.communities.gov.uk/housing/ – Housing pages for the UK Department of Communities and Local Government.

12 Transport, mobility and the city

Introduction

Transport and mobility, or the lack of it, are seen as fundamental concerns in terms of the organization and operation of the modern city. Much of life in cities is concerned with trying to get somewhere and the increasing pace, scope and complexity of journeys and communications has generated renewed interest in this area of study, linked to the wider 'mobilities turn' within the social sciences (Sheller and Urry 2006; Creswell 2006). As Creswell (2006: 3) notes, while mobility can be thought of as a 'brute fact', it also carries significant meaning as something that is practised, experienced and embodied. Scholars working within a mobilities paradigm have increasingly been interrogating the spaces of transport flows within the city to critical enquiry, illuminating the practices and powers involved in the everyday activity of moving around. Additionally, with rising levels of urbanization globally, the impacts of increasing inter- and intra-urban mobility, specifically the environmental impacts associated with increasing vehicle-based mobility, are contributing to key global urban challenges for the twenty-first century, namely, combating climate change and city sustainability (see also chapter 13).

The chapter first considers the links between transport, mobility and the city and then continues by exploring urban transport and mobility issues at different scales. At the global level the chapter considers urban connectivity, the growth in air transport and the contribution of urban transport to global concerns such as carbon emissions, climate change and pollution. At the intra-urban level, the chapter examines the 'governmentality' of transport and the challenges for urban transport policy and planning. The chapter concludes by looking at transport and mobility as an embodied practice, highlighting the importance of considering movement as a cultural practice, beyond an instrumental act of physical displacement. This scalar separation does not imply that there are separate transport and mobility concerns at different urban scales, but rather the chapter will demonstrate that these concerns interweave within the 'intransitive' or fluid and multi-scalar city (Hubbard 2006: 165).

Transport, mobility and the city

Within urban geography, discussion of transport has traditionally focused on the influence of changing transport technology on the morphology of the city. The size, form and internal organization of cities throughout history and across the world are a reflection of the prevalence of different forms of transport (Hart 2001). As we noted in chapter three, the form of the pre-industrial city was compact, based on walking as the primary form of urban transport and mobility. From the nineteenth century onwards, rapidly growing and industrializing cities in Europe and North America were characterized by waves of decentralization, linked to changing transport technologies. Developments such as the railway, the tram and latterly the motorcar were important determinants in the outward expansion of cities, and changes in mobility were physically expressed in the construction of dense networks of tramlines, suburban rail routes and motorways.

The urban sprawl of many contemporary cities reflects, and helps perpetuate, the rising importance of personalized motorized transport, specifically the private car and its associated transport infrastructure, to urban mobility. Within Europe and North America, the transition from walking to personalized motor transport as the predominant mode of transport within cities, and a change from access by proximity to access by movement, took place throughout the late-nineteenth and twentieth centuries (Rode et al. 2014). Within this 150-year period there was a transitional phase in the first half of the twentieth century where public mass transport grew in significance before being overtaken by the motorcar as the key transport mode (Hart 2001). In many fast-growing cities of the Global South this transition from the walking city to the car-based city has been different and extremely rapid, for example, 30 rather than 150 years, far from smooth and with little or no intervening 'mass transit' or public transport phase of development to ease the transition (Pacione 2009: 578–579). The result of this is that while the car (and in the case of many Asian cities, the motorcycle) accounts for a growing proportion of transport and mobility in these cities, walking and non-motorized vehicles have not declined as they have in cities of the North. These remain the main transport mode for the extensive urban poor in these cities, in the absence of significant public transit investment (UN-HABITAT 2013; Hickman and Banister 2014; Rode et al. 2014). As Rode et al. (2014: 4) highlight, cities around the world have therefore followed quite different 'urban accessibility pathways' which set certain parameters for their future transport development. The different pathways are reflected in the differing transport modal shares within cities in different regions of the world (see table 12.1). While these regional figures highlight the broad variations between different parts of the urban world in terms of transport development, they do mask some diversity amongst cities within these regions resulting from their specific urban development pathways (Hickman and Banister 2014).

Table 12.1 *Modal shares within different global regions*

Region	Total daily trips per capital, trips/person	Proportion of total trips by non-motorized modes, %	Proportion of total trips by motorized public modes, %	Proportion of total trips by motorized private modes, %
Western Europe	2.88	31.3	19.0	49.7
Eastern Europe	2.93	26.2	47.0	26.8
United States	3.81	8.1	3.4	88.5
Canada	2.88	10.4	9.1	80.5
Australia/ New Zealand	3.86	15.8	5.1	79.1
Middle East	2.09	26.6	17.6	55.9
Latin America	1.82	30.7	33.9	35.4
Africa	1.76	41.4	26.3	32.3
China	2.63	65.0	19.0	15.9
High income Asia	2.67	28.5	29.9	41.6
Low income Asia	1.98	32.4	31.8	35.9

Source: Adapted from Hickman and Banister (2014: 30)

The impact of the car on city morphologies is therefore not a simple division between Global North and South, and levels of car-based mobility in cities vary greatly around the world. This variability is linked to a range of economic, cultural and environmental factors underpinning the varying density of urban areas, including differences in land availability, planning systems, development industries, and cultural preferences for urban living (Hickman and Banister 2014). Generally, more densely populated cities tend to be characterized by lower levels of car use (fuel consumption), with more use of alternative modes of transport (Figure 12.1).

Population density is therefore a greater influence on levels of car use in cities than income levels, as many wealthy European cities that have high population densities are characterized by relatively low levels of car use (Hickman and Banister 2014, Rode et al. 2014). At similar wealth levels, sprawling Atlanta generates six times more private vehicle related carbon emissions than compact Barcelona. The same is true of wealthy Asian cities such as Tokyo or Hong Kong, with these higher density cities generating fewer emissions (Rode et al. 2014). Research into emissions in different Chinese cities of similar wealth indicates that those with lower densities and greater sprawl generate more transport related carbon emissions (Rode et al. 2014). This strong inverse relationship between car use and urban population density underpins much compact city advocacy as a route to the promotion of more sustainable urban mobility (UN-HABITAT 2013). However, current global trends indicate that

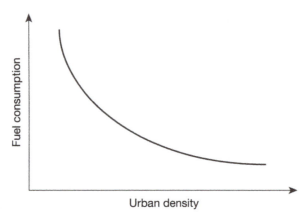

Figure 12.1 *The relationship between fuel consumption and urban density*

sprawl is increasing, with the amount of urban land expanding, most rapidly in cities of the developing world, with estimates suggesting that the total urban land area globally could triple between 2000 and 2030 (Rode et al. 2014). Equally, urban densities are decreasing, most notably in the developed world, although more recently there has been re-densification in some European and North American cities (Rode et al. 2014).

It is not simply the physical interrelationship between transport infrastructure and the city that is important, and mobilities research has increasingly sought to examine the co-production of urban technological and socio-cultural change. The 'mobilities paradigm' has been underpinned by the principle that mobility is something fundamental to contemporary culture, where the desirability of increasing speed and movement is central to the organization of modern society and life in cities more broadly (Sheller and Urry 2006). As Creswell (2006: 15) notes, the term 'modern' evokes images of technologically enhanced mobility – the car, the plane, the spaceship. These metaphors and images appear in futuristic views of the modern city in film, art and architecture (see also chapters 8 and 9). Subsequently, the idea of what constitutes the modern city and a successful life within it, for policy makers and urban populations alike, has been profoundly influenced by these metaphors of speed and the desire for faster, freer movement, underpinned by the automobile as the pre-eminent symbol of that freedom and mobility (Creswell 2006).

The central role of free and fast mobility to contemporary culture and social and economic organization has led urban theorists to think about the city differently. The city is increasingly described through a range of 'hydraulic' metaphors and is conceived of as a hyper-mobile global space of different flows of people, goods, services, capital, information, images and ideas linked to increasing global connectivity (Virilio 1986; Castells 1996; Massey 1999). Paul Virilio

(1986: 6) suggests that the city can be thought of as a system of 'habitable channels of communication and circulation', flowing through an urban environment of fixed and immobile infrastructure (of road and rail networks, ports, airports, etc.) that permit people (and things) to move. Virilio also argues that different channels of communication have their own speed limits and regulations and that a 'politics of speed' exists which is about regulating the flows through these channels, where some are expedited and others downgraded. This links to the idea of motility, or the capacity or potential to move, dependent on individual and collective capacities (Kaufmann 2002). Encompassing both social mobility and mobility capital, the notion of motility thus deals with questions of social inequality and the uneven distribution of mobility (Jensen 2011).

The idea that the contemporary mobile world has the potential to operate at different speeds, with some included and others excluded, underpins the idea of a 'politics of mobility' where some are afforded greater capacity to move in these flows than others (Creswell 2006; Jensen 2011). Those who control the channels of communication in the city represent powerful elites who influence the ability of others to move in step with these global, hypermobile flows or not. As Hubbard (2006: 173) notes, there is also a link to contemporary urban policy where 'the subtext … is that we need to create faster cities to keep up in these global times … the idea that the city needs to become more flexible and mobile has become something of a mantra among city governors'. These ideas underpin many current policy discourses concerning the promotion of 'wired cities', 'smart cities', '24-hour cities' and developments in airport and high-speed rail transport infrastructure to enhance global connectivity and economic competitiveness. However, if not planned carefully and inclusively these new developments may exacerbate the exclusion of those urban populations already confined to downgraded channels of communication, intensifying current inequalities in mobility.

Global transport connections and flows

Historically, while cities have been shaped internally by transport developments, they have also importantly acted as hubs in transport networks, as ports, marketing and distribution centres or as public transport termini. In the contemporary context, cities are seen as places where a range of increasingly globalized flows are embedded and entangled in a mesh of linkages, connections and relations across space. The role of cities as hubs within these webs of global connections was important to our discussion of world cities in chapter four, and international accessibility and connectivity were seen as key markers in defining world city status (Short and Kim 1999).

Fuller and Harley (2004: 140) discuss the emergence of the 'aviopolis', the networked and dispersed city of the air, which turns mobility and connection into a productive force that produces value in the global economy and in the process reshapes a city and its infrastructure. The growth in international air transport, both passenger numbers and freight movement, makes the point that urban networks are becoming more globally dispersed, and illustrate air transport's role in shaping the global urban system as the preferred mode of intercity movement for the transnational business class, migrants, tourists and high-value, low bulk goods:

> Quick, cheap regular air travel is one of the major enablers of space-time distanciation, the process whereby more and more social relations are routinely maintained at a distance, so that, for example, intercontinental business travel and transaction are regarded as commonplace.
>
> (Hubbard 2006: 166)

The impact of air transport on urban systems mirrors that of economic globalization more broadly, with the sophisticated and expensive infrastructure required for air networks concentrated within major cities in the global urban network, reinforcing their advantage. Short and Kim (1999) use air passenger and route data to analyse the increasing air service connectivity underpinning the development of the current global urban hierarchy of air route hubs. Their analysis highlights the distinction between cities that act as global or regional airport hubs. In their analysis, Heathrow, Paris and Frankfurt operate as key international hubs as they have more international flight connections and volumes. Below this, a second tier of cities act as centres for onward travel from international flights within their regions, such as Miami, Los Angeles and Amsterdam. Within their third tier are key regional cities linking neighbouring cities in a hub and spoke model, with each region having distinct geographical orientations. Within the US, for example, cities with some of the largest volumes of passenger air traffic, such as Atlanta, Chicago and Dallas, act as key regional centres in a national hub and spoke model (see also table 12.2).

Recent International Civil Aviation Organization (ICAO) data on passenger and freight numbers demonstrate both the growth in the volume of air traffic and the evolving global dynamic in flows, with the increasing significance of cities within the Asia/Pacific and Middle East regions in global air markets (ICAO 2015). In terms of aircraft departures for specific cities, eight of the top fifteen airports were US airports, four were in Europe, two were in Asia/Pacific and one in the Middle East (Istanbul) (table 12.2). Atlanta ranked first in terms of departures and passenger numbers, highlighting its continued role as a key regional airport hub. Shanghai achieved the highest growth of 2015 and entered the top fifteen, linked to strong domestic traffic demand in China. Globally the growth in freight traffic has been slower, although again Asia-Pacific and the Middle East have seen the strongest growth recently, with Hong Kong ranked

Table 12.2 *Top 15 airports 2015 (ranked by departures, passengers and freight)*

Airport	Number of Departures	Airport	Number of passengers	Airport	Tonnes of Freight
Atlanta (US)	441,249	Atlanta (US)	50,744,944	Hong Kong (CN)	4,379,762
Chicago (US)	437,568	Beijing (CN)	44,969,314	Memphis (US)	4,289,377
Dallas/Fort Worth (US)	340,622	Dubai (AE)	39,005,133	Shanghai (CN)	3,178,985
Los Angeles (US)	327,782	Chicago (US)	38,471,247	Anchorage (US)	2,624,312
Beijing (CN)	295,085	Tokyo (JP)	37,658,359	Dubai (AE)	2,505,507
Denver (US)	270,607	London (GB)	37,494,957	Incheon (KR)	2,489,662
Charlotte (US)	270,607	Los Angeles (US)	37,352,061	Louisville (US)	2,262,650
Las Vegas (US)	265,165	Hong Kong (CN)	34,165,405	Tokyo (JP)	2,085,275
Houston (US)	251,422	Paris (FR)	32,885,644	Taipei (CN)	2,008,703
Paris (FR)	237,888	Dallas/Fort Worth (US)	32,036,234	Frankfurt (DE)	1,993,467
London (GB)	237, 051	Istanbul (TR)	30,918,391	Miami (US)	1,970,616
Frankfurt (DE)	234,077	Frankfurt (DE)	30,516,011	Beijing (CN)	1,889,830
Amsterdam (NL)	232,761	Shanghai (CN)	30,026,694	Paris (FR)	1,861,197
Istanbul (TR)	232,433	Amsterdam (NL)	29,142,424	Singapore (SG)	1,853,000
Shanghai (CN)	224.107	New York (US)	28,422,625	Los Angeles (US)	1,846,010

Source: ICAO (2015)

first in terms of freight traffic (ICAO 2015) (table 12.2). Table 12.2 also highlights the absence of African cities from the top international airport rankings in terms of passenger and freight volumes. O.R. Tambo International Airport near Johannesburg is the busiest airport in Africa with just over 19.6 million passengers in 2014–15 (Airports Company South Africa 2016), below the 20th ranked airport globally (ACI 2015). However, we must be cautious about what these flight frequency, passenger number and freight tonnage data actually represent in terms of the nature of urban connectivity and influence, as these bold figures reveal little about who, or what, is actually travelling and why.

McNeill (2014) notes that the development of mutually supportive partnerships between the state, airport authorities and airline carriers can produce significant benefits in terms of the positioning and branding of the city-region as a global business space (see case study). Airport and container port infrastructures are important mechanisms through which people, organizations and firms are able to manage extra territorial relations and extend their influence regionally and globally, and are a visible manifestation

Case study: international airport development in Hong Kong

Hong Kong International Airport (HKIA) illustrates the pivotal role that airport development plays in maintaining the competitive advantage of a city. The development of a new airport was fast-tracked in order to secure Hong Kong's future economic development following the handing back of the territory to China by Britain in 1997. The site chosen for the new HKIA provided room for a large airport complex and future growth, but was remote from the existing downtown business hubs. To address this issue, development was underpinned by an infrastructure programme, which included building new road and rail links to the site from the existing downtown, reclaiming land for new port and industrial facilities and establishing new settlements near the airport.

Hong Kong's business and political elite were heavily involved in the development and management of HKIA, which boosted the profile of established companies, assisted the development of a multi-modal port to provide leverage in global trade, and which facilitated the emergence of Hong Kong's new transnational world city subjectivity and a business elite of 'entrepreneurial astronauts' shuttling around the world. More recently, the airport authority has sought to enhance HKIA's commercial operations in order to compete with other airports in the region, developing new convention, hotel, cargo, leisure and retail facilities. Additionally, HKIA has developed its links with the growing Pearl River Delta Region and with Shenzhen on the Chinese mainland, with ambitions to create a super-metropolis and air hub.

However, we must recognize the peculiar model of highly concentrated and centralized Hong Kong capitalism and political control, which drove the airport's early development and limited dissenting voices. Now there is an increasingly mobilized set of activist and amenity groups, which oppose major developments in terms of the impacts on the environment, culture and heritage. In this respect, the current development of HKIA mirrors the more complex and controversial situations underpinning airport development in other cities around the world.

Source: McNeill (2014)

of world city interactions (McNeill 2014). The development of air travel has consequently had a significant impact on the territorial organization of many cities, with major international airports becoming an important aspect of contemporary urban life and key gateways to the city (McNeill 2014) (figure 12.2).

The development of an airport represents a substantial investment in terms of sunk costs, and as such 'represent long-term accumulations of finance, technology, know-how, and organizational and geopolitical power' (Graham and Marvin, 2001: 12). Much of the sunk costs associated with new airport development tend to be spent on improved surface accessibility, through the provision of high-speed public rail access and increased road capacity, and the

Figure 12.2 *Hong Kong International Airport*
Source: Shutterstock

development of the wider airport platform, including the mix of aeronautical and non-aeronautical supporting activities. Increasingly, the airport can be seen not as a singular entity, but rather as an 'airport territory' composed of a variety of different physical and jurisdictional spaces, overseen by a web of private and state bodies operating at different territorial scales '… which position the airport as [both] polity and urban spatial formation' (McNeill 2014: 2999).

Despite the apparent success of these airport territorial complexes, there are a number of issues associated with these developments and vulnerabilities based on their increasing connections and impacts. The territorial entanglements of airports are clearly vulnerable in times of geopolitical crisis, when access to both maritime and air ports and corridors can be blockaded. Cresswell (2012) considers the idea of 'stuckness', and reflects on the personal impacts of being stuck when infrastructural mobilities break down, and being stranded in Washington, DC when European airspace was shut down due to the ash cloud from the eruption of the Icelandic volcano Eyjafjallajökull. Similarly, increasing aeromobility and the promise of cheap high-speed air travel for new social groups raises significant concerns in terms of increasing environmental damage and climate change impact, illustrating the need for mobilities research to bridge multiple scales (Jensen and Lassen 2011). As Charles et al. (2007: 1013) note: 'hydrocarbon-based fuel is integral to our present concept of air travel. Thus, a future for maritime transport without oil-based fuel is readily conceivable, but such a future for air travel is somewhat more problematic'.

The Airports Council International (ACI) world airport traffic forecasts indicate the continued growth of both passenger and cargo traffic to 2040. Passenger traffic is set to double globally to 14 billion by 2029, with international passenger traffic surpassing domestic traffic after 2028 (ACI 2016). In terms of percentage market share by region, Asia-Pacific will increasingly dominate global markets, rising from 30 per cent in 2015 to 47 per cent of market share in 2040 (ACI 2016). The growth in air travel is connected to increasingly problematic impacts of carbon emissions in the upper atmosphere (Jensen and Lassen 2011). Aviation is responsible for about three per cent of global fossil fuel consumption and twelve per cent of transportation-related carbon emissions, and emissions from aircraft are estimated to contribute 3.5–4.9 per cent of anthropogenic climate change impacts (Lee et al. 2010; Dessens et al. 2014). Yet, international action on emissions reduction has proved difficult when set against current growth, and consideration of emissions from international aircraft and marine travel is not covered in international protocols that look to mitigate climate change impacts (Hickman and Banister 2014). However, in 2016 the ICAO did gain agreement that from 2020, any increase in airline carbon emissions will be offset by activities such as tree planting, which absorbs carbon.

Increases in air travel are only part of the issue in terms of their contribution to global environmental concerns, and emissions from road travel, and specifically urban road transport, remain the key contributor (Hickman and Banister 2014). About 80 per cent of the increase in global transport emissions since 1970 has been due to road vehicles (IPCC 2014a). Overall, the transport sector produces around 23 per cent of global energy-related carbon emissions, equivalent to 6.7 gigatonnes of carbon in 2010 (IPCC 2014a). Emissions are growing more rapidly in the transport sector than in any other sector and are projected to increase by 50 per cent by 2035 and almost double by 2050 under a business-as-usual scenario (IPCC 2014a). While transport emissions per capita in developing countries are relatively low on an absolute basis compared to wealthier countries, almost all of the predicted increase in global transport-related carbon emissions is expected to occur in these countries. If developing countries expand their infrastructure (transport and others) to current global average levels using currently available technologies, around 470 gigatonnes of cumulated carbon emissions would be emitted (IPCC 2014b). It is widely accepted that future growth in average surface temperatures needs to be limited to below 2°C if the impacts to which many major cities are vulnerable, such as sea level rise, are to be minimized (Hickman and Banister 2014) (see also chapter 13). This requires a 50 per cent reduction in 1990 carbon emission levels. Many wealthier nations have set themselves 80 per cent reduction targets to allow for the need for convergence with the much less profligate patterns of fuel consumption in the world beyond the Global North (May 2013; Rode et al. 2014). However, transport remains a challenging and complex sector within which to act in tacking climate change, and behaviours and travel trajectories remain persistent and difficult to alter, at the international, national and the city scale (Hickman and Banister 2014).

Intra-urban transport and sustainability

> ... when the Secretary-General of the United Nations launched his '5-year action agenda' in January 2012, he identified sustainable transportation as one of the major building blocks of sustainable development. In particular, he stressed the need for urgent action to develop more sustainable urban 'transport systems that can address rising congestion and pollution'. He noted that action was required by a range of actors, including 'aviation, marine, ferry, rail, road and urban public transport providers, along with Governments and investors.
>
> (UN-HABITAT 2013: vii)

Urban travel currently constitutes more than 60 per cent of all kilometres travelled globally and, as a result, urban transport is currently the largest single source of global transport-related carbon emissions and the largest local source of urban air pollution (Rode et al. 2014). Between 1960 and 2010, the number of registered cars globally increased from about 100 million to 700 million, and registered trucks and buses from 30 million to 300 million, and by 2035 the total registered vehicle fleet is estimated to grow to 1.7 billion (Rode et al. 2014). In developing countries, motorized two-wheelers also account for a substantial proportion of vehicles, with 114 million added to global totals in 2013 alone (Rode et al. 2014). All regions of the world are forecast to continue their growth in vehicle ownership per capita. Although the levels in Asia and Africa will remain comparatively lower than in Europe, North America and Australia, these regions are predicted to experience significant growth in vehicle ownership levels and distances travelled throughout the first half of the twenty-first century. However, some caveats must be put on predicted increases, as transport projections have often been proved wrong, and the 'peak-car' hypothesis suggests there will be some decreases in ownership in some developed economies (Rode et al. 2014). While this increased mobility will bring economic benefits to many, it will impose increasing and interconnected burdens on wider urban societies. This will contribute to further rises in urban transport emissions globally in the longer term and contribute to rising levels of air pollution in the growing megacities of the Global South in the short term (UN-HABITAT 2013; Hickman and Banister 2014). These trends are inconsistent with the objectives that cities and governments are attempting to achieve, and it remains a significant challenge for cities around the world to achieve these substantial reductions in greenhouse gas emissions, while also achieving the economic and social objectives of sustainable urban development (May 2013; Hickman and Banister 2014) (see also chapter 13).

The dominance of the car has brought with it a range of social, economic and environmental problems. These revolve around the negative impacts generated by the car on the environment (toxic emissions, noise and vibration), people (road accidents, health problems, the loss of public space to roads and parking and the severing of communities by road development) and on economies (congestion and the costs associated with managing car-based mobility) (Hall, 2003; Rode et al. 2014). The rising congestion associated with increasing car use imposes significant economic costs to cities, with New York City estimated to lose US$13 billion annually to congestion (Rode et al. 2014). For middle income developing cities the costs can be higher, with Mexico City and Buenos Aires estimated to lose 2.6 and 3.4 per cent of their gross domestic product to congestion, respectively (Rode et al. 2014). Additionally, vehicles stuck in traffic generate more emissions posing greater environmental and health problems through pollution, while the rise in car numbers imposes greater safety risks on pedestrians, cyclists and powered two-wheel users (May 2013). However, these negative impacts are not evenly felt and fall most heavily on non-car users, a disproportionate number of whom are from low-income groups, and who occupy areas of heavier traffic within cities.

Rode et al. (2014) note that the key strategies available to urban policy makers to tackle the problems of motor vehicle congestion are to avoid (reducing travel through physical proximity), shift (to non-motorized transport modes), or improve (the efficiency of transport). In assessing the contribution of a selection of common policy instruments, May notes that no single instrument performs best against all sustainable transport objectives (Table 12.3), highlighting that effective strategies are based on a combination of different types of approach (May 2013). Traditionally, urban transport planning has looked to technological

Table 12.3 *The contribution of different types of intervention to policy objectives*

	Technology	Land use	Infrastructure	Management	Information	Pricing
Reducing greenhouse gas	✓✓✓	✓✓	✓	✓	✓✓	✓✓✓
Improving air quality/noise	✓✓	✓	✓	✓✓	✓✓	✓✓
Improving safety	✓	✓	✓✓	✓✓✓	✓✓	✓
Improving access/reducing exclusion		✓✓	✓	✓✓✓	✓	
Reducing congestion	✓	✓✓	✓✓	✓✓	✓✓	✓✓✓
Enhancing wealth/economy	✓	✓✓✓	✓✓✓	✓✓✓	✓	✓✓

Key: ✓✓✓ = high contribution; ✓✓ = moderate contribution; ✓ = low contribution
Source: Adapted from May (2013)

solutions and infrastructure provision and management to address transport problems. Improved vehicle technology has provided a key contribution to reducing emissions and improving air quality, through the provision of clean engine technologies and substitution of petrol/diesel vehicles by electric vehicles (May 2013). Improved vehicle design has also contributed to casualty reduction for drivers and passengers, and to some extent for pedestrians, although deaths from vehicle accidents in cities remain high (Rode et al. 2014). However, this technology does not impact on the number of journeys made, and gains are offset by increasing vehicle passenger miles travelled in cities around the world. Technological and infrastructure investments have also been used to try to improve the flow of traffic within cities, looking to reduce congestion and speed up journey times. Many cities have looked to invest in new road infrastructure to improve capacity, although the impact of these improvements can be short lived as new roads fill up due to what is known as the 'triple convergence' (Downs 1992), where new provision paradoxically encourages more car use at peak times. As a result of the cost and limited impact of new road infrastructure provision, more cities have looked to technological and pricing instruments to manage flows and capacity on existing infrastructure. Initiatives include the provision of smart highway technologies, dedicated car-share lanes, Bus Rapid Transit lanes, road pricing/congestion charging, or the use of fiscal instruments such as vehicle tax and fuel pricing. However, the use of instruments to limit or increase the cost of car use can prove unpopular with the public (Pacione 2009; May 2013).

There is now recognition that 'technological fixes' to existing motorized transport and infrastructure alone will not tackle wider sustainability issues and the transition to post-carbon economies and cities. The UN Report (2013) into sustainable urban mobility argues that the development of sustainable urban transport systems requires a conceptual leap in terms of urban planning and management thinking, where greater emphasis is given to promoting accessible cities through design and investing in non-motorized transport options. Cities have policy pathway choices leading towards either more sprawling, car-dependent urban development, or alternatively more compact, public transport-oriented cities, which has substantial implications, both for their economic and social performance and for their carbon emissions (Hickman and Banister 2014; Rode et al. 2014) (Figure 12.3).

There is a growing consensus that more compact urban growth and mixed-use developments, integrated with increased provision of public transport, walking and cycling infrastructure and services, and pro-active support for travel behaviour change and non-motorized transport use, is likely to deliver substantial net economic, social and environmental benefits (Hickman and Banister 2014; Rode et al. 2014). Singapore is a well-documented innovator, where car ownership is about a third of comparator cities, with car ownership

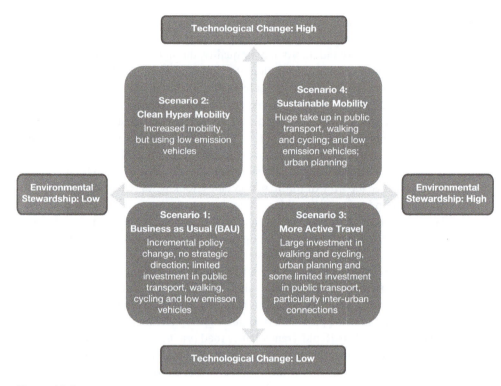

Figure 12.3 *Urban transport scenarios*
Source: Hickman and Banister (2014: 325)

control combined with public transport enhancement, and land use planned in conjunction with transport strategies (May 2004). Similarly, European cities such as Freiburg, Vienna and Zurich have effected reductions in car use by using a combination of demand management, public transport improvements and land use planning (Buehler and Pucher 2011). London has pursued a progressive policy to reduce greenhouse gas emissions through urban development built around multi-modal public transport provision, cycle hire schemes and the introduction of congestion charging in the central area and inner urban low emission zones (Hickman and Banister 2014) (Figure 12.4). This has led to declining levels of car ownership, increases in cycling and a rise in passenger transport numbers in the city (Rode et al. 2014). These more holistic approaches which work with the existing urban framework, acknowledge the limitations of earlier compact city arguments for extensive reurbanization as a means of achieving sustainable urban development and reductions in car travel, critiqued by Michael Breheny in terms of their false assumptions about compact and decentralized urban forms, their failure to consider wider socio-cultural factors which influence travel behaviour, and

Figure 12.4 *Multi-modal transport in London*
Source: Author's photograph

their failure to acknowledge the social and physical costs associated with the extensive replanning of cities (Breheny 1995).

Rather than focusing solely on the extensive replanning of cities to achieve travel change, policy makers have also looked to try to influence people's travel choices. Initiatives to promote sustainable urban travel behaviour change are either based around the promotion of 'substitution' or 'switching' (Marshall et al. 1997) (see exercise). Substitution can involve making linked trips, rather than multiple single journeys, using technology such as electronic communication to replace physical travel or trip modification, where a trip is modified by type, such as replacing a shopping trip with mobile goods delivery. Switching can either involve mode switching, such as walking rather than using the car, destination switching, shopping locally rather than a distant location, or time switching, travelling at off-peak, less congested times. Information is important to promoting behaviour change and developing public support for projects, including real time information for public transport passengers and awareness raising in terms of travel options (May 2013). However, to be successful, solutions proposed need to include effective consultation and participation in the policy formation stage, so that people can participate in developing their own 'travel futures' (Hickman and Banister 2014).

Exercise

Think about your journeys around the city/town where you live:

What types of journey do you normally make and how (to university/college, shopping, work, leisure, hobbies/activities)?

Do these journeys have any particular transport constraints (do you have access to a car, or are there issues with train/bus routes and times, for example)?

Could your journeys have been substituted in any way (linking, technology, modification) or switched (mode, destination, time)?

What are the challenges to modifying your journeys (substituting or switching)?

What does your experience tell you about the challenges of sustainable transport planning for cities?

If you can, do this exercise as a group, and discuss and compare your responses to the questions. Even in a small group your journey types, times and transport modes should illustrate the complexity of daily mobility patterns and reveal your attitudes to your own mobility. If you scale up your small 'survey' to the city-scale you can perhaps begin to appreciate the complexities and challenges for transport planners in tackling issues and changing behaviours.

While there is some consensus in terms of what policy instruments are needed to tackle car-based mobility and promote sustainable urban transport, there remain significant barriers to effecting change. Central to these challenges are the problems associated with the translation of transport policies to different cities. Barriers to delivering sustainable urban transport initiatives included poor integration and coordination across different policy areas, counterproductive institutional roles between different arms and levels of government, unsupportive regulatory frameworks, weaknesses in financing and pricing for transport initiatives, poor data quality and quantity for monitoring impacts, limited public support, and lack of political resolve for change (European Conference of Ministers of Transport 2006) (see also chapter 13). Transport systems in existing cities represent a considerable investment and can take a long time to change, with cities becoming locked in to particular urban accessibility pathways. Equally, many industry sectors, such as automotive, construction and real estate, remain highly dependent on a 'business as usual' model (Rode et al. 2014). There is a need for long-term political approaches and the development of a 'culture of policy learning' by city administrations, which is often hampered by short-termism and conflict in urban politics and tensions between the local and the national state (May 2013; Hickman and Banister 2014). Policy transfer from innovator cities is typically challenging, given that contexts and pathways to sustainable mobility are often very different, and that awareness of the various policy measures and empirical evidence of their impacts are not widely available (Hickman and Banister 2014).

The debates around the need for better policy transfer methods for encouraging sustainable mobility often mask deeper issues concerning the uneven distribution of benefits of transport policies and practices, paying insufficient attention to the political-economic dimensions that underpin policy development and implementation. Jensen (2011) considers the 'governmentality' of mobility, highlighting how particular knowledge, practices and imaginaries around mobility become integrated into the rationalities of policies and planning. The rationalities of governing bodies are socio-historically produced, and frame governing practices, which become enmeshed with people's daily mobility practices, informing them how to behave, perform and shape their identities in ways that align with taken-for-granted knowledge and accepted 'truths' about mobility in the city, which both allow and restrict particular mobility practices. Plans and visions for urban mobility often seem to produce certain 'imaginary mobile citizens' who are understood to perform in particular ways and engage with, and benefit from, the implemented policies (Jensen and Lassen 2011; Jensen 2013). Yet, urban transport policies frequently have social exclusions and differential justices bound up with their implementation. For example, Elsheshtawy (2012) highlights the unequal impact of Metro developments in Dubai, and the exclusion of poorer urban populations and areas of the city from new transit developments due to cost and network coverage (see also case study in chapter three). This exclusion is also evident in the development of the Metro system in Delhi, which involved the demolition of informal settlements, while only serving a limited elite in the city, also principally due to its limited network and high cost (Hickman and Banister 2014). This takes us back to the politics of mobility introduced previously, and control over the development and operation of channels of communication in the city, enabling some to move and participate in these developments, while on the other hand exacerbating the exclusion of other urban populations. As Jensen (2014) notes, further critical research is needed to explore how the layout and physical design of transport infrastructure and transit spaces enable, obstruct or prevent particular mobile practices and experiences.

Mobility as embodied practice

The discussion above of the problems in effecting travel behaviour change and addressing mobility inequalities reminds us that mobility in cities carries significant meaning as something that is practised, experienced and embodied. Jensen and Lassen (2011: 10) suggest that urban mobilities are constitutive for the structures that frame social life where cultural actions and identities are produced and reproduced, but also that social structures (economic, political and spatial) are constitutive for the ways mobilities develop. Jensen (2013) considers how actual and concrete mobilities are 'staged' through an array of physical, social, technical and cultural conditions that influence their 'staging' from above and below. The physical setting, material spaces and design aspects of mobilities contribute to staging from above, while social interactions and embodied performances stage from below. Consequently, we need to consider the embodied nature of mobile subjects, their social and psychological engagement with lived space, and how these intertwine with the material systems of the city and its policy frameworks.

As we noted previously, discourses about mobility are powerful co-players in the shaping of modernity, and are a part of the fabric upon which the city and social relations are built. These powerful discourses mould mobility practices, supporting certain social aspects of mobility while simultaneously silencing others, and contributing to the shaping of our emotions, desires and wants towards mobility (Jensen 2011). As Jensen (2011: 268) notes:

> Focus on the formation of our modern selves through mobility related rationalities, on emotions and the feeling of mobile technologies and places, on ambiences and atmospheres thus suggests that power in relation to mobility…also works in ways that connect just as much to what we do as to what is put into words …

So, as Jensen argues, while written policy and plans have an important role in shaping mobility practices, our unnoticed and taken-for-granted actions and emotional responses to mobility have an equally powerful place in reinforcing particular mobility discourses, inducing pleasures and working on our desires as much as they coerce, discipline and normalize practices (Jensen 2011). In this respect, ideas around automobility and freedom of movement are particularly powerful in imprinting their logics and desires on the making of our modern selves (Thrift 2004; Urry 2007).

As Harada and Waitt (2012: 145) note: 'the challenge of changing driving behaviour becomes evident when this practice is conceived of as a bodily habit co-constituted within an automobile assemblage'. The motions and emotions related to automobility are not only a kinaesthetic bodily and sensory experience, but are also entangled in cultural, social and family practices.

Doughty and Murray (2016) examine how key discourses on mobility permeate society and embed in everyday mobile practices. They identify five key institutional discourses of 'western' mobility that have underpinned transport policy; technocratic, rights to mobility, mobile riskiness, speedy connectivity and sustainable mobility. Their ethnographic study of mobile experiences and everyday practices of families living in Brighton, UK examines mobile practices as social texts embedding or resisting these wider discourses (Doughty and Murray 2016). The contradictions of these institutional discourses are evident in the everyday mobile practices of the people surveyed, as they negotiate sustainable mobility concerns alongside discourses that simultaneously position mobilities (especially automobilities) as central to the exercise of individual rights of freedom, economic success and citizenship. For many respondents, the car was narrated as symbolic of a transition to adulthood and working life, and thus central to personal success and independent mobility, offering freedom, possibilities and fun. Equally, the car was viewed positively as a controlled environment of one's own that could be shaped and managed, unlike the public realm which could be threatening and challenging to negotiate. For those encumbered with parenting responsibilities, the car was a way of managing modern life efficiently and effectively, undertaking activities which would be more difficult and complex to accomplish using other forms of transport, such as accessing buses and trains with young children or making multiple linked trips. Themes of legitimacy and freedom were set against perceptions of moral responsibility and guilt for younger participants who were more aware of sustainability discourses through their school learning. However, concerns about the need to engage with more sustainable mobilities were frequently resisted at the micro-level of everyday embodied engagements:

> … because it is easily overshadowed by mundane social and material constraints and affordances; and by bodily dispositions and disabilities. It is the often overlooked and obscured mundane sensate relationships that people have with mobilities and mobile spaces that hold most significance in constituting mobile behaviours.
>
> (Doughty and Murray 2016: 319)

The power of these discourses is therefore mediated through varying emotional experiences and cultural differences, which are productive of particular mobile emotions (ibid.).

Mobility is thus strongly intertwined with emotions, feelings and ambiences, which not only relate to the bodily kinaesthetic experience of movement but also to the technologies and materialities of the world by and through which movement is made. There is a growing interest in the affective and emotional aspects of urban everyday mobilities, and additionally how the rhythms of everyday movement shape the feel and tempo of cities. Different modes of travel, whether walking, biking or driving, or passengering in trains etc. can be included amongst the active corporeal engagements of human bodies with the sensed space (Jensen et al. 2015). However, Merriman (2015) notes that much of this research remains within the context of western mobility experiences, and there is a need to examine the diversity of mobile experiences and mobility cultures in cities across the world to develop our understanding. Recently, the journal *Transfers* has set itself the task of publishing trans-disciplinary and transnational research that goes beyond Western experiences and paradigms and bridges this gap in understanding (Mom and Kim 2013). Equally, much of the writing in this area focuses on individual embodiments and affects, rather than their collective circulation through social networks and relations to others. However, urban mobility is always a collective project and one person's mobility patterns can have a direct impact on another's capacity to be mobile. There is a need to consider mobile subjects as clusters of interacting agents, not simply singular and individuated actors:

> Different mobility practices and options are experienced and practised through the 'rhythms and relationships' of those mobilities, including the child who is driven to school, the bus driver who greets the elderly regular and the worker who handles the business traveller's luxury suitcase in the airport.
>
> (Jensen 2011: 266)

Within households, if one member changes his or her means of mobility, for example by deciding to ride a bicycle to work one day or to take the only car available another day, then the other household members must adjust to this choice (see case study).

Everyday family life and mobility in Copenhagen

The research explores the everyday mobility of eleven households with children in the multi-modal transport context of Copenhagen, Denmark. Data from household interviews demonstrate how everyday patterns of relational mobility are filtered through spatial affordances, affective ambience and temporalities of the life course to influence transport alternatives of route and modal choice.

Respondents in the study spoke differently about their specific emotions towards transport modes, which were reflected in the opportunities they afford. Participants used mode choice and route choice to manage stress or elicit positive feelings and to form familial relations. For some, car journeys to and from work offered the space for relaxation, a chance for 'family time', or a personal space for planning and work. Others thought driving stressful, with concerns about traffic and roadworks, preferring to use a bike and the train to get to work. Routes were chosen not always in a 'super logical' way, but were about making respondents feel better through the process of moving. Bike riding and routes

continued

were related with their stress-relieving properties and the idea of 'clearing the mind', and routes varied depending on time of day, relating to both changing emotions going to and from work and the desire to avoid negative feelings relating to the perceived security and safety of particular areas.

The rhythms of the household were also important in shaping mobility patterns, covering a whole range of differing temporalities. In daily journeys the rhythms of the household and the pulse or metabolism of the city often overlapped, with cycling respondents making adjustments to their routes in response to the rhythms of car traffic at different times of the day. Additionally, choreographies of household movement involved rhythms of synchronicity and asynchronicity in journeys, coordinating differing family commitments, such as shopping or family leisure activities, and times of being together and apart as a family.

What this small study demonstrates is the complex ways in which mobilities are negotiated, enacted and experienced within different households within the city.

Source: Jensen, Sheller, and Wind (2015)

Research into the relationship between walking and the city has also sought to illuminate the significance of social encounters to the ways in which pedestrian practices unfold in the public realm, and how everyday life in the city is negotiated through these encounters. Middleton (2016) uses diary and interview accounts of walking in London, UK to consider how people articulate their everyday pedestrian practices and negotiate the micro-politics of pedestrian encounters. Tensions and conflicts emerge between pedestrians and other urban walkers and cyclists, such as not obeying traffic signals and spaces appropriately, highlighting how mobility actively shapes citizen relations. This constitutes an emergent everyday politics of how people appropriate space on foot and illuminates wider concerns about the 'right to the city'. Whether driving, walking, bus riding, bicycling, or train passengering, each journey within the city has its own embodied dispositions, instinctual feelings, rhythms and affective resonances (Jensen et al. 2015). As people string together chains of activity, they are not only choosing routes, but also moving between different affective experiences of mobility and different emotions in relation to place and with others. All this helps to understand the complexity of transport and mobility in the city from below, and examine its lifeworlds and rhythms and the differential experiences of the urban world as embodied (see also chapter 9).

Middleton's (2016) research into walking also reflects a key contribution of this type of mobilities research in terms of experimentation with the use of 'mobile methods' that can capture, perform and intervene in processes of urban movement as they happen (Jensen et al. 2015) (see also case study in chapter 10 and exercise below). Mobile methods try to capture the complex and dynamic processes of moving via following-the-thing and participant-observation on the move, using methods such as walkalongs (walking-with, audio walks), mobile video ethnography, diaries, mapping, action research and arts-based interventions (Jensen et al. 2015). These embodied movements such as walking are also valued for their creative and expressive qualities (Merriman 2015). The most cited examples here are the work of the Situationists in Paris in the 1960s, as well as a later generation of scholars, artists and writers associated with the field of 'psychogeography' drawing on their peripatetic wanderings, usually in cities, including writers such as Iain Sinclair and Will Self (Merriman 2015). Movement and mobility are now part of a range of creative movements within Geography, and geographers no longer just write about art, film and poetry but are also experimenting with these techniques or collaborating with practitioners and scholars of such techniques (Merriman 2015).

Exercise

You might explore your own journeys around the town or city you live in, or your journey from where you live to your university/college, using the mobile methodologies outlined in this chapter (or Latham's diary-photo, diary-interview method outlined in chapter 10). Keep a diary recording your journeys over a week, not just noting down where you go and how, but also thinking about your emotions while travelling, the impact of others on your journeys and the feeling of the places you are travelling through (as in Jensen et al.'s (2015) study outlined above). Do any of the issues outlined by Jensen et al. resonate with your experiences? You might do this as a class exercise and compare your experiences to try and illuminate the complexity and diversity of journeys around your urban areas and the varying emotions they elicit.

Also, if you, or your academic department, have access to GPS technology, and/or a small video camera, you might film your movements and also track your routes through the city, building up a more detailed picture of your experiences and how these relate to the wider rhythms of the city, again comparing this to Jensen et al.'s study of Copenhagen.

Finally, you might also like to develop this as a creative exercise, and look to express the emotions of your movements in the city as prose, poetry, art or film as different ways in which geographers look to understand the city and its rhythms.

Summary

Transport and mobility are central to the functioning and development of the city, both in terms of its internal working, and the lives of people within the city, and also in terms of connectivity and the place of a city within global urban networks. The development and image of the modern city is strongly associated with transport technology and the idea of faster, freer movement, particularly embodied in the automobile and the aeroplane as symbols of this freedom and mobility. However, these contemporary urban mobility trajectories, or pathways, raise many important current issues for cities and their managers, most notably inequalities in mobility opportunities and choices, the unequal social and economic impacts on urban populations of increasing private motor vehicle mobility, and the global and local environmental impacts of increasing motorized mobility in terms of greenhouse gas emissions and pollution. The challenge for future sustainable urban transport planning is to take into account the complexities of mobility, particularly its embodied and emotional aspects in addition to its patterns and frequencies, and also acknowledge the politics of mobility and the wider discourses and power relations that shape our everyday mobility practices in the city.

Follow up activities

Essay title: Cutting the impact of private car travel is arguably the most important issue for urban transport planning in the twenty-first century. It is also one of the most difficult issues to tackle successfully. Discuss.

Commentary on essay title

A good answer would begin by considering the relationship between urban development and the car, and would also provide an overview of current trends in terms of rising car use in cities around the world and the environmental, economic and social impacts of this, drawing on recent statistical evidence and reference to wider reading. It would then examine the different ways in which urban transport planning has looked to tackle car-based mobility and promote sustainable urban travel, and the challenges in effecting change. Given the breadth of the issues in terms of different city problems and approaches, you may wish to focus down on some case studies of specific cities from which you can draw out key issues. An answer that managed to combine examples from across the global urban world, highlighting the diversity of issues and challenges present, could be highly effective. An excellent answer would also bring in theoretical perspectives on the issue, specifically considering the politics of mobility and the issues of power bound up with the implementation of urban transport policies, critically examining the inequalities in the impact of transport planning schemes. It would also consider the issue of mobility as an embodied practice and how this highlights the challenge for planning in recognizing the complexity of urban mobility practices and the problems in effecting travel behaviour change.

Project idea

Try and assess the 'urban accessibility pathway' that your urban area has followed in the last twenty years or so, depending on the information available to you. Firstly consider the different types of transport mode available and look to see if there is any statistical information available on the proportion of trips that people make by these different modes and if this has changed over recent time (your local municipal authority would be a good potential source for this travel information). If data are not available, you might consider conducting your own travel survey, devising a questionnaire to assess a sample of residents' travel behaviour.

Once you have some information on travel behaviour, you can then look at current and, if available, recent past policies for transport planning in your area. Critically assess these policies to consider what the key aims and objectives are and what ideas about transport planning underpin these. Is there an emphasis on sustainable travel options, or is planning for the car dominant? Is development focused on infrastructure delivery and technological solutions or changing travel behaviours? If there is not a current plan, this might be instructive in itself in highlighting a lack of political will in addressing transport issues explicitly (you might follow this up by setting up a discussion/interview with city leaders/planners). If specific plans are currently being implemented you again might carry out your own survey to find out what local residents make of these ideas, whether they have had any impact on their travel behaviours and whether any specific barriers exist to their effective implementation.

Further reading

Books

- Cidell, J. and Prytherch, D. (eds) (2015) *Transport, Mobility, and the Production of Urban Space*, Abingdon: Routledge
  Collection of writings with a diverse, international focus that explore the connections, in theory and practice, between transport geographies and 'new mobilities' in the production of urban space.

- Dennis, K. and Urry, J. (2009) *After the Car*, Cambridge: Polity Press
  An imaginative look at the future of motorized transport and its likely consequences for the reconfiguration of cities. Perhaps better titled: 'The car but not as we know it'.

- Grieco, M. and Urry, J. (eds) (2011) *Mobilities: New Perspectives on Transport and Society*, Farnham: Ashgate
  Not a specifically urban text, but a collection that draws together a wide range of recent perspectives on mobility and looks to bridge the earlier transport/mobility research divide.

- Hickman, R. and Banister, D. (2014) *Transport, Climate Change and the City,* Abingdon: Routledge
  Key text discussing the challenges of sustainability facing urban transport planning, which includes in-depth case studies of current problems and future scenarios from cities across the world.

- Jensen, O.B. (2013) *Staging Mobilities*, Abingdon: Routledge
  Important work bringing together Jensen's key research and which develops his conceptual idea of mobilities as 'staged'. A companion volume, *Designing Mobilities* (2014), looks to apply this theoretical perspective to urban planning and design.

- Kasarda, J. and Lindsey, G. (2011) *Aerotropolis: The Way We'll Live Next,* London: Allen Lane
  Popular scholarly work which looks at the increasing significance of airports to the city and their impact on business practices, and urban life more broadly, in an increasingly globalized world.

- UN-HABITAT (2013) *Planning and Design for Sustainable Urban Mobility: Global Report on Human Settlements*, Abingdon: Routledge
  Key global overview of the key issues and challenges facing sustainable urban transport planning.

Journal articles

- Freire-Medeiros, B. and Name, L. (2017) 'Does the future of the favela fit in an aerial cable car? Examining tourism mobilities and urban inequalities through a decolonial lens', *Canadian Journal of Latin American and Caribbean Studies*, 42(1): 1–16
  Provides a case discussion of the impact of aerial cable car systems as a transport solution to provide access to hillside favelas. Employed in a number of Latin American cities, questions are raised as to the extent to which these high-cost systems help ease the accessibility problems of the poor residents of these areas.

- Kenworthy, J. (2014) 'Total daily mobility patterns and their policy implications for forty-three global cities in 1995 and 2005', *World Transport Policy and Practice*, 20(1): 41–55
  Relatively rare comparative consideration of changing mobility patterns between cities, looking at car travel, walking and cycling and public transport use. Focused mainly on cities in the Global North, although also drawing comparison with Taipei and São Paulo.

- McLellan, A. and Collins, D. (2014) '"If you're just a bus community … you're second tier": Motivations for rapid mass transit (RMT) development into mid-sized cities', *Urban Policy and Research* 32(2): 203–217

Examines the motivations for rapid mass transit (RMT) development in two mid-sized cities (Auckland and Edmonton), relating this to issues of urban competition in addition to rationales linked to perceived local transport benefits.

- O'Connor, K. and Fuellhart, K. (2016) 'Airports and regional air transport markets: a new perspective', *Journal of Transport Geography*, 53: 78–82
  Provides a good overview of recent changes in global air transport markets looking beyond single airport hierarchies to consider airport regions, providing a broader perspective on recent growth. Also considers the issues associated with this air hub growth.

- Samanta, G. and Roy, S. (2013) 'Mobility in the margins: hand-pulled rickshaws in Kolkata', *Transfers*, 3(3): 62–78
  Drawing on both quantitative and qualitative research, the paper examines the often overlooked and marginal mobilities of non-motorized transport in Kolkata, India.

- Shirgaokar, M. (2014) 'Employment centres and travel behavior: exploring the work commute of Mumbai's rapidly motorizing middle class', *Journal of Transport Geography* 41: 249–258
  Considers the differences between commuting trips for those who commute to the city versus the exurbs for work, highlighting wealth inequalities in who commutes where and exploring why the growing middle class are increasing their car use.

- Skelton, T. (2013) 'Young people's urban im/mobilities: relationality and identity formation', *Urban Studies*, 50(3): 467–483
  Considers the local experiences of urban im/mobilities and everyday transportation of a group often neglected in urban research.

Websites

- http://www.carfree.com/
  Site of a pressure group promoting alternatives to car-based urban mobility. Contains a wealth of resources.

- http://www.wbcsd.org/Projects/smp2
  Website for the World Business Council for Sustainable Development's sustainable mobility project illustrating business-led approaches to tackling urban transport planning in cities, offers cases studies of demonstrator/trial city projects.

- https://www.gov.uk/government/organisations/department-for-transport/about/statistics
  Transport statistics for the UK. Equivalent sites for other countries available.

- http://www1.uwe.ac.uk/et/research/cts
  Centre for Transport and Society at the University of the West of England, Bristol. One of a number of university transport research centres. Based in an Environment/ Technology Faculty, while others are based in Engineering Faculties.

- http://www.lancaster.ac.uk/cemore/
  Centre for mobilities Research at Lancaster University. Pioneers of the mobilities paradigm.

13 Urban futures

The challenge of creating new and, one hopes, superior forms of settlement has occupied philosophers, architects, planners and urban theorists for centuries.

(Pacione 2009: 615–616)

As the global population becomes increasingly urbanized, cities have emerged as the dominant arenas to address the grand challenges facing humanity. Problems associated with climate change, economic under-development and social inequality are essentially urban in character. And so are their solutions.

(Evans et al. 2016: 1)

The complex nature of urbanization across the globe, and the seemingly insurmountable challenges of transforming urban futures require multi-disciplinary, multi-stakeholder research efforts across diverse geographies.

(Friend et al. 2016: 67)

Introduction

The world faces many imminent and interrelated challenges. These include those associated with population growth and migration, climate change, environmental degradation, food and resource pressures, economic inequalities and social divisions and diverse and dispersed geopolitical security concerns (Chen et al. 2013; Kilcullen 2013; Hou et al. 2015). In essence many of these concerns are urban in nature: growth in population is concentrating in cities; mobile, consumption oriented city lifestyles in the Global North are major contributors to climate change; cities draw upon food and other resources from across often global hinterlands in order to sustain themselves and generate waste which needs to be disposed of; and cities are frequently the sites of conflicts, terrorist attacks, organized criminal activities and other geopolitical flashpoints. Although demographic, environmental, resource, socio-economic and geopolitical issues have been intimately intertwined with processes of urbanization throughout history, the unprecedented global urban growth of the late twentieth and early twenty-first centuries has generated the sense of their present urgency (Pacione 2009: 605). The UN-HABITAT *World Cities Report 2016* summarizes eight 'persistent issues and emerging urban challenges due to increased urban population' as:

- urban growth;
- change in family patterns;
- increased residency in slums and informal settlements;
- challenges in providing urban services;
- climate change;
- exclusion and rising inequality;
- insecurity;
- upsurge in international migration.

(UN-HABITAT 2016: 2)

There is a growing recognition, therefore, across many realms of urban theory and practice, that questions of making a better, more sustainable world are, to a large extent, questions of making better, more sustainable cities (Caprotti 2014: 1285). As Friend et al. (2016: 67) argue, 'it is in the urban arena that much of the struggle to avoid a global climate catastrophe while achieving social development objectives will be played out'. These are questions that challenge urban residents, authorities, planners, developers, researchers and students alike (Kourtit et al. 2014: 1).

While these questions are global in nature they are also locally contingent at the same time, reflections of the complex mosaic of different urban patterns and processes across the world which were outlined and discussed in chapter 1. Cities have to both attend to their immediate local urban issues which, in Detroit, for example, might include the extensive abandonment of its neighbourhoods as its economy continues to decline and its population shrink (Chen et al. 2013: 326) and, in São Paulo, the persistence, growth and intractability of its informally housed favela populations (Smith 2008). At the same time these and other cities have to attempt to pursue paths of development that do not compromise the pressing global issues outlined above. Briefly, to recap some of the discussions of contemporary world urbanization patterns and processes from chapter one, it is apparent that the global urban landscape is being reshaped in a variety of ways in different regions and across different types of city. This includes: the continued urbanization of the developing world, both in large megacities and smaller settlements, the latter of which might lack the institutional capacities to successfully balance the immediate needs of their growing populations with the global impacts of this urbanization (Chen et al. 2013: 348); the massive current and future growth of the urban populations of the rising economies of China and India – China, for example, will add 221 cities with populations of more than one million by 2025 (the US by comparison currently has only nine such cities) (Chen et al. 2013: 336; Simon 2016); the expansion of several large US urban agglomerations to absorb surrounding smaller settlements producing extensive 'cities of cities', 'a continual string of cities, suburbs, absorbed CBDs, special purpose districts, parks, recreation areas, some empty areas – multi-nuclear, multi-nodal, multi-centric' (Brown 2014: 4–9); the likely continued reliance of automobile, private transport, albeit in smarter, perhaps driverless forms, across many cities of the world with the attendant pressures this brings to generate patterns of sprawling and/or multi-nodal urbanization (Chen et al. 2013: 343; Brown 2014: 8–9; see also chapter 12); and the aforementioned decline of those cities, such as Detroit, that grew until the mid to late twentieth century in the US and Europe around their development of extensive economies of manufacturing and heavy industry (Dicken 2011; Chen et al. 2013; Brown 2014: 6). This highly differentiated global urban landscape will engender the necessity for multiple, localized responses which remain attentive to a series of more broadly framed goals. The sheer diversity of urban environments around the world, though, and the range of pressing issues that different cities face, are undoubtedly major challenges to the transfer of discourses of sustainable future urban development (Pacione 2009: 608; Kourtit et al. 2014: 2). Indeed, in some instances the goals of immediate socio-economic relief may not be obviously compatible with the imagination of sustainable urban futures. The challenges of future urban development, therefore, are particularly complex in many urban settings in the Global South where Pacione (2009: 611) argued: 'in seeking to achieve higher levels of sustainability in Third World cities it is essential to develop linkages between socio-economic and environmental goals'.

This chapter takes you through a series of possible geographies of urban futures. It considers the challenges that face cities, challenges that cities have contributed to and that they have been recognized as key arenas in the resolution of. It explores some of the challenges to meeting these challenges. It then considers some possible future paths of urbanization that have emerged in response to the widespread recognition that urbanization, as it has largely occurred throughout history, is unsustainable. Here though it considers the limits of these emergent urban futures and the likely political, social, economic and environmental geographies that they might engender.

A word of caution though is appropriate here. Amidst the occasionally utopian rhetoric about building better urban futures, it should be remembered that the vast majority of the world's urban future will occur within existing urban settlements. New cities will be built without doubt; many new and significant cities in the case of China, for example (Chen et al. 2013: 336). However, the populations of the world's new cities will remain tiny compared to the populations of existing cities. Brown (2014: 4) is right to warn us then that 'The city of the future may more closely resemble the city of the present and the past than many would imagine, or hope for'. This presents a significant challenge. As we have already seen the cities that exist across the world today, for all of their wonderful diversity, are far from perfect in many ways. Building better urban futures then involves working, largely, within the constraints of the cities that we already have.

> Since it is rarely possible to make brand new cities … it is more realistic to talk about how to make existing cities better. Since any existing cities, especially those of a large scale, are entrenched and complex places, remaking them to achieve measurable improvements in either quality of life or the built environment can be a very long and challenging process.
>
> (Chen et al. 2013: 322)

Sustainable and unsustainable urbanization

> As urbanization and urban growth continue apace, a critical question will be whether cities can move closer to being sustainable, self-regulating systems, not only in their internal functioning but also in their relationship with the outside world.
>
> (Pacione 2009: 609–610)

Any discussion of future cities must start with the recognition that much urbanization that has occurred throughout history is environmentally unsustainable. The processes and patterns of urbanization that have prevailed across the world to date have been overwhelmingly based on the extensive acquisition and use of natural resources which has, over time, absolutely depleted the global reserves of these resources, created patterns of local and regional scarcity, generated problems of waste and pollution and generated additional and ongoing demands that are satisfied only by the further acquisition and use of these diminishing resources (Chen et al. 2013: 300–301; Simon 2016). It is now clear that 'The world's cities cannot continue to prosper if the aggregate impact of their economies' production and their inhabitants' consumption draws on global resources at unsustainable rates and deposits waste in global sinks at levels that lead to detrimental climatic change' (Pacione 2009: 606).

Unsustainability implies that at some point in the future, development will be compromised as environmental capacity is reached or environmental limits are breached. The question of what, and where, environmental limits and capacities are, is a not unproblematic thought. Views on where these limits lie depend on whether one adopts a robust or a precautionary stance towards the environment's capacities (Mohan 1999) and whether one holds that these limits are 'technical, cultural and social rather than environmental' (Rapoport 2014: 140). Environmental capacities are undoubtedly far from fixed and absolute. They may expand as new technologies come on-stream, genetically modified foodstuffs, being one example. However, many apparent technological panaceas might compromise environmental capacities in other, initially unforeseen, ways (Hinchliffe 2007).

The sheer scale of the impacts of cities on the environment through their use and disposal of resources of all kinds is conveyed well by the concepts of the ecological footprints and the global hinterlands of cities (Haughton and Hunter 1994; Blowers and Pain 1999; UN-HABITAT 2016). For cities to function they need to draw upon and, consequently, impact upon vast areas of land and water from beyond their own immediate geographical hinterlands. Cities draw resources, materials for construction and consumption, food, energy and so on, from all over the world. They also disperse pollution and waste locally, regionally and globally. The ecological connections stemming to and from cities can be visualized by the idea of the ecological footprint. Rees and Wackernagel's classic definition of the ecological footprints of cities comes in the form of a question: 'How large an area of productive land is needed to sustain a defined population indefinitely, wherever on Earth that land is located?' (2013: 159, originally published in 1996). The ecological footprint of the Canadian city of Calgary has been calculated at 100 times the area of its city limits (The City of Calgary 2007), while that of Santiago de Chile was in the 1990s calculated as being 16 times larger than its metropolitan area (Wackernagel 1998: 16). The idea of the ecological footprint, despite its complexity and its undoubted limitations, helps to visualize the sheer scale of the impacts of cities on the environment and also the ways in which cities in the Global North impact significantly more per capita than those in the Global South. Clearly, if sustainable urban development is to be achieved, significantly reducing the ecological footprints of cities is an imperative (Girardet 1996: 24–25; Gottdiener and Budd 2005: 158; Pacione 2009: 609–610; Simon 2016), and this is an imperative that falls most heavily on the cities of the Global North, although this does not always translate into action by these cities or their national governments.

The concept of sustainable urban development, typically based around principles of intergenerational equity, social justice, trans-frontier responsibility (Pacione 2009: 606), has been increasingly central to discourses of future urban development. The idea of sustainable urban development implies that there might be urban development trajectories that do not necessarily involve an inexorable march towards non-renewable resource depletion and environmental overload, but that can continue, in theory, indefinitely while remaining within, rather than breaching, environmental limits. The most commonly cited definition of sustainable development is that put forward by the Brundtland World Commission on Environment and Development (1987): 'development that meets the needs of the present without compromising the ability of future generations to meet their own needs'. The difficulty with this definition is that, beyond basic survival, the concept of need is socially constructed and is not absolute. Blowers and Pain (1999: 265) point out, for example, that 'what may be regarded as needs in the cities of the North would be luxuries in those of the South'.

Exercise

One of the key issues in the definition of sustainability is socio-economic and cross-cultural differences in the definition of needs, wants and luxuries. Can you think of a definition of 'needs' in the context of the cities of the Global North? What would you argue is a 'need' in this context? Indeed is it possible to come to a consensus about this? Does this definition of 'needs' hold up when transferred to the context of cities of the Global South? What are the implications of these cross-cultural differences for the definition and promotion of sustainable urban development? Will the urban societies of the Global North, or perhaps certain sections of these societies, be prepared to recognize that what they might have long regarded as needs are seen, viewed from another perspective, as wants or luxuries? Can lifestyle changes in these cities follow on from this? How is this change achieved and what are the barriers to this?

Increasingly, discussions about sustainable urban development have moved beyond the purely environmental concerns that typified many early contributions. More and more the notion of sustainability here has come to be deployed in ways that encompass social and economic sustainability which some have argued, necessitate the need to achieve greater degrees of democracy and equity in cities. This has been referred to as the 'triple bottom line' of sustainable development (Krueger and Savage 2007, cited in Cugurullo 2013: 24). However, the present seems to provide a particularly problematic period within which to consider paths to achieving this. A series of fundamental challenges have emerged from the late twentieth, and increasingly the early twenty-first, centuries, including a lack of international political consensus around the ways in which socio-economic sustainability might be achieved, as characterized by growing doubts from academic researchers, and increasingly from politicians and the wider public, around market-led economic approaches (Clement 2016; Harvey 2013; 2015) and the political reactions against economic globalization as evidenced by the withdrawal of the UK from the European Union and the elections of locally protectionist minded, seemingly anti-globalization politicians in Europe and the US; the uncertain and erratic economic growth patterns that have characterized the world economy since the end of the post-war 'long boom' in the mid-1970s (Harvey 1989); and the increasing informal provision of employment and housing in urban areas, particularly, but not exclusively, those in the Global South (Chen et al. 2013: 348; Vasudevan 2014; Chiodelli et al. 2017). The latter points are a particular challenge to sustainability as informal sectors lie largely beyond the control of the state and there is little evidence that those engaged in these sectors see sustainability as a priority. That is not to deny, of course, the considerable potentials of the informal sector to act as ways forward in terms of urban sustainability in areas such as urban design (Smedley 2013) and waste management (Wilson et al. 2006).

Thus, there is a broad consensus within the literature that there are multiple dimensions of sustainability (Gottdiener and Budd 2005: 158, 161; Chen et al. 2013: 342). This is the triple bottom line of environmental, social and economic sustainability mentioned above, although some writers' versions of this articulate five dimensions of sustainability, economic, social, natural, physical and political. These dimensions are not separate but, rather, are deeply interrelated. There is a growing consensus that they are unlikely to be achieved separately. Thus, Chen et al. (2013: 345) have argued: 'Economically deprived cities that are socially divided and exclusive are [even] more difficult to sustain'. The challenges of sustainable future cities then, are not just environmental challenges. David Simon (2016), and the other authors who contributed to the book *Rethinking Sustainable Cities*, see such cities as needing to be accessible and fair as well as green, a position echoed by others such as Friend et al. (2016: 68) who argue that future challenges for cities are 'ensuring wellbeing, social justice and equity for an ecologically viable future'. Elsewhere Hou et al. (2015: 4) have recognized that 'scientists, policy-makers, and academics increasingly acknowledge the interdependent nature of built and natural environments and the consequent challenges such relationships suggest in the advancement of urban sustainability, social equity and political empowerment'.

To the elements of sustainability outlined above we might also add connectivity. Cities function upon webs of various forms of connection. These include 'transportation, but also … electricity, waste disposal, water, digital, and other infrastructures' (Kourtit et al. 2014: 1) and these connections are both internal and external to cities. Many of these connections are directly related to the environmental sustainability of cities, such as the dominance of the private car as the preferred form of mobility in many cities globally, which was explored in chapter 12. However, equally, the connections associated with energy, water, waste, and digital infrastructures affect the use and disposal of non-renewable and renewable resources. Connectivity, therefore, can be thought of as an element of urban sustainability both in that it directly shapes the many aspects of cities that contribute to their ecological footprints, and in that it is key to the general functioning of cities which, in turn, shapes the possibilities of their pursuing more sustainable development trajectories. Connectivity of all kinds, then, is a major future challenge for cities and one which will shape the geographies of sustainable urbanism. Kourtit et al. (2014: 1) recognized this when they asked, 'whether the megalopolis can keep up with connectivity needs – will setbacks stifle urban growth and/or cast a pall over urban living?'.

Climate change and cities

> Given the current ubiquity of narratives concerning climate change, whether that be with respect to vulnerability and resilience, or to forms of low-carbon development and transition, within urban arenas it is hard to imagine that just a short decade ago such agendas were far from common place.
>
> (Bulkeley and Betsill 2013: 139)

That climate change poses major and multiple threats to cities and that it has become a priority for cities is now widely recognized (Bulkeley 2013; Bulkeley and Betsill 2013; Chen et al. 2013: 348). The river and littoral locations of many cities, particularly those in the Global South experiencing rapid population growth, make them environments increasingly susceptible to unpredictable, severe flooding events associated with rising sea and river levels and growing incidence of extreme storms. The uncertainties associated with these processes and their manifestations pose complex challenges for decision-makers and planners (Chen et al. 2013; Kourtit et al. 2014: 1; Friend et al. 2016: 68). Cities, particularly large cities, are typically the economic powerhouses of wider, sometimes, extensive, regions. The impacts of the devastation and disruption associated with flooding often go beyond the immediate damage and inconvenience associated with the flood events themselves and include considerable longer-term economic, not to say social, costs (see case study below).

The impacts of climate change on cities are mediated through their particular urban geographies. The physical environments of cities shape the short-term impacts of climate change associated events. Patterns of flooding, for example, are shaped by local topographies. However, these physical geographies are overlain by the social geographies of wealth and poverty which further mediate these impacts. The environmental problems in cities tend to impact most severely on the poorest and hence most vulnerable groups in urban society and upon the places in the city that they tend to occupy (Chen et al. 2013: 310). These groups are less able to insulate themselves from the impacts of environmental problems. They tend to occupy more marginal land, in shelter that might be informal or have access to only very basic infrastructures. The contrast between the environmental risks borne by poorer, vulnerable urban groups, for example, and the comparative lack of risks faced by the wealthy in cities of the Global South bears this out most clearly. Although to a lesser, but by no means insignificant, extent the same applies to the cities of the North. As Haughton and Hunter argue: 'Cities are the centre for the creation and redistribution of major environmental externalities. These are passed on unevenly, both within the city and outside' (1994: 52). This applies to both the immediate externalities generated by urbanization such as pollution and also their more long-term consequences such as the impacts of climate change.

Wealthy urban populations are typically more able to protect themselves from climate change impacts, either by investing in technological fixes such as air conditioning, which mitigate the immediate effects of extreme heat waves (while, ironically, indirectly contributing to their greater future occurrence through heavy energy consumption), or by choosing to live in, typically more expensive, locations which are likely to be less severely impacted. Flooding, for example, impacts most heavily upon the poorest in large cities, especially those in precarious informal habitation. Similarly, deaths during heat waves, which, while they attract less attention, are greater in the cities of the US than those associated with more dramatic earthquake, flood, and tornado events, disproportionately affect the urban poor who are less able to afford air conditioning units and may live in greater social isolation due to higher crime rates and a lack of public infrastructure within their neighbourhoods (Chen et al. 2013: 348). There is evidence also that the subsequent responses to disruptive environmental events within urban areas are not necessarily socially benign. It has been observed, in the case of Hurricane Katrina that devastated New Orleans in August 2005,

> not only that disasters disproportionately affect poor and minority communities but also that the rebuilding that follows disasters can perpetrate or exacerbate these inequalities ... the inequalities present before disasters tend to be amplified during and after these events, lending an important social dimension to these purportedly 'natural' catastrophes.
>
> (Chen et al. 2013: 306–307)

Further, the political geographies of cities, for example, their institutional capacity and resilience significantly affects their abilities to plan for and respond to the impacts of climate change. This is something that is explored in the case study of Lagos, Nigeria below.

Case Study: climate change and the megacity of Lagos, Nigeria

Lagos is a megacity of 18 million people which has experienced extreme population growth over the previous 50 years. Its annual population growth rate of six per cent will make it one of the world's largest megacities by 2025. Lagos is the industrial and commercial centre of Nigeria, albeit one with 70 per cent of its population living in poor slum communities. Further, the city's physical geography, its low-lying coastal location, makes it particularly vulnerable to climate change. The population growth and spatial expansion that Lagos has experienced has exceeded its administrative, management and infrastructural capacities, putting the city under multiple strains. Since the return of stable democratic governance in Nigeria in 1999 there have been a series of reforms aimed at addressing these inadequacies. Recently, these have begun to recognize the necessity of addressing issues related to

continued

climate change, both in terms of its causes and impacts on Lagos. In 2015, Peter Elias and Ademola Omojola, from the Department of Geography at the University of Lagos, published a critical assessment of the challenges that climate change poses for Lagos city and state, and the responses to them.

Lagos faces a number of risks from climate change including sea level rise, storm surge, flooding, high temperatures and high rainfall intensity. Its vulnerability to these hazards is exacerbated by social factors including the city's high population density and its large poor population. Rising temperatures, for example, have coincided with high population densities and poor drainage to worsen the impacts of disease in the city and surrounding areas. This has severe human health consequences, particularly amongst children and the elderly. Lagos's institutional weaknesses also limit its ability to respond effectively to such issues.

While Elias and Omojola acknowledge that engagement with the climate change agenda has taken place in Nigeria, they recognize that this has been somewhat uneven across different levels of government. This is particularly serious as others have recognized the importance of co-ordinated, multi-level 'urban' governance of climate change (Bulkeley and Betsill 2013). The federal government has long been active in climate change activities, but this has been less the case at the state and, especially, the local levels. Lagos State's government have begun to engage with climate change initiatives as evidenced by their joining the C40 Large Cities Climate Leadership network in 2007. This led to the formation of a climate change unit, part of the state's Ministry of Environment, which has

coordinated several adaptation initiatives, albeit somewhat ad hoc in nature. These have included a range of environmental advocacy programmes and annual climate change summits, adaptation strategies aimed at improving food security, maintaining and expanding provision of sea walls and storm water barriers, relocation of vulnerable populations and climate sensitive urban renewal and infrastructure provision programmes. The state government has also developed a climate change adaptation policy (2012–2014) which is subject to a cycle of three-yearly review and which aims to harmonize the state's various climate change activities. In addition, this is articulated with key international climate conventions, treaties and protocols. The state has also recognized the importance of strengthening institutional capacity within, for example, its emergency services and its Ministry of Agriculture and Cooperatives, to addressing climate change concerns. In addition to these adaptation initiatives, the state is also involved in many mitigation efforts such as waste reduction and management strategies, infrastructure provision, tree planting, promoting the green economy and improving public transport provision and infrastructures.

However, Elias and Omojola recognize arenas that the state has failed to fully engage within climate change initiatives, responses and decision-making processes. These include local government, non-governmental organizations, the private sector and vulnerable coastal and other communities, particularly those from informal settlements or those engaged in informal activities. This undoubtedly limits the effectiveness of adaptation programmes within the state. The authors offer the

continued

damning judgement that Lagos's responses to climate change have been 'haphazard, largely top-down, uncoordinated and fragmented' (2015: 74). Certainly, there is much at the neighbourhood level in terms of risk and adaptation potentials that remains unknown and untapped. Lagos's vulnerability to climate change, a function of its physical, social and political geographies, has become increasingly apparent through the extensive damage which has resulted from severe weather events of various kinds. While this has prompted action across various levels of government this has been variable and incomplete in its reach, and there remains much that needs to be done to strengthen and expand the responses of this vulnerable, increasingly populous urban region.

Source: Elias and Omojola (2015)

The relationships between climate change and cities do not flow one way. While climate change has major impacts on cities, cities are also major contributors to processes of climate change. While occupying only around two per cent of the world's land surface, cities now contain more than one half of the world's population. Cities also consume three quarters of the world's resources and generate a majority of its waste (Blowers and Pain 1999: 249). Curbing this voraciousness is the key goal of the urban sustainability agenda discussed above. The environmental demands of city dwellers vary enormously. For example, city dwellers in the Global North typically generate up to twice as much waste per day as those in the South (Haughton and Hunter 1994: 11). Whereas, for much of history, the consequences of these problems were primarily local, the scale of contemporary urbanization ensures that they are now felt across multiple scales including globally. Cities, then, are both the arenas which contribute most to the causes of climate change and also, if sustainable ways of urban living can be found and operationalized across a range of urban environments, the arenas that can contribute most to its management and mitigation (Castán Broto and Bulkeley 2013: 92).

Since the early 2000s cities have increasingly sought to integrate climate change concerns into their pursuit of wider urban goals, priorities and agendas. This has often been driven by elected officials who have been responsible for the emergence and growth of a number of transnational urban networks, which are either explicitly organized around addressing climate change or which have come to recognize the importance of climate change issues to cities and of cities to these climate change issues. These networks include the Climate Protection Agreement, signed by over 1,000 city mayors by 2011, the European Covenant of Mayors, which had over 2,000 members by 2011 and the C40 Cities Climate Leadership Group, a network of 40 global cities (Bulkeley and Betsill 2013: 140–141). Evidence from an extensive, global review of urban climate change experiments, which is discussed below, suggests that membership of these city networks is strongly associated with the willingness of cities to undertake their own urban climate change experiments (Castán Broto and Bulkeley 2013: 97).

Two main climate change policy responses have emerged since the early 1990s; mitigation and adaptation. While mitigation is designed to address the causes of climate change, for example, by reducing the emission of greenhouse gasses, adaptation is designed to reduce the vulnerability of cities to the impacts of climate change. This is often framed as increasing the resilience of cities to climate change (Clark 2012). The foci of many urban climate change networks, such as the ones listed above, and hence the cities they represent, have tended to shift more towards adaptation responses over time, although this is not to suggest by any means that cities have ceased to recognize the importance of and to promote mitigation measures (Bulkeley and Betsill 2013: 141). Indeed, the majority of urban climate change experiments analysed in the review discussed below were mitigation, rather than adaptation, measures, which only accounted for 12.1 per cent of initiatives reviewed (see table 13.1).

Table 13.1 *Types of climate change schemes in cities*

Objective in relation to climate change	Sector	Types of schemes
Mitigation	Urban infrastructure	Alternative energy supply (renewable or low carbon)
		Landfill gas capture
		Alternative water supply
		Collection of waste for recycling and reuse
		Energy and water conservation measures
		Network demand reduction measures
	Built environment	Use of energy-efficient materials
		Energy-efficient design
		Building-integrated alternative energy supply
		Building-integrated alternative water supply
		New-built energy and water-efficient technologies
		Retrofitting energy and water-efficient technologies
		Energy and water-efficient appliances
		Building-integrated demand reduction measures
	Urban form	Urban expansion and suburban development
		New urban development
		Reuse of brownfield land
		Neighbourhood and small-scale urban renewal
	Transport	New low-carbon transport infrastructure
		Low-carbon infrastructure renewal
		Fleet replacement

Table 13.1 *(continued)*

Objective in relation to climate change	Sector	Types of schemes
		Fuel switching
		Enhancing energy efficiency
		Mobility demand reduction measures (reducing travel)
		Mobility demand enhancement measures (alternative means of travel)
	Carbon sequestration	Urban capture and storage
		Urban tree-planting programmes
		Restoration of carbon sinks
		Preservation and conservation of carbon sinks
		Carbon offset schemes
Adaptation		Cooling services and designs
		Measures securing energy and water supply
		Flood protection
		Bushfire protection
		Relocation and zoning policies
		Blue and green infrastructure
		Building codes for extreme weather
		Early warning systems
		Behaviour-based measures

Source: Castán Broto and Bulkeley (2013: 95), adapted from UN-HABITAT (2011)

Vanesa Castán Broto and Harriet Bulkeley's (2013) valuable comparative review of 627 climate change experiments in 100 cities in both the Global North and Global South revealed the diversity of climate change experiments and the actors involved across a wide range of international urban spaces. They found urban climate change experiments occurring in cities in all world regions and a clear growth in the numbers of these experiments occurring after 2005, showing the extent to which climate change has become embedded within urban governance and policy agendas globally. Experiments in urban infrastructure, built environment and transport dominated the cases reviewed. Across these sectors, measures were most heavily focused on optimizing the production, distribution and consumption of energy. Castán Broto and Bulkeley's evidence showed that urban infrastructure experiments are particularly important in Asia, while transport projects are popular in Central and South America. This, they attribute to rapid urban growth in Asia which has created the opportunities and capital from private sector actors for urban

infrastructure experiments and, in Central and South America, the influence
of extensive, innovative, high-profile urban transport projects in cities
such as Curitiba in Brazil and Bogotá in Colombia. One further critical
finding to emerge from this work was that only two per cent of the experiments
that Castán Broto and Bulkeley surveyed were located in the world's least
developed nations. This is particularly significant as it is widely recognized that
these nations are likely to be the most severely impacted by climate change
(Caprotti 2014: 1289).

The authors show that local governments play important leadership roles in
many climate change experiments but also that there are a wide range of other
actors involved in the experiments reviewed, including non-state actors such as
community groups, private corporations, NGOs and international foundations.
Further, Castán Broto and Bulkeley found that private actors are particularly
important in urban climate change experiments in Asian cities, more so than is
the case in other regions. One important observation from this review and
other literature on urban climate change experimentation is that social justice
concerns do not emerge strongly as desired outcomes of urban climate change
experimentation. As Castán Broto and Bulkeley (2013: 100) observe 'it appears
that the distribution of climate change responsibilities and vulnerabilities is
often parallel to existing patterns of urban inequality' (see also Satterthwaite
2008). This chimes closely with the observation made at many points in this
book that prevailing forms of urban change of many kinds within neoliberal
urban environments either perpetuate existing patterns of inequality or
significantly worsen them. It also connects directly with the discussion
of the wider politics of future urbanization discussed later in this chapter.

Addressing future challenges

The difficulties of achieving sustainable trajectories of future urbanization are
almost ubiquitous within academic discussions of the topic. The complexities
and uncertainties associated with all aspects of predicting, planning and
delivering sustainable urban futures have been acknowledged extensively within
the opening sections of this chapter. One of the key challenges, or perhaps more
accurately a series of related challenges, involves co-ordination. Even if it were
possible to construct widely accepted visions of what sustainable future cities
might be, and much debate, disagreement and diversity, surrounds these visions
and their associated policies (Gottdiener and Budd 2005: 161), there remain the
questions of how these might infuse the multiple spaces, scales and actors
involved in processes of urbanization. Cities are complex, a point that should be
more than apparent by this point in the book. They are made and remade by
combinations of powerful, external macro forces of many kinds and locally
grounded individual and community contributions (Chen et al. 2013: 334).

The questions of at which scale, or rather scales, templates of sustainable urban development should be constructed are problems to which urban geographers should be particularly attuned. Actually, achieving sustainable urban development across these multiple scales and with the diversity of actors involved are challenges demanding immense coordination: 'regionalists (or metropolitanists) ... argue that pressing environmental, social, and governance problems cannot be solved by independent jurisdictions acting alone [;] ... cross-jurisdictional problems demand cross-jurisdictional solutions' (Katz 2000: 3 cited in Brown 2014: 9). This is a particularly pressing challenge in the face of planning systems that are increasingly focused on short-term goals (Malecki 2014 in Kourtit et al. 2014: 2) and the rescaling of urban decision making which includes, in some ways, a greater decentralization (Chen et al. 2013: 348), potentially rendering questions of co-ordination even more challenging.

The economic and political uncertainty that has characterized much of the twenty-first century to date adds another layer of difficulty to these challenges (Hou 2015: 4). In the face of recent global economic difficulties and shifting political priorities, the immediate extent to which urban agendas in the coming years will be shaped by the concerns outlined in this chapter is perhaps less certain than it has been for some time. On a more optimistic note, rapidly developing technologies offer multiple potentials in the spheres of smart, sustainable urbanism and yet there are further uncertainties and challenges associated with this (Chen et al. 2013: 322), some of which are explored in more detail later in the chapter.

Finally, competing and differently powered notions of global environmental justice have come to frame debates about sustainable urban futures. Increasingly, the growing contributions of the cities of the South to global environmental problems have been recognized (Chen et al. 2013: 315). This has tended to take the form of blaming these cities as the sites of production for a range of environmental externalities that are associated with their growing manufacturing economies. This has discursively framed approaches that have sought to impose regulation on these cities to reduce the environmental impacts stemming from this manufacture. An alternative view, though, would be to tie these outputs to the consumers of the products manufactured here, who predominantly reside in the cities of the Global North, thus relocating the cause of these externalities beyond the sites of their immediate manufacture. This might potentially open up alternative questions of how these issues might be tackled, perhaps putting the emphasis more on the reform of damaging consumption habits in the Global North in addition to the cleaning up of manufacture in the Global South. This, however, is not a discourse that has gained much traction politically in environmental, and other, forums shaped, largely, by Northern concerns. What follows from this is a concern that prevailing discourses of sustainable development are in danger of further perpetuating existing patterns of socially and environmentally harmful global inequalities at a global scale. Caprotti argues:

> The fact that the increasingly environmentally polluting role of emerging economies is intimately and directly tied to increasing levels of consumption in the 'clean' and ecologically modernising countries of the North is not often explicitly stated. As a result, leading emerging economies are highlighted as the new culprits of human-induced climate change.
>
> (Caprotti 2014: 1288)

The politics of future urbanization

This section explores the emergent models of urbanization that purport to shape urban futures in more sustainable ways. It looks particularly at two of the most prominent of these models, the overlapping concepts of the eco-city and smart urbanism. This section adopts a critical perspective on these concepts, highlighting the multiple concerns that a number of authors have raised concerning their environment and social justice credentials. It will begin, though, by situating these models within the economic and political discourse, the idea of ecological modernization, that they emerged from within.

Rather than seeing the widespread recognition of the ecological consequences of modern urbanization and industrialization as an argument to restructure the prevailing regime of accumulation, the response of governments internationally, and especially those of the Global North, has been to argue that rising levels of economic development and personal consumption are not incompatible with sustainable development. This is a stance referred to as ecological modernization within the literature. Governments have sought to instigate measures that aim to secure sustainability without compromising levels of economic development. The stance of governments has broadly been to encourage the 'greening' of capitalism without fundamentally affecting its operation. Measures adopted or proposed by governments have included encouraging, or forcing, the internalization of the costs of environmental externalities by incorporating them into the costs of industrial production. Although in many ways welcome, forcing companies to realize that production has an environmental cost, such measures are severely limited. It is possible with measures such as this for companies to continue to generate environmental externalities simply by paying for the 'right' to pollute (Haughton and Hunter 1994). Such measures might not even severely affect profit margins if these increased costs can be passed on to the consumer. Further, critics argue, that by not fundamentally disrupting the operations of neoliberal, global capitalism, ecological modernization does nothing to alter the prevailing conditions that led to emergence of the range of challenges that cities globally now face and which have been outlined above.

Key parts of these attempts to green capitalism have been attempts to green processes of urbanization through the development of more sustainable future cities. As Rapoport argues: 'translated into the urban arena, ecological modernization promises that technological and procedural innovation can solve urban environmental problems ... Accordingly, many contemporary eco-cities rely heavily on technology as a means for achieving their sustainability objectives' (2014: 138). There is huge diversity to urban projects under the banner of ecological modernization in terms of their scale, the actors involved and their specific foci. However, the most prominent discourse of future urbanization is the concept of the eco-city. Within this there is also a powerful technocratic discourse of smart urbanism. These should be regarded, given their diversity, as no more than two umbrella concepts. Joss (2011: 280) defines eco-cities as 'a development of substantial scale, occurring across multiple sectors, which is supported by policy processes' (cited in Rapoport 2014: 139). Smart urbanism is specifically concerned with the potentials of new technologies and interactive infrastructures to shape cities in ways that potentially offer 'a futuristic solution brought to the present to deal with a broad multiplicity of urban maladies, including issues of transport congestion, resource limitation, climate change and even the need to expand democratic access, amongst others' (Luque-Ayala and Marvin 2015: 2106–2107). There is an extensive literature that offers guides to the emerging geographies of eco and smart urbanization (see, for example, Joss 2011; Chang and Sheppard 2013; Cugurullo 2013; Caprotti 2014; Rapoport 2014; Luque-Ayala and Marvin 2015). One of the key themes to emerge within this critical literature is the warning that these developments do not necessarily offer panaceas to present and future crises and that may reproduce, albeit in different contexts, existing patterns of urban inequality and injustice.

There are four main arguments that have emerged within the critical literatures of eco and smart urbanism in recent years. The first is that environmental concerns are not nearly as central to these modes of urbanization as their advocates claim (Rapoport 2014: 138–142; Luque-Ayala and Marvin 2015: 2110 and see case study below) indeed, it has been argued, eco-ness might be attached to some of these projects largely for marketing purposes and to distinguish them from other rival projects of urban development (Chang and Sheppard 2013 in Rapoport 2014: 142). While this is not to dismiss the environmental claims and ambitions of these projects entirely, they should be seen more accurately, critics would argue, as part of a series of objectives, prime of which is capital accumulation through the attraction of outside investment. The context of these eco-cities, as part of international networks of competitive urban spaces, should not be dismissed within readings of their ambitions. Rapoport argues that 'most eco-city projects claim to incorporate environmental, economic and social aspects of sustainability. However, a

Case study: eco-city myths? Masdar City, Abu Dhabi

Since the late 1980s there has been much prominent rhetoric concerning sustainable urban development, a significant proportion of which covers the apparent potentials of the eco-city model. The three dimensions of sustainability that have emerged from this literature are the economic, the social and the environmental. What has been somewhat lacking, until recently, are empirical evaluations of the substance of sustainable urban and eco-city rhetoric and development. Federico Cugurullo, a geographer based at King's College London, aimed to address this by evaluating the degrees to which these three dimensions were present, and the relations between them, in the Masdar City eco-city project in a paper published in the *Journal of Urban Technology* in 2013. Masdar City, a master-planned city, intended initially to be zero-carbon and zero-waste, was launched in 2006 and is the most developed eco-city project available to study to date. By 2010–2011 when the fieldwork for Cugurullo's paper took place, while it was still under construction, the core of the city was complete and the first residents had moved in. Cugurullo argued that Masdar City offered not just a model of future sustainable development but of urban planning more generally making it an important site to understand the shaping of urban futures around the world. His analysis of Masdar City referenced the triple-bottom line of sustainable development: environmental, social and economic sustainability. However, the findings of his analysis were damning and he was able to argue that while the economic dimension has been central to the development of Masdar City, social and environmental concerns were present only at more superficial levels and were far from central to the project.

Masdar City was planned by Foster and Partners, a globally-known architectural practice. It is a 6km^2 development which utilizes the latest green technologies such as smart utility grids, concentrated solar power and electric personal rapid transport, as well as technologies designed to reduce energy consumption. The early prominence of the project attracted the involvement of companies such as General Electric, Schneider and Siemens as well as support from a number of international environmental organizations. The initial optimism in the potentials of Masdar City was shaken by the credit crunch of 2008 and this resulted in the reduction of some of the project's ambitions. For example, rather than being zero-carbon, the city became carbon-neutral and the electric, personal rapid transport (PRT) system, which was originally planned to cover all of the development, was scaled back to cover only ten per cent of its total area. The financial backing from the Abu Dhabi government has been reduced and the deadline for completion of the project pushed back from initially 2016 to 2020, then 2025 and since possibly to 2030.

Despite the undoubted impact of the recession on the development of Masdar City, Cugurullo was keen to stress that this did not derail the eco-visions of the city, rather it revealed more starkly than would otherwise have been the case, its inherently commercial orientations. Cugurullo's analysis reveals the city as a site which combines the characteristics of a laboratory, in which green tech products from the companies backing the development are

continued

tested in a real-world setting, and a permanent showroom for these products which, if they are successful, are incorporated into the fabric and landscape of the city. These technologies are further promoted through Masdar's web-portal The Future Build and two major annual renewable energy events organized by the Masdar Initiative: The World Future Energy Summit and the European Future Energy Forum, which Cugurullo, who undertook fieldwork at these events, described as more akin to trade fairs than forums or summits.

There is a requirement, imposed on the development by the Abu Dhabi Planning Council, for 20 per cent of the housing in the development to be reserved for low income workers. However, concerns emerged within Cugurullo's research that this remains underdeveloped with little apparent idea of how it will be realized. This would appear to be a largely symbolic attempt at realizing social justice within Masdar City. There is no evidence elsewhere in the city of how the growth generated by the technological developments might be shared equitably between its citizens.

Despite these problems, Masdar City still argues that it is one of the world's most sustainable places. However, Cugurullo counters this by suggesting that the idea of Masdar City as an eco-city was something that was created discursively prior to its construction. Since this image was created for the development, Cugurullo argues that much of the plan has changed, with only ten per cent of the development having been realized to date. Further, Masdar City had not, at the time of Cugurullo's research, released any data relating to its environmental performance, thus, it is impossible to verify any claims as to its environmental performance. In summary, Cugurullo is able to claim that Masdar City is primarily a project of capital accumulation rather than of sustainable social and environmental development. Further, given its vulnerability to downturns in the economic cycle, he argues, that Masdar City has all of the resilience of a 'sandcastle'.

Source: Cugurullo (2013)

number of recent publications drawing on detailed case studies of eco-city projects have found that economic concerns consistently take priority over environmental ones' (2014: 141–142). Luque-Ayala and Marvin (2015: 2110) have found that such criticisms are equally applicable to smart city projects.

A second criticism of these models of development is that they tend to offer broadly standard, universal solutions that may not be applicable across the range of urban settings to which they are applied, particularly where these settings lie in the Global South. Eco-cities, and especially their smart manifestations, offer very technocratic forms of address rooted in on-going capitalist processes to which new technologies are central. Eco- and smart cities, then, seem unable to imagine solutions to environmental or urban problems that are not similarly rooted within these technocratic approaches. Rapoport (2014: 143) and others, however, have contested the claim that models of eco-city development have universal applicability. They point to alternative, indigenous, low-tech solutions that might be more appropriate within some, particularly Global South, contexts

Figure 13.1 *Masdar City, Abu Dhabi*
Source: LMspencer / Shutterstock.com

but which are excluded from eco and smart city thinking and doing because of their limited potential to be turned into future revenue streams for corporate developers. This may, then, have the dual effect of compromising the potentially positive ecological outcomes of this city building and also of excluding those who are unable to engage with the technologies of eco and smart cities either for reasons of poverty or for reasons of culture or practice. These exclusions, Rapoport argues, may apply to the majority of the world's urban dwellers and point to a third, related, criticism of eco and smart city development, namely that they may intensify, rather than reduce, patterns of inequality and exclusion within and between cities (Caprotti 2014: 1296; Luque-Ayala and Marvin 2015: 2108). One of the most direct ways in which this occurs, but by no means the only way, is through the exploitation of migrant workers who provide the labour for the production of eco-cities. Eco-cities have been described as a form of enclave urbanism (Wong 2011; Caprotti 2014: 1293), which is little connected to wider urban spaces, networks and systems and thus does little by way of engagement with, or address of, the broader urban and environmental problems that they purport to offer solutions to. Caprotti points out that 'these exclusive developments provide environmental 'goods' to those who can afford to live within the eco-city – while little attention is paid to those who built it, or to those who live in its shadow or on its fringes' (2014: 1293).

Finally, Caprotti (2014: 1289–1290) questions the discourses of crisis that are often mobilized to justify eco and smart city models of development.

He does not argue that ecological change stemming from urbanization and industrialization is not real or that its impacts are not severe and socially and economically mediated. Rather, he suggests that these processes are presented in particular ways, most notably through narratives of crisis, that appear to both render them addressable, primarily or exclusively, through high-tech driven eco and smart urbanism, and to suggest that this course of action is a natural response to these crises, rather than a political position rooted within sets of specific economic interests. These crisis discourses then posit the actors behind eco and smart city development, typically powerful corporate interests and national governments, as those most empowered to address them. Such discursive boundary drawing is an important component in the justification of the emergence of modes of future urbanization which do not fundamentally disrupt prevailing economic relations and are active in the marginalization of more radical, disruptive alternatives. Caprotti sums up this criticism by arguing that 'economic and technological interventions are justified and rationalized not for what they frequently are – speculations, for profit investments, and projects devoid of socio-political equity – but as shining examples of high-tech responses to crisis' (2014: 1297). Having said all this, however, the potential contributions of eco and smart cities should not be dismissed entirely. Some authors recognize that these models may increase awareness of sustainable urbanization and provide experimental contexts within which new ideas might be tested (Rapoport 2014: 144; Evans et al. 2016), while it has been recognized that smart cities offer the potential, in some contexts, for 'more "community", "civic" or "metropolitan" forms of service provision and urban life' (Luque-Ayala and Marvin 2015: 2108).

Future urban priorities

It has been widely argued, both within this chapter and across the literature it draws on, that to shape urban futures in ways that are more sustainable is only achievable if approaches encompass economic, social and environmental dimensions in ways that are balanced. This points towards approaches grounded more in areas such as resilience theories and social ecological systems perspectives (Friend et al. 2016: 67) and, perhaps, less in the technocratic and commercial world views that have proven so influential within eco and smart city practice. Inclusive, more democratic cities, to follow this logic, cities where communities rather than corporations, have control over the decisions that shape their spaces and resources, are more sustainable cities (Harvey 2008; Chen et al. 2013: 345–346). Achieving this degree of democratic control of city spaces and resources is a major challenge in that it goes against prevailing neoliberal market logics. As the discussion of eco and smart cities above shows, powerful corporate and political interests are adept at negotiating potential crises, indeed in presenting them to their advantage, to ensure the reproduction of their hegemonic positions. The narration of ecological crisis discussed in the previous section, and the discursive boundary drawing that goes with this, certainly locates such ambitions outside of hegemonic modes of future urbanization which seem to be squarely built around the ideas of eco and smart urbanism discussed earlier. David Harvey outlines the fundamental nature of the transformation towards truly inclusive, democratic cities in his discussion of the right to the city in ways which are difficult to imagine being incorporated into eco- and smart city practices and discourses.

> The right to the city is far more than the individual liberty to access urban resources: it is a right to change ourselves by changing the city. It is, moreover, a common rather than an individual right since this transformation inevitably depends upon the exercise of a collective power to reshape the processes of urbanization. The freedom to make and remake our cities and ourselves is, I want to argue, one of the most precious yet most neglected of our human rights.
>
> (Harvey 2008: 23)

The journey towards more inclusive, democratic and genuinely sustainable cities is a long and complex one and involves multiple dimensions. Indeed, perhaps it should be more accurately considered as many interconnected journeys rather than a single, uniform one. Primarily, the present and future challenges that this chapter has discussed pose questions for urban governance where questions of social justice, equity and ecology (Friend et al. 2016: 68) must become more central to its values, strategies and operations. Below the governance of cities, these rights questions play into the approaches of those responsible for the management and production of the urban environment, such as architects and planners (Chen et al. 2013: 342). Gottdiener and Budd, for example, discuss the work of Portney (2002) who outlines some considerations for democratic planning:

> Enabling a community to identify what it values and then prioritizing the values.
> Holding individuals and larger groups accountable for achieving results.
> Building democratic mechanisms for decision-making and participation.
> Allowing community residents to measure what is important and get that to planning.
>
> (Gottdiener and Budd 2005: 162)

However, the nature of the problems that city professionals, at all levels, face seem to demand not just that they become more attuned to the issues and considerations outlined above but that they become prepared to approach

problems in entirely new ways, and ways that may not fully, indeed at all, accord
to their traditional practices. Friend et al., for example, discuss climate change as
a wicked problem for cities, a wicked problem being one that is characterized by
complexity, uncertainty and incomplete knowledge. Such a problem, they argue,
'requires plural solutions, clumsy governance and flexible, learning-oriented,
and adaptive institutions [which] stands in direct contrast to the core foundations
of public administration theory and practice, which require efficiency,
transparency, and accountability' (Friend, et al. 2016: 69). As part of this they go
on to argue that responses to these problems should be networked and involve
connections between cities, often cities distant in space, rather than territorially
defined in the ways that they have traditionally been. This recognizes that the
challenges that this chapter has discussed, exceed individual administrative
units and are characterized by interconnections between often distant spaces
and formerly discrete sectors of society and the economy. Further, achieving
the global transformations necessary to address current and future challenges,
would appear to require political action across multiple scales from the local to
the global (Friend et al. 2016: 69). While the potential limitations of existing
forms of urban governance, planning and production, in the faces of these
challenges, are becoming more widely recognized, we are only at the very
earliest stages of transforming them in ways that future challenges demand.
We cannot be sure that they will fully achieve these transformations given the
challenge of marrying the long-term visions that these challenges require with
more immediate and local priorities. Within the broad realm of future urban
challenges there are a number of specific dimensions which are major in
themselves. These include engaging with the cities of the Global South, and
recognizing their experiences and perspectives, through democratic practices
rather than by imposing Northern perspectives that may not be appropriate or
through unequal, even exploitative relations (Friend et al. 2016: 70). Further, the
challenges of achieving sustainable mobility, within and between cities, as the
previous chapter showed, will be central to achieving desirable global urban
futures (Pacione 2009; 614; Chen et al. 2013: 316).

Given the range, complexities, interconnections and uncertainties associated
with the challenges facing the urban world now and into the future, perhaps
the starting point is to acknowledge that we do not have anything approaching
complete knowledge or even any degree of certainty of how specifically
we should approach these challenges. Increasingly our approaches to these
challenges are likely to be characterized by experimentation and reflective
learning. Perhaps then, as Evans et al. argue, we should see the global urban
future as one in which we are required to embrace the potentials of our cities
as sites for experimentation.

> Experimentation forms a common thread running through otherwise disparate
> contemporary urban trends, from corporatised attempts to create smart, low

carbon cities to grassroots civic movements to make neighbourhoods more socially cohesive ... While assuming many forms, urban experimentation can be distinguished conceptually from conventional urban development or policy by an explicit emphasis on learning from real-world interventions.

(Evans et al. 2016: 2)

Summary

As this chapter has argued, it is apparent that historic processes of urbanization are unsustainable and have been major contributors to a series of pressing interrelated global challenges. Cities then are increasingly being required to develop new policies and modes of planning, development and governance that combine these long-term global challenges with their urgent short-term and local priorities. Building better, more sustainable, cities then is an issue of global significance. These are processes that governments and corporate interests are becoming increasingly engaged in, but, as we have seen above, in ways which do not necessarily balance the environmental, social and economic dimensions of sustainable development. Thus, the challenges of urban futures are as much political challenges of the redistribution of power between different actors as they are questions of finding technological solutions to scientific environmental problems. They appear to require thought and action that is multidisciplinary as well as multi-scale in nature. Within all of this it should be clear that the city itself remains a key research site, or laboratory, within which to explore our urban futures.

Follow-up activities

Essay title: There are three dimensions of sustainable urban development, sometimes called the 'triple bottom-line': economic, social and environmental. To what extent do projects of sustainable urban development balance these three dimensions?

Commentary on essay title

An effective essay would outline a little of the history of the idea of sustainable urban development, noting the advocacy of a blended approach combining the three dimensions above. However, it would also recognize the critical academic literature, such as those papers by Caprotti and Cugurullo which are discussed in this chapter, which have emerged within discussions of sustainable urban development particularly in recent years. Much of the subsequent discussion is likely to be evaluative and based upon the review of other analyses of sustainable urban development projects from a variety of geographical contexts (no specific geographical region is noted in the essay title so it is safe to assume that case studies can be drawn from any region). If this was a research project rather than an essay then you might conduct your own original analysis of a sustainable urban development project, or a number of different projects. A really effective answer would also contain some discernible nuance. For example, rather than trying to arrive at a single conclusion you might look for subtleties that emerge from your analysis. For example, are sustainable urban development projects of one type, or those found within one region more balanced in their incorporation of the three dimensions listed in the essay title than others? You should take the opportunity for the conclusion to be more than just a summary of the essay, although this should be part of this section. For example, are there any limitations in the sustainable urban development literature that restrict what you are able to conclude or which point towards future research agendas?

Project idea

Many universities now have research groups that explore the relationships between climate change and cities and are concerned with finding ways in which future urbanization can help address the causes of climate change and making cities more resilient to its impacts. The majority of these research groups are located in the relatively well-resourced universities of the Global North. Can you find any such units located in the Global South? If so what is the nature of the research they are doing and in what cities is this research based? Look at the projects listed on their websites or at the publications of their researchers.

How does this compare to the climate change research being done on cities located in the Global North by researchers based there? Look at the websites of a selection of research groups based in Universities in the Global North – a few are listed below. Do they tend to research issues of climate change and cities by looking at cities in the Global North or in the Global South? Can you find any examples of research on climate change and the cities of the Global South by the research groups based in universities in the Global North? How does the geography of urban climate change research mirror that of the urban climate change experiments that were explored in Castán Broto and Bulkeley's (2013) paper, which found that only two per cent of the climate change experiments that they reviewed were taking place in the world's least developed nations? On the basis of your research do you think urban climate change researchers are doing enough to support the world's populations that are most vulnerable to climate change?

Further reading

Books

- Bulkeley, H. (2013) *Cities and Climate Change*, Abingdon: Routledge
  A key resource for a topic at the heart of questions of urban futures. It covers the science, policy, governance and responsive aspects of the relationship between cities and climate change in a highly accessible, authoritative discussion.

- Chen, X., Orum, A.M., Paulsen, K.E. (2013) *An Introduction to Cities: How Place and Space Shape Human Experience*, Chichester: Wiley-Blackwell
  Part V 'Challenges of today and the metropolis of the future', offers a comprehensive global perspective that is rich in international examples. It includes some excellent summary discussions of many of the complex issues surrounding questions of the challenges facing cities into the future and the ways in which they might respond to them.

- Davis, M. and Bertrand Monk, D. (eds) (2007) *Evil Paradises: Dreamworlds of Neoliberalism*, New York: The New Press
  A collection of lively, accessible essays exploring the global proliferation of luxury enclaves for the wealthy, the very wealthy and the super-rich. A series of case studies that demonstrate well the forms of exclusive urbanization that are springing up around the world, often in some unlikely locations, the desire of their inhabitants to be screened off from what they perceive as the problems of the world around them and their impacts on patterns of inequality.

- Evans, J., Karvonen, A. and Raven, R. (2016) *The Experimental City*, Abingdon: Routledge
  An edited book exploring the idea that the city (and city governance) rather than the state is the key agent for global change. Rejecting the idea of universal solutions, it considers the city as the site for policy, planning and governance experimentation for sustainable development. The perspective is multidisciplinary and considers the increasing involvement of experts and researchers from the natural sciences in issues around urban development.

- Fry, T. (2014) *City Futures in the Age of a Changing Climate*, Abingdon: Routledge
  Along with the works cited here by Bulkeley and Giradet this is a key resource. It is distinguished by its historical discussions and focus on the likely growing significance of the movement of people in response to climate changes and the plethora of fragmented urban forms that may emerge under these conditions.

- Girardet, H. (2008) *Cities, People, Planet: Urban Development and Climate Change*, 2nd edn, Hoboken, NJ: Wiley
  Girardet has long been a leading author who has made a number of key interventions into discussions about sustainable urban futures. This is the second edition of a classic text.

- Simon, D. (ed.) (2016) *Rethinking Sustainable Cities: Accessible, Green and Fair*, Bristol: Policy Press
  There is now a huge literature consisting of various takes on the concept of sustainable cities, far too many to list here. This is a concise edited collection exploring various dimensions of urban sustainability and how future cities might be planned to achieve desirable outcomes. The aspirations for cities that the book lays out are that they should be sustainable in form, economically and socially fair and respectful of environmental limits. A strength of this book is the extent to which its discussions are rooted in examples and case studies.

- UN-HABITAT (2016) *Urbanization and Development: Emerging Futures. World Cities Report 2016*, Nairobi: United Nations Human Settlements Programme (UN-HABITAT)
  The latest (at time of writing), annual *World Cities Report* from the United Nations. This report looks back over twenty years of world urbanization and forwards to the futures that these processes are likely to produce. It recognizes the key global challenges that face cities and that cities contribute to and outlines an urban agenda to address these. Rich in data and examples and wide ranging in its scope this is an essential source book for the serious student of urban futures.

- Wheeler, S.M. and Beatley, T. (2014) (eds) *The Sustainable Urban Development Reader*, 3rd edn, Abingdon: Routledge
  A collection of over 100 classic writings and contemporary case studies on the many dimensions of sustainable urban development. While the oldest piece (by Ebenezer Howard), dating back to 1898, shows how long questions of sustainability have preoccupied urban thinkers, the plethora of more recent extracts show the extent to which these concerns are becoming ever more central to urban geography and other disciplines concerned with the city. Essential.

Journal articles

- Brown, L.A. (2014) 'The city in 2050: a kaleidoscopic perspective', *Applied Geography*, 49: 4–11
  Part of the *Applied Geography* special collection on urban futures mentioned below. A timely reminder that the city of the future is not built on a blank stage but reflects past and current urbanization trends as much, if not more, than it does new phases and innovations. Very US-focused, it has little to say about the Global South, for example, but a concise reflection on the issues likely to face these cities in decades to come.

- Buhaug, H. and Urdal, H. (2013) 'An urbanization bomb? Population growth and social disorder in cities', *Global Environmental Change*, 23(1): 1–10
  A paper that explores the challenge of massive urban population growth, exploring its relation to social disorder through the analysis of riot data.

The paper finds that it is not population growth that is related to the outbreaks of riots but rather weak political institutions, economic shocks and civil conflict. In doing so it highlights a series of other challenges facing future cities.

- Caprotti, F. (2014) 'Eco-urbanism and the eco-city, or, denying the right to the city?', *Antipode*, 46(5): 1285–1303
  A highly critical intervention that punctures the utopian myth-making surrounding the development of eco-cities as technological solutions to ecological crises. The article highlights the roles of eco-cities in perpetuating inequalities in the transition to low-carbon, green capitalism. It examines the discourses of crisis and solution upon which eco-city development is often advanced and the fates of the low-paid migrant workers upon whose labour these developments are often built but who remain largely hidden from view and little considered in academic accounts.

- Castán Broto, V. and Bulkeley, H. (2013) 'A survey of urban climate change experiments in 100 cities', *Global Environmental Change*, 23(1): 92–102
  The majority of research literature on urban climate change responses had been based around single case studies, the majority of which were located in the cities of the Global North. Recognizing the limitations of this literature, this is an attempt to extend our knowledge of what experiments are occurring across a significantly wider, global terrain. It is a key reference. While its discussion of urban climate change experiments inevitably lacks the depth of individual case study approaches, it offers an unprecedented breadth in its perspective.

- Kourtit, K., Nijkamp, P. and Reid, N. (2014) 'The new urban world: challenges and policy', *Applied Geography*, 49: 1–3
  The introductory paper for a special collection of research articles from the journal *Applied Geography* that address the ways in which cities might respond positively to the future challenges they face. While this introductory paper provides an excellent concise overview, the papers that follow delve into the specifics. It is worth reading all of the papers in this collection which include discussions of urban connectivity, liveability, the inertia inherent in cities, entrepreneurialism, new data, and infrastructure. Brown's paper above is another in this special collection.

- Luque-Ayala, A. and Marvin, S. (2015) 'Developing a critical understanding of smart urbanism?', *Urban Studies*, 52(12): 2105–2116
  Smart urbanism (technologically and digitally driven) has been hailed as a panacea to many of the ills of the contemporary and future city. This short review reveals that we actually know very little about it from a critical perspective. The literature to date has been focused predominantly on technological issues rather than social questions. The paper outlines a research agenda to address this deficit which is interdisciplinary and comparative in its approach.

Websites

- Cabot Institute based at the University of Bristol www.bristol.ac.uk/cabot/research/futurecities.html) – A research group looking at a number of themes cutting across the promotion of sustainable urban development. There are a wealth of other research groups looking at similar themes based in universities around the world. A couple are listed below but you are advised to explore others as many contain valuable resources that may be of use in your studies.

- The C40 Cities network (www.c40.org/) – A network of 40 megacities seeking to explore the contributions of cities to addressing climate change.

- Masdar City (http://masdar.ae/en/masdar-city/live-work-play) – The website for Masdar City, one of the case studies discussed earlier in this chapter.

- The Tyndall Centre for Climate Change Research (http://www.tyndall.ac.uk/) – an interdisciplinary climate change research group based in the UK.

- United Nations Cities and Climate Change Initiative (http://unhabitat.org/urban-initiatives/initiatives-programmes/cities-and-climate-change-initiative/)

- The Urban Climate Change Research Network (http://uccrn.org/) – a research group based at Columbia University in New York. The site contains a number of valuable case studies.

- The Urban Institute (http://urbaninstitute.group.shef.ac.uk/) – a research group based at the University of Sheffield in the UK. Their focus is on the promotion of innovative thinking about contemporary cities and its application to shaping future cities.

Glossary

Arts and Crafts: An international design movement that originated in England and flourished between 1880 and 1910, continuing its influence up to the 1930s. The movement advocated truth to materials and traditional craftsmanship using simple forms and often medieval, romantic or folk styles of decoration. It also proposed economic and social reform and has been seen as essentially anti-industrial.

Breakthrough street: A street constructed to link two or more existing streets. These were particularly common in the early nineteenth century: the era of transport innovations. A breakthrough street may involve the demolition of building fabric and considerable transformation of plot layouts.

Burgage cycle: The progressive filling-in with buildings of the back lands of burgage plots, culminating in the clearing of buildings and a period of urban fallow, or plot vacancy, prior to the initiation of a redevelopment cycle. See *burgage plot* below.

Burgage plot: An urban strip-plot held by a burgess in an English medieval town or city and charged with a fixed annual rent. A unit of land ownership.

Carceral: from the Latin word *carcer* meaning prison, used in the case of the city to imply a city where physical boundaries are used to control urban space. In these 'public' spaces, gatherings of strangers to the area are discouraged, and barricades of various forms can prevent people from entering or passing through.

Central place theory: Central place theory seeks to explain the number, size and location of human settlements in an urban system. It assumes that marketing is the dominant function in the urban system and seeks to explain how services are distributed and why there are distinct patterns in this distribution; organized by hexagons to eliminate unserved or overlapping market areas.

Cognition: The processes by which the brain analyses environmental information it receives through the senses.

Collective consumption: The provision and management of public goods and services such as municipal housing, health and education services that are distributed through non-market mechanisms. Central to the analysis of the city as a bureaucratic entity in the work of Castells.

Dystopic: Pertaining to, or characterized by, dystopia – a state in which the conditions of life are extremely bad, such as from deprivation, oppression or terror. Dystopic imagery has figured prominently in modern depictions of the urban landscape where the city is often portrayed as a terrifying world of darkness, crisis and catastrophe.

Embeddedness: The idea that economic behaviour is not determined by universal factors that are unvariable (as in *neo-classical economics*) but is deeply related to cultural values that are specific in time and space.

Enlightenment: The broad trend in western intellectual thought, beginning in the Renaissance period in the fourteenth and fifteenth centuries, which sought to analyse and control society through the principles of scientific analysis and rational thought.

Fordist: Named after the car manufacturer Henry Ford, it is associated with mass production in large factories and controlled production lines, inflexibly geared towards the production of a standardized commodity. This organizational structure was the dominant form of manufacturing within the old industrial cores of Europe and North America throughout much of the twentieth century.

Fringe belts: Belt-like zones originating from the temporarily stationary or very slowly advancing fringe of a town and composed of a characteristic mixture of land-use units initially seeking peripheral location.

Incorporation: An American phenomenon in the current age. Concerned with the formation of new cities from areas that were previously separate administrative units.

Karl Marx: Marx was a nineteenth-century philosopher, political economist and revolutionary. Often called the father of communism, he was both a scholar and a political activist and is best known for his analysis of capitalism. Marx believed that capitalism, like previous socioeconomic systems, will produce internal tensions which will lead to its destruction and replacement by communism. Marxian geography uses the analytical insights, concepts and theoretical frameworks from Marx's writings to address geographical questions.

Keynesian: Based on the ideas of the economist John Maynard Keynes – a key idea of his was to manage economies by countering the lack of demand in recessions through government spending. These ideas were applied to a set of policies that underpinned the management of cities in the 1950s and 1960s which involved a significant role for both the national and local state. This approach was undermined by global economic crises in the 1970s.

Masculinism: An approach that privileges and represents as normal the activities of men. A masculinist approach in urban geography was seen to have failed to consider the differences of urban experiences due to gender and therefore ignored women's experiences of the city.

Materiality: The material qualities of phenomena. Within urban geography this refers to the built form of the city. A growth in interest in the material qualities of the city has been interpreted as a response to the preoccupation, particularly within cultural geography, with signification and representation.

Mode of production: The mode of production consists of both the forces of production and the relations of production. The forces of production are the productive capacities of the economy, significantly linked to levels of technological development. The relations of production are the social relations between different economic actors, including property relations and work relations.

Mode of regulation: The various types of norm, rule and regulation manifest in various institutions and legislation that seek to mediate the conflicts within a capitalist society.

Morphogenetic: The creation of physical forms viewed as a developmental or evolutionary set of processes that create and reshape the physical fabric of urban form.

Neo-classical economics: Attempts to update the ideas of classical economists of the late eighteenth and early nineteenth centuries and is characterized by a belief in market mechanisms. It looks for universal, unchanging principles of human economic behaviour and tends to ignore the social context of economic activity. Human actions are explained as a consequence of rational economic choices. Contrast with *embeddedness.*

New international division of labour (NIDL): A form of the division of labour in industrial processes and service operations that is associated with the internationalization of production and the spread of industrialization in a number of newly industrializing countries of the Global South as capital seeks to maintain its levels of profit.

New Right: A political movement that reflects a belief in the market as the most efficient mechanism for the allocation of resources. Tends to see the state as a bureaucratic impediment to the supposed efficiency of the market. In cities, it is most associated with policies such as Urban Development Corporations in the UK. Rose to political prominence from the 1970s associated with governments such as Margaret Thatcher's in the UK and Ronald Reagan's in the US. Reflective of the broader neoliberal perspective.

Nomothetic: Concerned with the pursuit of universal laws, patterns and regularities.

Patriarchy: Social arrangements such as in the form of institutional arrangements and prevailing social attitudes that enable men to dominate women.

Positivist: A perspective that posits scientific methods and forms of knowledge. It is concerned with the measurement and understanding of what exists, in contrast to the normative approach which is concerned with what ought to be. A central plank of spatial science in geography.

Post-Fordist: Modes of industrial organization that involve greater flexibility of production for an increasing range of products with shorter runs, delivered 'just in time' to customers rather than having large stock holdings. For companies this has increased sub-contracting of parts of the production process, greater vertical integration within firms and a more 'flexible' use of labour. This has allowed firms to respond more quickly to market volatility, permits high rates of productivity growth and reduces employment costs.

Regime of accumulation: Refers to the dominant form of securing and accumulating profit. From time to time in capitalist societies there emerge stable sets of social, economic and institutional arrangements that serve to link production and consumption.

Semiotics: The study of signs and their meanings. Linked to the idea that culture involves signs, symbols and material activities that convey meaning and which provide a way of interpreting and understanding the world.

Spatiality: Refers to the inherently spatial dimensions or qualities of phenomena.

Structuralism: A theoretical perspective which involves looking beneath the surface appearance of human activity to examine the underlying economic, political and cultural structures that affect human behaviour.

Sub-cultures: Groups who define themselves as in some way different to the dominant or majority culture. There may be many bases upon which subcultures are defined such as lifestyle, ethnicity, sexuality or religion, among others. Patterns of subcultural differentiation are often reflected to some extent in patterns of residential differentiation. Long recognized in many sociological accounts of the city and increasingly so in post-modern accounts that stress difference in the city.

Utopian: An ideal community or society possessing a perfect socio-politicolegal system. Originally a Greek term, it was used by Sir Thomas More for his 1516 book *Utopia*, describing a fictional island in the Atlantic Ocean. The term has been used to describe both intentional communities that attempted to create an ideal society, and fictional societies portrayed in literature.

 **Bibliography**

Abel, C. (2004) *Architecture, Technology and Process*, Oxford: Architectural Press.

Acuto, M. (2010) 'High-rise Dubai urban entrepreneurialism and the technology of symbolic power', *Cities*, 27(4): 272–284.

Airports Company South Africa (2017) *Passenger Statistics, O.R. Tambo International Airport*, http://www.acsamedia.co.za/stats.htm?airport=ORTIA (Accessed 6 March 2017).

Airports Council International (2017) *2016 World Airports Traffic Forecasts Infographic*, www.aci.aero/Data-Centre/Airport-Statistics-Infographics (Accessed 6 March 2017).

Albrechts, L. (2013) 'Reframing strategic spatial planning by using a coproduction perspective', *Planning Theory*, 12(1): 46–63.

Aldrich, R. (2004) 'Homosexuality and the city: an historical overview', *Urban Studies*, 41(9): 1719–1737.

Alonso, W. (1964) *Location and Land Use*, Cambridge, MA: Harvard University Press.

Ambrose, P. (1994) *Urban Processes and Power*, London: Routledge.

Amin, A. (2002a) 'Spatialities of globalisation', *Environment and Planning A*, 34(3): 385–399.

Amin, A. (2002b) *Ethnicity and the Multicultural City: Living with Diversity*, Report for Department of Transport, Local Government and the Regions, Durham: University of Durham.

Amin, A. and Thrift, N. (2002) *Cities: Reimagining the Urban*, Cambridge: Polity Press.

Amin, A. and Thrift, N. (2007) 'Cultural-economy and cities', *Progress in Human Geography*, 31(2): 143–161.

Amin, A., Massey, D. and Thrift, N. (2000) *Cities for the Many Not the Few*, Bristol: The Policy Press.

Anderson, J. (2010) *Understanding Cultural Geography: Places and Traces*, Abingdon: Routledge.

Andersson, I. (2014) 'Placing place branding: an analysis of an emerging research field in human geography', *Geografisk Tidsskrift: Danish Journal of Geography*, 114(2): 143–155.

Appleyard, D. (1981) *Livable Streets*, Berkeley, CA: University of California Press.

Arlidge, J. (1994) 'Blob on the landscape', *The Independent*, 17 November: 24.

Ashworth, G.J. (1989) 'Urban tourism: an imbalance in attention', in Cooper, C.P. (ed) *Progress in Tourism, Recreation, and Hospitality Management*, vol. 1, London: Belhaven Press.

Ashworth, G.J. and Tunbridge, J.E. (2000) *The Tourist-Historic City: Retrospect and Prospect of Managing the Heritage City*, 2nd edn, Oxford: Pergamon Press.

Atkinson, A. (2007a) 'Cities after oil – 1: "sustainable development" and energy futures', *City*, 11(2): 201–213.

Atkinson, A. (2007b) 'Cities after oil – 2: background to the collapse of "modern" civilisation', *City*, 11(3): 293–312.

Atkinson, A. (2008) 'Cities after oil – 3: collapse and the fate of cities', *City*, 12(1): 79–106.

Attfield, J. (1995) 'Inside pram town: a case study of Harlow house interiors', in Attfield, J. and Kirkham, P. (eds) *A View from the Interior: Women and Design*, London: Women's Press.

Audirac, I. (2003) 'Information age landscapes outside the developed world', *Journal of the American Planning Association*, 69(1): 16–32.

Bachrach, P. and Baratz, M.S. (1962) 'Two faces of power', *American Political Science Review*, 54(4): 947–952.

Bailey, J.T. (1989) *Marketing Cities in the 1980s and Beyond*, Chicago, IL: American Economic Development Council.

Bairoch, P. (1988) *Cities and Economic Development: From the Dawn of History to the Present*, London: Mansell.

Ballard, J.G. (1996) *Cocaine Nights*, London: Flamingo.

Ballard, J.G. (2000) *Super-Cannes*, London: Flamingo.

Bandarin, F. and van Oers, R. (2012) *The Historic Urban Landscape: Managing Heritage in an Urban Century*, Chichester: Wiley Blackwell.

Banfield, E.C. (1961) *Political Influence*, Glencoe: Free Press.

Barau, A.S., Maconachie, R., Ludin, A.N.M. and Abdulhamid, A. (2015) 'Urban morphology dynamics and environmental change in Kano, Nigeria', *Land Use Policy* 42, 307–317.

Barke, M. and Harrop, K. (1994) 'Selling the industrial town: image, identity and illusion', in Gold, J.R. and Ward, S.V. (eds) *Place Promotion: The Use of Publicity and Marketing to Sell Towns and Regions*, Chichester: John Wiley & Sons.

Barnes, T. (2003) 'The '90s show: culture leaves the farm and hits the streets', *Urban Geography*, 24(6): 479–492.

Barnett, S. (1991) 'Selling us short: cities, culture and economic development', in Fisher, M. and Owen, U. (eds) *Whose Cities?*, Harmondsworth: Penguin.

Bassens, D. and Van Meeteren, M. (2015) 'World cities under conditions of financialized globalization: Towards an augmented world city hypothesis', *Progress in Human Geography*, 39: 752–775.

Bassett, K. and Harloe, M. (1990) 'Swindon: the rise and decline of a growth coalition', in Harloe, M., Pickvance, C. and Urry, J. (eds) *Place, Policy and Politics: Do Localities Matter?*, London: Unwin Hyman.

Bates, C., Imrie, R. and Kulman, K. (eds) (2017) *Care and Design: Bodies, Buildings, Cities*, Chichester: Wiley-Blackwell.

Baum, H. (2015) 'Planning with half a mind: why planners resist emotion', *Planning Theory and Practice*, 16(4): 498–516.

Baum, S., Van Gellecum, Y. and Yigitcanlar, T. (2004) 'Wired communities in the city: Sydney, Australia', *Australian Geographical Studies*, 42: 175–192.

Beal, V. and Pinson, G. (2014) 'When mayors go global: international strategies, urban governance and leadership', *International Journal of Urban and Regional Research*, 38(1): 302–317.

Beaverstock, J.V., Hubbard, P.J. and Short, J.R. (2004) 'Getting away with it: exposing the geographies of the super-rich', *Geoforum*, 35(4): 401–407.

Bell, D. (2007) 'The hospitable city: social relations in commercial spaces', *Progress in Human Geography*, 31(1): 7–22.

Bell, D. and Jayne, M. (eds) (2004) *City of Quarters: Urban Villages in the Contemporary City*, Aldershot: Ashgate.

Bender, B. (ed) (1993) *Landscape: Politics and Perspectives*, Oxford: Berg.

Benton-Short, L. and Short, J.R. (2008) *Cities and Nature*, Abingdon: Routledge.

Beriatos, E. and Gospodini, A. (2004) '"Glocalising" urban landscapes: Athens and the 2004 Olympics', *Cities*, 21(3): 187–202.

Berman, M. (1988) *All That is Solid Melts into Air: The Experience of Modernity*, New York: Penguin.

Berry, B.J.L. (1964) 'Cities as systems of cities within systems of cities', *Papers of the Regional Science Association*, 13: 147–163.

Berry, B.J.L. (1967) *Geography of Market Centers and Retail Distribution*, Englewood Cliffs, NJ: Prentice-Hall.

Bianca, S. (2000) *Urban Form in the Arab World*, London: Thames & Hudson.

Bianchini, F., Dawson, J. and Evans, R. (1992) 'Flagship projects in urban regeneration', in Healey, P., Davoudi, S., O'Toole, M., Tavsanoglu, S. and Usher, D. (eds) *Rebuilding the City: Property-led Urban Regeneration*, London: E & F.N. Spon.

Binns, T. and Lynch, K. (1999) 'Feeding Africa's growing cities into the 21st century: the potential of urban agriculture', *Journal of International Development*, 10(6): 777–793.

Blomley, N.K. (1996) '"I'd like to dress her all over": masculinity, power and retail capital', in Wrigley, N. and Lowe, M.S. (eds) *Retailing Consumption and Capital: Towards the New Retail Geography*, Harlow: Addison-Wesley Longman.

Blowers, A. (1997) 'Society and sustainability', in Blowers, A. and Evans, B. (eds) *Town Planning into the 21st Century*, London: Routledge.

Blowers, A. (ed) (1993) *Planning for a Sustainable Environment*, London: Earthscan.

Blowers, A. and Pain, R. (1999) 'The unsustainable city', in Pile, S., Brook, C. and Mooney, G. (eds) *Unruly Cities? Order/Disorder*, London: Routledge/Open University.

Boland, P., Murtagh, B. and Shirlow, P. (2016) 'Fashioning a city of culture: "life and place changing", or "12-month party"', *International Journal of Cultural Policy*, http://dx.doi.org/10.1080/10286632.2016.1231181

Bondi, L. (1999) 'Gender, class and gentrification: enriching the debate', *Environment and Planning D: Society and Space*, 17(3): 261–282.

Borden, I. (2001) *Skateboarding, Space and the City: Architecture and the Body*, Oxford: Berg.

Borden, I. (2004) 'A performative critique of the city: the urban practice of skateboarding, 1958–98', in Miles, M., Hall, T. and Borden, I. (eds) *The City Cultures Reader*, 2nd edn, London: Routledge.

Borer, M.I. (2006) 'The location of culture: the urban culturalist perspective', *City and Community*, 5(2): 173–197.

Bosco, F.J. (2006) 'Actor-network theory, networks and relational approaches in human geography', in Aitken, S. and Valentine, G. (eds) *Approaches to Human Geography*, London: Sage.

Bourne, L.S. (1981) *The Geography of Housing*, London: Edward Arnold.

Box, J. and Baker, G. (1998) 'Delivering sustainability through local nature reserves', *Town and Country Planning* (December): 360–363.

Boyle, M. and Hughes, G. (1991) 'The politics of representation of the "real": discourses from the left on Glasgow's role as European City of Culture', *Area*, 23(3): 217–228.

Boyle, M. and Hughes, G. (1995) 'The politics of urban entrepreneurialism in Glasgow', *Geoforum*, 25(4): 453–470.

Bradley, A. and Hall, T. (2006) 'The festival phenomenon: festivals, events and the promotion of small urban areas', in Bell, D. and Jayne, M. (eds) *Small Cities: Urban Experience Beyond the Metropolis*, Abingdon: Routledge.

Brand, P. with Thomas, M.J. (2005) *Urban Environmentalism: Global Change and the Mediation of Local Conflict*, Abingdon: Routledge.

Brand, S. (1995) *How Buildings Learn: What Happens After They're Built*, New York: Penguin Books.

Breheny, M. (1995) 'The compact city and transport energy consumption', *Transactions of the Institute of British Geographers* (NS), 20(1): 81–101.

Breheny, M. and Hall, P. (1996) 'Four million households – where will they go?', *Town and Country Planning* (February): 39–41.

Brenner, N. and Keil, R. (eds) (2006) *The Global Cities Reader*, London: Routledge.

Bromley, R. and Thomas, C. (1993) 'The retail revolution, the carless shopper and disadvantage', *Transactions of the Institute of British Geographers*, 18(2): 222–236.

Brooks Pfeiffer, B. and Nordland, G. (eds) (1987) *Frank Lloyd Wright: In the Realm of Ideas*, Carbondale: Southern Illinois University Press.

Brosseau, M. (1995) 'The city in textual form: *Manhattan Transfer*'s New York', *Ecumene*, 2(1): 89–114.

Brotchie, J. (1992) 'The changing structure of cities', *Urban Futures* (February): 13–23.

Brown, L.A. (2014) 'The city in 2050: a kaleidoscopic perspective', *Applied Geography*, 49(1): 4–11.

Brown, D. and McGranahan, G. (2016) 'The urban informal economy, local inclusion and achieving a global green transformation', *Habitat International*, 53: 97–105.

Brunn, S.D. (2006) 'Gated lives and gated minds as worlds of exclusion and fear', *GeoJournal*, 66(1–2): 5–13.

Bryson, J., Taylor, M. and Daniels, P.W. (2008) 'Commercializing "creative" expertise: business and professional services and regional economic development in the West Midlands, United Kingdom', *Politics and Policy*, 36(2): 306–328.

Bryson, J.R. (2007) 'A "second" global shift? The offshoring or global sourcing of corporate services and the rise of distanciated emotional labour', *Geografiska Annaler*, 89B (S1): 31–43.

Buckley, J. (2010) *The Construction of Collective and Individual Identities within the Space of the School Environment: A Study of Two Rural Secondary Schools*, Unpublished dissertation, Department of Natural and Social Sciences, University of Gloucestershire.

Buehler, R. and Pucher, J (2011) 'Sustainable transport in Freiburg: lessons from Germany's environmental capital', *International Journal of Sustainable Transportation* 5(1):43–70.

Buhang, H. and Urdal, H. (2013) 'An urbanization bomb? Population growth and social disorder in cities', *Global Environmental Change*, 23(1): 1–10.

Bulkeley, H. (2013) *Cities and Climate Change*, Abingdon: Routledge.

Bulkeley, H., and Betsill, M. (2013) 'Revisiting the urban politics of climate change', *Environmental Politics*, 22(1): 136–154.

Burgess, E. (1925) 'The growth of the city', in Park, R. and Burgess, E. (eds) *The City*, Chicago: University of Chicago Press.

Burgess, J. (1985) 'News from nowhere: the press, the riots and the myth of the inner city', in Burgess, J. and Gold, J.R. (eds) *Geography, the Media and Popular Culture*, London: Croom Helm.

Burgess, J. (1998) 'Not worth taking the risk? Negotiating access to urban woodland', in Ainley, R. (ed) *New Frontiers of Space, Bodies and Gender*, London: Routledge.

Burtenshaw, D., Bateman, M. and Ashworth, G.J. (1991) *The City in Western Europe*, 2nd edn, Chichester: John Wiley & Sons.

Butler, R. and Bowlby, S. (1997) 'Bodies and spaces: an exploration of disabled people's experiences of public space', *Environment and Planning D: Society and Space*, 15(4): 411–433.

Butler, R. and Parr, H. (eds) (1999) *Mind and Body Spaces: Geographies of Illness, Impairment and Disability*, London: Routledge.

Butler, T. and Lees, L. (2006) 'Supergentrification in Barnsbury, London: globalization and gentrifying global cities at the neighbourhood level', *Transactions of the Institute of British Geographers* 31(4): 467–487.

Byrne, D. (2001) *Understanding the Urban*, London: Palgrave.

Calhoun, C. (2002) 'The class consciousness of frequent travelers: towards a critique of actually existing cosmopolitanism', in Vertovec, S. and Cohen, R. (eds) *Conceiving Cosmopolitanism: Theory, Context and Practice*, Oxford: Oxford University Press.

Caprotti, F. (2014) 'Eco-urbanism and the eco-city, or, denying the right to the city?', *Antipode*, 46(5): 1285–1303.

Carley, M. (1991) 'Business in urban regeneration partnerships: a case study of Birmingham', *Local Economy*, 6(2): 100–115.

Carley, M. (2000) 'Urban partnerships, governance and the regeneration of Britain's cities', *International Planning Studies*, 5(3): 273–297.

Carter, H. (1995) *The Study of Urban Geography*, 4th edn, London: Arnold.

Castán Broto, V. and Bulkeley, H. (2013) 'A survey of urban climate change experiments in 100 cities', *Global Environmental Change*, 23(1): 92–102.

Castells, M. (1977) *The Urban Question*, London: Arnold.

Castells, M. (1978) *City, Class and Power*, London: Macmillan.

Castells, M. (1983) *The City and the Grassroots: A Cross-cultural Theory of Urban Social Movements*, London: Arnold.

Castells, M. (1996) *The Information Age: Economy, Society and Culture, Volume 1: The Rise of the Network Society*, Oxford: Blackwell.

Castells, M. and Hall, P. (1994) *Technopoles of the World: The Making of 21st Century Industrial Complexes*, London: Routledge.

Castree, N. and Gregory, K. (eds) (2006) *David Harvey: A Critical Reader*, Oxford: Blackwell.

Cervero, R. (1995) 'Changing live-work spatial relationships: implications for metropolitan structure and mobility', in Brotchie, J., Batty, M., Blakely, E., Hall, P. and Newton, P. (eds) *Cities in Competition: Productive and Competitive Cities for the Twenty-first Century*, Melbourne: Longman.

Chance, T. (2009) 'Towards sustainable residential communities: the Beddington Zero Energy Development (BedZED) and beyond', *Environment and Urbanization*, 21 (2): 527–544.

Chang, I.-C.C. and Sheppard, E. (2013) 'China's eco-cities as variegated urban sustainability: Dongtan eco-city and Chongming eco-island', *Journal of Urban Technology*, 20(1): 57–75.

Chant, C. (2008) 'Science and technology', in Hall, T., Hubbard, P. and Short, J.R. (eds) *The Sage Companion to the City*, London: Sage, 47–66.

Charles, M.B., Barnes, P., Ryan, N.F. (2007) 'Airport futures: Towards a critique of the aerotropolis model', *Futures* 39(4): 1009–1028.

Chen, X., Orum, A.M., Paulsen, K.E. (2013) *An Introduction to Cities: How Place and Space Shape Human Experience*, Chichester: Wiley- Blackwell.

Ching, F.D.K., Jarzombek, M.M. and Prakash, V. (2006) *Global History of Architecture*, Chichester, John Wiley & Sons.

Chiodelli, F., Hall, T. and Hudson, R. (eds) (2017) *The Illicit and Illegal in the Development and Governance of Cities and Regions: Corrupt Places*, Abingdon: Routledge.

Choi, N. (2016) 'Metro Manila through the gentrification lens: disparities in urban planning and displacement risks', *Urban Studies*, 53(3): 577–592.

Chouinard, V., Hall, E. and Wilton, R. (eds) (2010) *Towards Enabling Geographies: 'Disabled' Bodies and Minds in Society and Space*, Farnham: Ashgate.

Chu, C.L. (2015) 'Spectacular Macau: visioning futures for a World Heritage City', *Geoforum*, 65: 440–450.

Cidell, J. and Prytherch, D. (eds) (2015) *Transport, Mobility, and the Production of Urban Space*, Abingdon: Routledge.

Clark, D. (2003) *Urban World /Global City*, London: Routledge.

Clark, D. (2012) 'What is climate change adaptation?', *The Guardian*, 27 February: www.theguardian.com/environment/2012/feb/27/climate-change-adaptation (Accessed 25 November 2016).

Clark, W.A.V., Deurloo, M.C. and Deileman, F.M. (2006) 'Residential mobility and neighbourhood outcomes', *Housing Studies*, 21(3): 323–342.

Clement, M. (2016) *A People's History of Riot, Protest and the Law: The Sound of the Crowd*, Basingstoke: Palgrave Macmillan.

Cloke, P., Philo, C. and Sadler, D. (1991) *Approaching Human Geography*, London: Paul Chapman.

Coaffee, J. (2016) *Terrorism, Risk and the Global City: Towards Urban Resilience*, Abingdon: Routledge.

Coaffee, J. and Murakami Wood, D. (2008) 'Terror and surveillance', in Hall, T., Hubbard, P. and Short, J.R. (eds) *The Sage Companion to the City*, London: Sage, 352–372.

Cochrane, A. (1999) 'Administered cities', in Pile, S., Brook, C. and Mooney, G. (eds) *Unruly Cities? Order/Disorder*, London: Routledge/Open University.

Cochrane, A. (2007) *Understanding Urban Policy. A Critical Approach*, Oxford: Blackwell.

Coe, N.M., Kelly, P.F. and Yeung, H.W.C. (2007) *Economic Geography: A Contemporary Introduction*, Oxford: Blackwell.

Cohen, N. (1993) 'The renaissance that never was', *The Independent on Sunday*, 10 October: 11.

Colls, R. and Dodd, P. (1986) *Englishness: Politics and Culture 1880–1920*, London: Croom Helm.

Commission of the European Communities (1990) *Green Paper on the Urban Environment*, Brussels: European Commission.

Conzen, M.P. (2001) 'The study of urban form in the United States', *Urban Morphology*, 5(1): 3–14.

Conzen, M.P. (2009) 'How cities internalize their former fringe belts: a cross cultural comparison', *Urban Morphology*, 13(1): 29–54.

Conzen, M.P., Whitehand, J.W.R. and Gu, K. (2012) 'Comparing traditional urban form in China and Europe: a fringe-belt approach' *Urban Geography*, 33(1): 22–45.

Conzen, M.R.G. (1960) *Alnwick, Northumberland: A Study in Town Plan Analysis*, Institute of British Geographers Publication 27, London: George Philip.

Conzen, M.R.G. (1981[1966]) 'Historical townscapes in Britain: a problem in applied geography', in Whitehand, J.W.R. (ed) *The Urban Landscape: Historical Development and Management: Papers by M.R.G. Conzen*, Institute of British Geographers Special Publication 13, London: Academic Press, 55–74.

Conzen, M.R.G. (1981[1978]) 'The morphology of towns in Britain during the industrial era', in Whitehand, J.W.R. (ed) *The Urban Landscape: Historical Development and Management: Papers by M.R.G. Conzen*, Institute of British Geographers Special Publication 13, London: Academic Press, 87–126.

Cosgrove, D. (1989) 'Geography is everywhere: culture and symbolism in human landscapes', in Gregory, D. and Walford, R. (eds) *Horizons in Human Geography*, Basingstoke: Macmillan.

Cosgrove, D. and Daniels, S. (eds) (1988) *The Iconography of Landscape: Essays on Symbolic Representation, Design and Use of Past Environments*, Cambridge: Cambridge University Press.

Couch, C. (2016) *Urban Planning: An Introduction*, London: Palgrave Macmillan.

Cowen, H., McNab, A., Harrison, S. and Howes, L. (1989) 'Affluence amidst recession: Cheltenham', in Cooke, P. (ed) *Localities: The Changing Face of Urban Britain*, London: Unwin Hyman.

Cox, K. and Mair, A. (1987) 'Levels of abstraction in locality studies', *Antipode*, 21(2): 121–132.

Cox, K. and Mair, A. (1988) 'Locality and community in the politics of local economic development', *Annals of the Association of American Geographers*, 78(2): 307–325.

Craggs, R. and Neate, H. (2017) 'Post-colonial careering and urban policy mobility: between Britain and Nigeria, 1945-1990', *Transactions of the Institute of British Geographers* 42(1): 44–57.

Crang, M. (1999) *Cultural Geography*, London: Routledge.

Crang, M. (2000) 'Public space, urban space and electronic space: would the real city please stand up?', *Urban Studies*, 37(2): 301–317.

Cresswell, T. (1996) *In Place/Out of Place: Geography, Ideology and Transgression*, Minneapolis, MN: University of Minnesota Press.

Cresswell, T. (2006) *On the Move: Mobility in the Modern Western World*, Abingdon, Routledge.

Cresswell, T. (2012) 'Mobilities II: still', *Progress in Human Geography*, 36(5): 645–653.

Crewe, L., Gregson, N. and Brooks, K. (2003) 'Alternative retail spaces', in Leyshon, A. and Lee, R. (eds) *Alternative Economic Geographies*, London: Sage.

Crouch, D. and Ward, C. (1997) *The Allotment: Its Landscape and Culture*, Nottingham: Five Leaves Publications.

Crowder, K. and Krysan, M. (2016) 'Moving beyond the big three: a call for new approaches to studying racial residential segregation', *City & Community*, 15(1):18–22.

Cruickshank, D. (ed) (1996) *Sir Bannister-Fletcher's A History of Architecture*, 20th edn, Oxford: Architectural Press.

Cugurullo, F. (2013) 'How to build a sandcastle: an analysis of the genesis and development of Masdar City', *Journal of Urban Technology*, 20(1): 23–37.

Cullingworth, J.B. and Nadin, V. (2006) *Town and Country Planning in the UK*, London: Routledge.

Dahl, R. (1961) *Who Governs?* New Haven, CT: Yale University Press.

Daly, G. (1998) 'Homelessness and the street: observations from Britain and the US', in Fyfe, N. (ed) *Images of the Street: Planning, Identity and Control*, London: Routledge.

Daly, G. (2008) 'Housing and homelessness', in Hall, T., Hubbard, P. and Short, J.R. (eds) *The Sage Companion to the City*, London: Sage, 267–281.

Daniels, P.W. (2004) 'Urban challenges: the formal and informal economies in mega-cities', *Cities*, 21(6): 501–511.

Datta, A. (2006) 'From tenements to flats: gender, class and "modernization" in Bethnal Green Estate', *Social and Cultural Geography*, 7(5): 789–805.

Davidson, J. (2000) '"…the world is getting smaller": women, agoraphobia and bodily boundaries', *Area*, 32(1): 31–40.

Davidson, J., Bondi, L. and Smith, M. (eds) (2016) *Emotional Geographies*, Abingdon: Routledge.

Davidson, M. and Martin, D. (eds) (2014) *Urban Politics: Critical Approaches*, London: Sage.

Davies, J.S. and Imbroscio, D. (eds) (2009) *Theories of Urban Politics*, London: Sage.

Davis, M. (1990) *City of Quartz: Excavating the Future in Los Angeles*, London: Verso.

Davis, M. (2000a) 'Fortress LA', in Miles, M., Hall, T. and Borden, I. (eds) *The City Cultures Reader*, London: Routledge.

Davis, M. (2000b) *The Ecology of Fear: Los Angeles and the Imagination of Disaster*, New York: Vintage.

Davis, M. (2006) *Planet of Slums*, London: Verso.

Davis, M. and Bertrand Monk, D. (eds) (2007) *Evil Paradises: Dreamworlds of Neoliberalism*, New York: The New Press.

Dear, M. (2000) *The Postmodern Urban Condition*, Oxford: Blackwell.

Dear, M. and Flusty, S. (1998) 'Postmodern urbanism', *Annals of the Association of American Geographers*, 88: 50–72.

Dear, M. and Flusty, S. (2005) 'Postmodern urbanism', in Fyfe, N. and Kenny, J. (eds) *The Urban Geography Reader*, London: Routledge.

Dear, M. and Wolch, J. (1987) *Landscapes of Despair*, Princeton, NJ: Princeton University Press.

Deffner, A. (2005) 'Culture', in Caves, R.W. (ed) *Encyclopedia of the City*, London: Routledge.

Degen, M. (2003) 'Fighting for the global catwalk: formalizing public life in Castlefield (Manchester) and el Raval (Barcelona)', *International Journal of Urban and Regional Research*, 27(4): 867–880.

Degen, M. (2004) 'Barcelona's games: the Olympics, urban design and global tourism', in Sheller, M. and Urry, J. (eds) *Tourism Mobilities: Places to Play, Places in Play*, London: Routledge.

Dennis, K. and Urry, J. (2009) *After the Car*, Cambridge: Polity Press.

Department of the Environment (1993) *Strategy for Sustainable Development*, London: HMSO.

Department of the Environment, Transport and the Regions (1998) *New Deal for Transport: Better for Everyone*, London: HMSO.

Derudder, B., Taylor, P., Ni, P., De Vos, A., Hoyler, M., Hanssens, H., Bassens, D. and Huang, J. (2010) 'Pathways of change: shifting connectivities in the world city network, 2000–08', *Urban Studies*, 47(9): 1861–1877.

Dessens, O., Köhler, M.O., Rogers, H.L., Jones, R.L. and Pyle, J.A. (2014) 'Aviation and climate change', *Transport Policy*, 34: 14–20.

DeVerteuil, G., Lee, W. and Wolch, J. (2002) 'New spaces for the local welfare state? The case of general relief in Los Angeles County', *Social and Cultural Geography*, 3(3): 229–246.

DeVerteuil, G., May, J. and Von Mahs, J. (2009) 'Complexity not collapse: recasting the geographies of homelessness in a "punitive" age', *Progress in Human Geography*, 33(5): 646–666.

Dicken, P. (1986) *Global Shift: Industrial Change in a Turbulent World*, London: Harper & Row.

Dicken, P. (2007) *Global Shift: Mapping the Changing Contours of the World Economy*, 5th edn, London: Sage.

Dicken, P. (2011) *Global Shift: Mapping the Changing Contours of the World Economy*, 6th edn, London: Sage.

Dicken, P. (2015) *Global Shift: Mapping the Changing Contours of the World Economy*, 7th edn, New York: Guilford Press.

Dicken, P., Kelly, P.F., Olds, K. and Yeung, H.W-C. (2001) 'Chains and networks, territories and scales: towards a relational framework for analysing the global economy', *Global Networks*, 1(2): 89–112.

DiGaento, A. and Klemanski, J.S. (1993) 'Urban regimes in comparative perspective – the politics of urban development in Britain', *Urban Affairs Quarterly*, 29(1): 54–83.

Dikec, M. (2006) 'Two decades of French urban policy: from social development of neighbourhoods to the republican penal state', *Antipode*, 38(1): 59–81.

Dodd, D. (2003) 'Barcelona: the making of a cultural city', in Miles, M., Hall, T. and Borden, I. (eds) *The City Cultures Reader*, 2nd edn, London: Routledge.

Domosh, M. (1989) 'A method for interpreting landscape: a case study', *Area*, 21(4): 347–355.

Domosh, M. (1996) *Invented Cities: The Creation of Landscape in Nineteenth-Century New York and Boston*, New Haven, CT: Yale University Press.

Dos Passos, J. (1925) *Manhattan Transfer*, New York: Harper.

Doughty, K. and Murray, L. (2016) 'Discourses of mobility: institutions, everyday lives and embodiment', *Mobilities*, 11(2): 303–322.

Dowell, D.E. and Monkkonen, P. (2007) 'Consequences of the *Plano Piloto*: The urban development and land markets of Brasilia', *Urban Studies*, 44(10): 1871–1887.

Downs, A. (1992) *Stuck in Traffic: Coping with Peak-Hour Traffic Congestion*, Washington, DC: The Brookings Institution.

Drakakis-Smith, D. (2000) *The Third World City*, 2nd edn, London: Routledge.

Duncan, J. (1999) 'Elite landscapes as cultural (re)productions: the case of Shaughnessy Heights', in Anderson, K. and Gale, F. (eds) *Cultural Geographies*, Melbourne: Addison-Wesley Longman.

Duncan, S. and Goodwin, M. (1988) *The Local State and Uneven Development*, Cambridge: Polity Press.

Dunster, B., Simmons, C. and Gilbert, B. (2007) *The ZEDbook: Solutions for a Shrinking World*, Abingdon: Taylor and Francis.

Durning, A.T. (1996) *The Car and the City: 24 Steps to Safe Streets and Healthy Communities*, Seattle, WA: Northwest Environmental Watch.

Dwyer, C. (2005) 'Diasporas', in Cloke, P., Crang, P. and Goodwin, M. (eds) *Introducing Human Geographies*, 2nd edn, London: Hodder Arnold.

Eade, J. and Mele, C. (eds) (2002) *Understanding the City: Contemporary and Future Perspectives*, Oxford: Blackwell.

ECOTEC (1993) *Reducing Transport Emissions Through Planning*, London: HMSO.

Edensor, T. (2005) 'The ghosts of industrial ruins: ordering and disordering memory in excessive space', *Environment and Planning D: Society and Space*, 23(6): 829–849.

Edensor, T. (2008) 'Mundane hauntings: commuting through the phantasmagoric working-class spaces of Manchester, England', *Cultural Geographies*, 15(3): 313–333.

Edensor, T. and Jayne, M. (eds) (2012) *Urban Theory Beyond the West: A World of Cities*, Abingdon: Routledge.

Edwards, C. and Imrie, R. (2015) *The Short Guide to Urban Policy*, Bristol: Policy Press.

Elias, P. and Omojola, A. (2015) 'Case study: the challenges of climate change for Lagos', *Current Opinion in Environmental Sustainability*, 13: 74–78.

Elkin, T., McLaren, D. and Hillman, M. (1991) *Reviving the City: Towards Sustainable Urban Development*, London: Friends of the Earth.

Elsheshtawy, Y. (2012) 'Urban (im)mobility. Public encounters in Dubai', in Edensor, T. and Jayne, M. (eds) *Urban Theory Beyond the West: A World of Cities,* Abingdon: Routledge.

Engels, F. (1887[1844]) *The Condition of the Working Class in England*, London: Swan Sonnenschein.

English Heritage (2010) *Heritage Counts 2010*, London: English Heritage.

European Conference of Ministers of Transport (2006) *Sustainable Urban Travel: Implementing Sustainable Urban Travel Policies: Applying the 2001 Key Messages*, Paris: ECMT.

Evans, G. (2003) 'Hard branding the cultural city: from Prado to Prada', *International Journal of Urban and Regional Research*, 27(2): 217–240.

Evans, J., Karvonen, A. and Raven, R. (2016) *The Experimental City*, Abingdon: Routledge.

Fainstein, S.S. and DeFilippis, J. (eds) (2016) *Readings in Planning Theory*, 4th edn, Chichester: John Wiley & Sons.

Fainstein, S.S. and Servon, L.J. (eds) (2005) *Gender and Planning: A Reader*, New Brunswick, NJ: Rutgers University Press.

Ferenčuhová, S. (2016) 'Accounts from behind the Curtain: history and geography in the critical analysis of urban theory', *International Journal of Urban and Regional Research*, 40(1): 113–131.

Fischer, B. and Pollard, B. (1998) 'Exclusion, risk and social control: reflections on community policing and public health', *Geoforum*, 29(2): 187–197.

Florida, R. (2002) *The Rise of the Creative Class and How It's Transforming Work, Leisure and Everyday Life*, New York: Basic Books.

Florida, R. (2004) *Cities and the Creative Class*, London: Routledge.

Freeman, D.B. (1991) *A City of Farmers: Informal Agriculture in the Open Spaces of Nairobi, Kenya*, Montreal: McGill University Press.

Freire-Medeiros, B. and Name, L. (2017) 'Does the future of the favela fit in an aerial cable car? Examining tourism mobilities and urban inequalities through a decolonial lens', *Canadian Journal of Latin American and Caribbean Studies*, 42(1): 1–16.

Fretter, A.D. (1993) 'Place marketing: a local authority perspective', in Kearns, G. and Philo, C. (eds) *Selling Places: The City as Cultural Capital, Past and Present*, Oxford: Pergamon Press.

Friedmann, J. (1986) 'The world city hypothesis', *Development and Change*, 17(1): 69–83.

Friend, R.M., Anwar, N.H., Dixit, A., Hutanuwatr, K., Jayaraman, T., McGregor, J.A., Menon, M.R., Moench, M., Pelling, M. and Roberts, D. (2016) 'Re-imagining inclusive urban futures for transformation', *Current Opinion in Environmental Sustainability*, 20: 67–72.

Fry, T. (2014) *City Futures in the Age of a Changing Climate*, Abingdon: Routledge.

Fuller, G. and Harley, R. (2004) *Aviopolis: A Book About Airports*, London: Black Dog.

Fuller, M., Healey, M., Bradley, A. and Hall, T. (2004) 'Barriers to learning: a systematic study of the experience of disabled students in one university', *Studies in Higher Education*, 29(3): 303–318.

Fyfe, N.R. and Kenny, J.T. (eds) (2005) *The Urban Geography Reader*, London: Routledge.

Gallent, N. (2006) 'The rural-urban fringe: a new priority for planning policy?', *Planning Practice and Research*, 21(3): 383–393.

Gallent, N. and Anderson, J. (2007) 'Representing England's rural-urban fringe', *Landscape Research*, 32(1): 1–21.

Gallent, N., Bianconi, M. and Anderson, J. (2006) 'Planning on the edge: England's rural-urban fringe and the spatial-planning agenda', *Environment and Planning B: Planning and Design*, 33 (3): 457–476.

García Lamarca, M. and Kaika, M. (2016) 'Mortgaged lives': the biopolitics of debt and housing financialisation', *Transactions of the Institute of British Geographers*, 41(3): 313–327.

Garreau, J. (1991) *Edge City: Life on the New Urban Frontier*, New York, NY: Doubleday.

Geddes, P. (1915) *Cities in Evolution*, London: Williams and Norgate.

Gelder, K. (ed) (2005) *The Subcultures Reader*, London: Routledge.

Gelder, K. (2007) *Subcultures: Cultural Histories and Social Practice*, London: Routledge.

Ghertner, D.A. (2015) 'Why gentrification theory fails in "much of the world"', *City*, 19(4): 552–563.

Gibson, L. and Pendlebury, J. (2009) (eds) *Valuing Historic Environments*, Farnham: Ashgate.

Gilbert, A.G. (2000) 'Housing in Third World cities: the critical issues', *Geography*, 85(2): 145–155.

Gildenbloom, J. (2008) *Invisible City*, Austin: University of Texas Press.

Girard, L.F., Forte, B., Cerreta, M., De Toro, P. and Forte, F. (eds) (2003) *The Human Sustainable City: Challenges and Perspectives from the Habitat Agenda*, Aldershot: Ashgate.

Girardet, H. (1996) *The Gaia Atlas of Cities: New Directions for Sustainable Urban Living*, London: Gaia Books.

Girardet, H. (2008) *Cities, People, Planet: Urban Development and Climate Change*, 2nd edn, Hoboken, NJ: Wiley.

Gleeson, B. (1996) 'A geography for disabled people', *Transactions of the Institute of British Geographers*, 21(2): 387–396.

Glenny, M. (2008) *McMafia: Crime Without Frontiers*, London: Bodley Head.

Gold, J.R. (2001) 'Under darkened skies: the city in science-fiction film', *Geography*, 86(4): 337–345.

Gold, J.R. and Gold, M.M. (1994) '"Home at last!": building societies, home ownership and the imagery of English suburban promotion in the interwar years', in Gold, J.R. and Ward, S.V. (eds) *Place Promotion: The Use of Publicity and Marketing to Sell Towns and Regions*, Chichester: John Wiley & Sons.

Gold, J.R. and Revill, G. (2004) *Representing the Environment*, London: Routledge.

Goldfrank, B. and Schrank, A. (2009) 'Municipal neoliberalism and municipal socialism: urban political economy in Latin America', *International Journal of Urban and Regional Research*, 33(2): 443–462.

Golledge, R. and Stimson, R.J. (1997) *Spatial Behaviour: A Geographic Perspective*, New York, NY: Guilford Press.

Goodey, B. and Gold, J.R. (1985) 'Behavioural and perceptual geography: from retrospect to prospect', *Progress in Human Geography*, 9(5): 585–595.

Goodwin, M. (1992) 'The changing local state', in Cloke, P. (ed) *Policy and Change in Thatcher's Britain*, Oxford: Pergamon Press.

Goodwin, M. (1993) 'The city as commodity: the contested spaces of urban development', in Kearns, G. and Philo, C. (eds) *Selling Places*, Oxford: Pergamon Press.

Gordon, P., Richardson, H.W. and Jun, M.J. (1991) 'The commuting paradox: evidence from the top twenty', *Journal of the American Planning Association*, 57(4): 416–420.

Goss, J. (1988) 'The built environment and social theory: towards an architectural geography', *Professional Geographer*, 40(4): 392–403.

Goss, J. (1993) 'The "Magic of the Mall": An analysis of form, function, and meaning in the contemporary retail built environment', *Annals of the Association of American Geographers*, 83(1): 18–47.

Goss, J. (1999) 'Once-upon-a-time in the commodity world: an unofficial guide to the mall of America', *Annals of the Association of American Geographers*, 89: 45–75.

Gottdiener, M. and Budd, L. (2005) *Key Concepts in Urban Studies*, London: Sage.

Gottdiener, M. and Lagopoulos, A.Ph. (eds) (1986) *The City and the Sign: An Introduction to Urban Semiotics*, New York, NY: Columbia University Press.

Graham, S. (2016) *Vertical: The City from Satellites to Bunkers*, London: Verso.

Graham, B., Ashworth, G.J. and Tunbridge, J.E. (2000) *The Geography of Heritage: Power, Culture and Economy*, London: Arnold.

Graham, S. and Marvin, S. (1996) *Telecommunications and the City: Electronic Spaces, Urban Places*, London: Routledge.

Graham, S. and Marvin, S. (2001) *Splintering Urbanism*, London, Routledge.

Grant, R. and Nijman, J. (2002) 'Globalization and the corporate geography of cities in the less developed world', *Annals of the Association of American Geographers*, 92(2): 320–340.

Greed, C. (1994) *Women and Planning: Creating Gendered Realities*, London: Routledge.

Greed, C. (2000) 'Women in the construction professions: achieving critical mass', *Gender, Work and Organisation*, 7(3): 181–197.

Greed, C. (2006) 'Making the divided city whole: mainstreaming gender into planning in the United Kingdom', *Tijdschrift voor Economische en Sociale Geografie*, 97(3): 267–280.

Green, A., Grace, D. and Perkins, H. (2016) 'City branding, research and practice: an integrative review', *Journal of Brand Management*, 23(3): 252–272.

Grieco, M. (1995) 'Time pressures and low income families: the implications for "social" transport policy in Europe', *Community Development Journal*, 30(4): 347–363.

Grieco, M. and Urry, J. (eds) (2011) *Mobilities: New Perspectives on Transport and Society*, Farnham: Ashgate.

Grieco, M., Turner, J. and Hine, J. (2000) 'Transport, employment and social exclusion: changing the contours through information technology', *Local Work*, 26: np.

Gruffydd Jones, B. (2012) '"Bankable slums": The global politics of slum upgrading', *Third World Quarterly*, 33(5): 769–789.

Gwilliam, K. (2003) 'Urban transport in developing countries', *Transport Reviews*, 23(2): 197–216.

Habraken, N.J. (2005) *Palladio's Children*, Abingdon: Taylor and Francis.

Hägerstrand, T. (1982) 'Diorama, path and project', *Tijdschrift voor Economische en Sociale Geografie*, 73(6): 323–339.

Hall, P. (1998) *Cities in Civilization*, London: Weidenfeld & Nicolson.

Hall, P. (2002) *Cities of Tomorrow: An Intellectual History of Urban Planning and Design in the Twentieth Century*, Oxford: Blackwell.

Hall, P. (2014) *Cities of Tomorrow: An Intellectual History of Urban Planning and Design since 1880*, 4th edn, Chichester: John Wiley & Sons.

Hall, P. and Pfeiffer, U. (2000) *Urban Future 21: A Global Agenda for 21st Century Cities*, London: E. & F.N. Spon.

Hall, P. and Preston, P. (1988) *The Carrier Wave: New Information Technology and the Geography of Innovation 1846–2003*, London: Unwin Hyman.

Hall, P.S. (1966) *The World Cities*, London: Weidenfeld & Nicolson.

Hall, T. (1997) 'Images of industry in the postindustrial city: Raymond Mason and Birmingham', *Ecumene*, 4(1): 46–68.

Hall, T. (2003) 'Art and urban change', in Blunt, A., Gruffudd, P., May, J., Ogborn, M. and Pinder, D. (eds) *Cultural Geography in Practice*, London: Edward Arnold.

Hall, T. (2007) *Everyday Geographies*, Sheffield: Geographical Association.

Hall, T. (2008) 'Contesting the urban renaissance: journalism and the post-industrial city', *Geography*, 93(3): 148–157.

Hall, T. (2009a) 'Tourism, Urban', in Kitchin, R. and Thrift, N. (eds) *International Encyclopedia of Human Geography*, Vol. 11, Oxford: Elsevier, 318–323.

Hall, T. (2009b) 'The camera never lies: photographic research methods in human geography', *Journal of Geography in Higher Education*, 33(3): 453–462.

Hall, T. (2010a) 'Economic geography and organized crime: a critical review', *Geoforum*, 41: 841–845.

Hall, T. (2010b) 'Where the money is: the geographies of organised crime', *Geography*, 95(1): 4–13.

Hall, T. and Hubbard, P. (1996) 'The entrepreneurial city: new urban politics, new urban geographies?', *Progress in Human Geography*, 20(2): 153–174.

Hall, T. and Hubbard, P. (eds) (1998) *The Entrepreneurial City: Geographies of Politics, Regime and Representation*, Chichester: John Wiley & Sons.

Hall, T., Healey, M. and Harrison, M. (2002) 'Fieldwork and disabled students: discourses of exclusion and inclusion', *Transactions of the Institute of British Geographers* NS, 27(2): 213–231.

Hambleton, R. (1991) 'American dreams, urban realities', *The Planner*, 77(23): 6–9.

Hamilton, K., Ryley Hoyle, S. and Jenkins, L. (1999) *The Public Transport Gender Audit: The Research Report*, London: Transport Studies, University of East London.

Hamilton, N. and Chinchilla, N.S. (2001) *Seeking Community in a Global City: Guatemalans and Salvadorans in Los Angeles, 2001*, Philadelphia: Temple University Press.

Hannigan, J. (1998) *Fantasy City: Pleasure and Profit in the Postmodern Metropolis*, London: Routledge.

Harada, T. and Waitt, G. (2012) 'Researching transport choices: the possibilities of 'mobile methodologies' to study life-on-the-move', *Geographical Research* 51(2): 145–152.

Harris, A. and Moore, S. (2013) 'Planning histories and practices of circulating urban knowledge', *International Journal of Urban and Regional Research*, 37(5): 1499–1509.

Harris, C. and Ullman, E.L. (1945) 'The Nature of Cities', *The Annals of the American Academy of Political and Social Science*, 242: 7–17.

Harris, R. and Larkham, P.J. (1999) *Changing Suburbs: Foundation, Form and Function*, London: E & F.N. Spon.

Hart, T. (2001) 'Transport and the city', in Paddison, R. (ed) *The Urban Studies Handbook*, London: Sage.

Hartshorn, T. and Muller, P.O. (1989) 'Suburban downtowns and the transformation of metropolitan Atlanta's business landscape', *Urban Geography*, 10: 375–395.

Harvey, D. (1973) *Social Justice and the City*, London: Arnold.

Harvey, D. (1982) *The Limits to Capital*, Oxford: Blackwell.

Harvey, D. (1985) 'The geopolitics of capitalism', in Gregory, D. and Urry, J. (eds) *Social Relations and Spatial Structures*, London: Macmillan.

Harvey, D. (1988) 'Voodoo cities', *New Statesman and Society*, 30 September: 33–35.

Harvey, D. (1989a) *The Condition of Postmodernity: An Enquiry into the Origins of Cultural Change*, Oxford: Blackwell.

Harvey, D. (1989b) *The Urban Experience*, Oxford: Blackwell.

Harvey, D. (1996) 'The environment of justice', in Merrifield, A. and Swyngedouw, E. (eds) *The Urbanization of Injustice*, London: Lawrence & Wishart.

Harvey, D. (2001) 'Heritage pasts and heritage presents', *International Journal of Heritage Studies*, 7(4): 319–338.

Harvey, D. (2008) 'The right to the city', *New Left Review*, 53, September-October: 23–40.

Harvey, D. (2013) *Rebel Cities: From the Right to the City to the Urban Revolution*, London: Verso.

Harvey, D. (2015) *Seventeen Contradictions and the End of Capitalism*, London: Profile Books.

Hastings, A. (2004) 'Stigma and social housing estates: beyond pathological explanations', *Journal of Housing and the Built Environment*, 19(3): 233–254.

Haughton, G. and Hunter, C. (1994) *Sustainable Cities*, London: Regional Studies Association.

Haughton, G. and Hunter, C. (2003) *Sustainable Cities*, London: Routledge.

Hawkins, R. (1986) 'A road not taken: sociology and the neglect of the automobile', *California Sociologist*, 9: 61–79.

Hayden, D. (1981) *Grand Domestic Revolution*, Cambridge, MA: MIT Press.

Hayden, D. (1995) *The Power of Place: Urban Landscapes as Public History*, Cambridge, MA: MIT Press.

Haynes, J. (1997) *Democracy and Civil Society in the Third World: Politics and New Political Movements*, Cambridge: Cambridge University Press.

Healey, P. (2004) 'Towards a "social democratic" policy agenda for cities', in Johnstone, C. and Whitehead, M. (eds) *New Horizons in British Urban Policy: Perspectives on New Labour's Urban Renaissance*, Aldershot: Ashgate.

Healey, P., Magalhaes, C. de, Madanipour, A. and Pendlebury, J. (2002) *Shaping City Centre Futures: Conservation, Regeneration and Institutional Capacity*, Newcastle upon Tyne, Centre for Research in European Urban Environments, University of Newcastle.

Hebdige, D. (1979) *Subculture: The Meaning of Style*, New York: Methuen.

Henderson, J.C. (2008) 'Managing ethnic urban heritage: Little India in Singapore', *International Journal of Heritage Studies*, 14(4): 332–346.

Henderson, S., Bowlby, S. and Raco, M. (2007) 'Refashioning local government and inner-city regeneration: the Salford experience', *Urban Studies*, 44(8): 1441–1463.

Herbert, D.T. and Thomas, C.J. (1982) *Urban Geography. A First Approach*, Chichester: John Wiley & Sons.

Hernandez, J. (2009) 'Redlining revisited: mortgage lending patterns in Sacramento 1930–2004', *International Journal of Urban and Regional Research*, 33(2): 291–313.

Herschel, T. and Newman, P. (2002) *Governance of Europe's City Regions*, London: Routledge.

Hersey, G. (1999) *The Monumental Impulse: Architecture's Biological Roots*, Cambridge, MA: MIT Press.

Hewitt, L. and Graham, S. (2015) 'Vertical cities: representations of urban verticality in twentieth-century science-fiction literature', *Urban Studies*, 52(5): 923–937.

Hickman, R. and Banister, D. (2014) *Transport, Climate Change and the City*, Abingdon: Routledge.

Highmore, B. (2002a) *Everyday Life and Cultural Theory: An Introduction*, London: Routledge.

Highmore, B. (2002b) *Everyday Life Reader*, London: Routledge.

Highmore, B. (2010) *Ordinary Lives: Studies in the Everyday*, Abingdon: Routledge.

Hill, J. (ed) (1998) *Occupying Architecture*, London: Routledge.

Hinchliffe, S. (2007) *Geographies of Nature: Societies, Environments, Ecologies*, London: Sage.

Hine, J. and Mitchell, F. (2001) *The Role of Transport in Social Exclusion in Urban Scotland*, Edinburgh: Scottish Executive Central Research Unit.

Hober, G. (2013) 'Surviving the era of deindustrialization: the new economic geography of the urban rust belt', *Journal of Urban Affairs*, 35(4): 417–434.

Hodge, D.C. (1995) 'My fair share: equity issues in urban transportation', in Hanson, S. (ed) *The Geography of Urban Transportation*, New York: Guilford Press.

Holcomb, B. (1994) 'City make-overs: marketing the post-industrial city', in Gold, J.R. and Ward, S.V. (eds) *Place Promotion: The Use of Publicity and Marketing to Sell Towns and Regions*, Chichester: John Wiley & Sons.

Holloway, L. and Hubbard, P. (2001) *People and Place: The Extraordinary Geographies of Everyday Life*, Harlow: Prentice Hall.

Hopkins, J. (1990) 'West Edmonton Mall: landscape of myths and elsewhereness', *Canadian Geographer*, 34(1): 2–17.

Hopkins, P. (2008) 'Critical geographies of body size', *Geography Compass*, 2(6): 2111–2126.

Hopkins, R. (2008) *The Transition Handbook: From Oil Dependency to Local Resilience*, Totnes, Devon: Green Books.

Hou, J., Spencer, B., Way, T. and Yocom, K. (eds) (2015) *Now Urbanism: The Future City is Here*, Abingdon: Routledge.

Hough, M. (1995) *Cities and Natural Processes*, London: Routledge.

Howard, E. (1898) *Tomorrow, A Peaceful Path to Real Reform*, London: Swan Sonnenschein & Co.

Howard, E. (1902) *Garden Cities of Tomorrow*, London: Swan Sonnenschein & Co.

Hoyt, H. (1939) *The Structure and Growth of Residential Neighborhoods in American Cities*, Washington DC: Federal Housing Administration.

Hubbard, P. (2006) *City*, Abingdon: Routledge.

Hubbard, P., Kitchin, R., Bartley, B. and Fuller, D. (2002) *Thinking Geographically: Space, Theory and Contemporary Human Geography*, London: Continuum.

Huby, M. and Burkitt, N. (2000) 'Is the "New Deal for Transport" really "Better for Everyone"? The social policy implications of the UK's 1998 Transport White Paper', *Environment and Planning C: Government and Policy*, 18(4): 379–392.

Hudson, S., Cárdenas, D., Meng, F. and Thal, K. (2017) 'Building a place brand from the bottom up: a case study from the United States', *Journal of Vacation Marketing*, DOI: 10.1177/1356766716649228

Hughes, M. (1992) 'Regional economics and edge-cities', in Federal Highway Administration (ed) *Edge City and ISTEA – Examining the Transport Implications of Suburban Development Patterns*, Washington DC: Federal Highway Administration.

Hunter, F. (1953) *Community Power Structure*, Chapel Hill, NC: University of North Carolina Press.

Imrie, R. (1996) *Disability and the City*, London: Paul Chapman.

Imrie, R. and Edwards, C. (2007) 'The geographies of disability: reflections on the emergence of a sub-discipline', *Geography Compass*, 1(3): 623–640.

Imrie, R. and Raco, M. (2003) *Urban Renaissance? New Labour, Community and Urban Policy*, Bristol: Policy Press.

Imrie, R. and Thomas, H. (1999) *British Urban Policy: An Evaluation of Urban Development Corporations*, London: Sage.

Ingersoll, R. and Kostof, S. (2013) *World Architecture: A Cross-Cultural History*, Oxford: Oxford University Press.

Intergovernmental Panel on Climate Change (2014a). *Climate Change 2014: Mitigation of Climate Change – Transport. Working Group III: Mitigation of Climate Change*, Potsdam: Intergovernmental Panel on Climate Change.

Intergovernmental Panel on Climate Change (2014b). *Climate Change 2014: Mitigation of Climate Change – Human Settlements. Working Group III: Mitigation of Climate Change*, Potsdam: Intergovernmental Panel on Climate Change.

International Civil Aviation Organization (ICAO) (2015) *Air Transport Yearly Monitor 2015*, ICAO, www.icao.int/sustainability/Pages/Air-Traffic-Monitor.aspx (Accessed 24 February 2017).

Jackson, J.B. (1984) *Discovering the Vernacular Landscape*, London: Yale University Press.

Jackson, P. (1989) *Maps of Meaning: An Introduction to Cultural Geography*, London: Unwin Hyman.

Jackson, P. and Holbrook, B. (1995) 'Multiple meanings: shopping and the cultural politics of identity', *Environment and Planning A*, 27(12): 1913–1930.

Jacobs, J. (1961) *The Death and Life of Great American Cities*, New York: Random House.

Jacobs, J.M. (1992) 'Cultures of the past and urban transformation: the Spitalfields Market redevelopment in East London', in Anderson, K. and Gale, F. (eds) *Inventing Places: Studies in Cultural Geography*, Melbourne: Longman Cheshire.

Jacobs, J.M. (1996) *Edge of Empire: Postcolonialism and the City*, London: Routledge.

Jacobs, J.M. (1999) 'Cultures of the past and urban transformation: the Spitalfields Market redevelopment in East London', in Anderson, K. and Gale, F. (eds) *Cultural Geographies*, Melbourne: Addison-Wesley Longman.

Jacobs, J.M., Cairns, S. and Strebel, I. (2012) 'Doing building work: methods at the interface of geography and architecture', *Geographical Research*, 50(2): 126–140.

Jacobs, M. (ed) (1997) *Greening the Millennium: The New Politics of the Environment*, Oxford: Blackwell.

Jarvis, H., Pratt, A.C. and Cheng-Chong, P. (2001) *The Secret Life of Cities: The Social Reproduction of Everyday Life*, Harlow: Prentice Hall.

Jayne, M. (2005) 'Creative cities: the regional dimension', *Environment and Planning C: Government and Policy*, 23(4): 537–556.

Jayne, M., Holloway, S. and Valentine, G. (2006) 'Drunk and disorderly: alcohol, urban life and public space', *Progress in Human Geography*, 30(4): 451–468.

Jencks, C. (1991) *The Language of Post-Modern Architecture*, London: Academy Editions.

Jensen, A. (2011) 'Mobility, space and power: on the multiplicities of seeing mobility', *Mobilities*, 6(2): 255–271.

Jensen, O.B. (2007) 'Culture stories: understanding cultural urban branding', *Planning Theory*, 6(3): 211–236.

Jensen, O.B. (2013) *Staging Mobilities*, Abingdon: Routledge.

Jensen, O.B. (2014) *Designing Mobilities*, Aalborg: Aalborg University Press.

Jensen, O.B. and Lassen, C. (2011) 'Mobility challenges', *Danish Journal of Geoinfomatics and Land Management*, 46(1): 9–21.

Jensen, O.B., Sheller, M. and Wind, S. (2015) 'Together and apart: affective ambiences and negotiation in families' everyday life and mobility', *Mobilities*, 10(3): 363–382.

Johnston, R.J. and Sidaway, J.D. (2004) *Geography and Geographers: Anglo-American Human Geography Since 1945*, London: Hodder Arnold.

Jonas, A.E.G., McCann, E. and Thomas, M. (2015) *Urban Geography: A Critical Introduction*, Chichester: Wiley-Blackwell.

Jones, P. (2009) 'Putting architecture in its social place: a cultural political economy of architecture', *Urban Studies*, 46(12): 2519–2536.

Jones, P. and Evans, J. (2008) *Urban Regeneration in the UK*, London: Sage.

Joss, S. (2011) 'Eco-cities: the mainstreaming of urban sustainability; key characteristics and driving factors', *International Journal of Sustainable Development and Planning*, 6(3): 268–285.

Judd, D. and Parkinson, M. (1990) 'Urban leadership and regeneration', in Parkinson, M. and Judd, D. (eds) *Leadership and Urban Regeneration*, London: Sage.

Judd, D. and Swanstrom, T. (1998) *City Politics*, London: Macmillan.

Judd, D. and Swanstrom, T. (2009) *City Politics: The Political Economy of Urban America*, 7th edn, Harlow: Longman.

Judge, J., Stoker, G. and Wolman, H. (eds) (1995) *Theories of Urban Politics*, London: Sage.

Kaika, M. (2010) 'Architecture and crisis: reinventing the icon, re-imag(in)ing London and re-branding the City', *Transactions, Institute of British Geographers*, NS, 35: 453–474.

Kasarda, J. and Lindsey, G. (2011) *Aerotropolis: The Way We'll Live Next*, London: Allen Lane.

Katz, B. (ed) (2000) *Reflections on Regionalism*, Washington: Brookings Institution Press.

Kaufmann, V. (2002) *Rethinking Mobility*, Aldershot: Ashgate.

Keating, M. (1991) *Comparative Urban Politics, Power and the City in the US, Canada, Britain and France*, Aldershot: Edward Elgar.

Kenworthy, J. (2014) 'Total daily mobility patterns and their policy implications for forty-three global cities in 1995 and 2005', *World Transport Policy and Practice*, 20(1): 41–55.

Kiel, R. (1998) 'Globalisation makes states', *Review of International Political Economy*, 5(4): 616–646.

Kilcullen, D. (2013) *Out of the Mountains: The Coming Age of the Urban Guerrilla*, London: C. Hurst.

Kim, Y-H. (2008) 'Global-local', in Hall, T., Hubbard, P. and Short, J.R. (eds) *The Sage Companion to the City*, London, Sage.

King, A. (1984) *The Bungalow: The Production of a Global Culture*, London: Routledge.

King, A. (1990) *Global Cities: Post-Imperialism and the Internationalization of London*, London: Routledge.

King, A. (2004) *Spaces of Global Cultures*, London: Routledge.

Kitchin, R. (1994) 'Cognitive maps: what are they and why study them?', *Journal of Environmental Psychology*, 14(1): 1–19.

Kitchin, R. (2000) *Disability, Space and Society*, Sheffield: Geographical Association.

Kniffen, F. (1965) 'Folk housing: key to diffusion', *Annals of the Association of American Geographers*, 55: 549–577.

Knopp, L. (1990) 'Some theoretical implications of gay involvement in an urban land market', *Political Geography Quarterly*, 9(4): 337–352.

Knowles, C. and Sweetman, P. (eds) (2004) *Picturing the Social Landscape: Visual Methods and the Sociological Imagination*, London: Routledge.

Knowles, R. and Wareing, J. (1976) *Economic and Social Geography Made Simple*, London: Heinemann.

Knowles, R., Shaw, J. and Docherty, I. (eds) (2007) *Transport Geographies: Mobilities, Flows and Spaces*, Oxford: Blackwell.

Knox, P. (1987) 'The social production of the built environment: architects, architecture and the postmodern city', *Progress in Human Geography*, 11(3): 353–377.

Knox, P. (1991) 'The restless urban landscape: economic and sociocultural change and the transformation of metropolitan Washington D.C.', *Annals of the Association of American Geographers*, 8(1): 181–209.

Knox, P. (1995) 'World cities and the organisation of global space', in Johnston, R., Taylor, P. and Watts, M. (eds) *Geographies of Global Change*, Oxford: Blackwell.

Knox, P. (ed) (1993) *The Restless Urban Landscape*, Englewood Cliffs, NJ: Prentice Hall.

Knox, P. and Pinch, S. (2010) *Urban Social Geography: An Introduction*, 6th edn, Harlow: Pearson.

Knox, P., Agnew, J. and McCarthy, L. (2008) *The Geography of the World Economy*, 5th edn, London: Hodder Education.

Kolb, A. (1984) *Experiential Learning: Experience as the Source of Learning and Development*, Upper Saddle River, NJ: Prentice Hall.

Kostof, S. (1995) *A History of Architecture: Settings and Rituals*, 2nd edn, Oxford: Oxford University Press.

Kourtit, K., Nijkamp, P. and Reid, N. (2014) 'The new urban world: challenges and policy', *Applied Geography*, 49: 1–3.

Krätke, S. (2002) 'Network analysis of production clusters, the Potsdam/Babelsberg film industry as an example', *European Planning Studies*, 10(1): 27–54.

Kropf, K. (2016) *The Handbook of Urban Morphology*, London: John Wiley & Sons.

Krueger, R. and Savage, L. (2007) 'City-regions and social reproduction: a "place" for sustainable development?', *International Journal of Urban and Regional Research*, 31(1): 215–223.

Lacy, S. (1994) *Mapping the Terrain: New Genre Public Art*, San Francisco, CA: Bay Press.

Landry, C. (2000) *The Creative City*, London: Earthscan.

Langegger, S. and Koester, S. (2016) 'Invisible homelessness: anonymity, exposure, and the right to the city', *Urban Geography*, 37(7): 1030–1048.

Larkham, P.J. (1996) *Conservation and the City*, London: Routledge.

Larkham, P.J. (2006) 'The study of urban form in Great Britain', *Urban Morphology*, 10(2): 117–141.

Larkham, P.J. and Barrett, H.J. (1998) 'Conservation of the built environment under the Conservatives', in Allmendinger, P. and Thomas, H. (eds) *Urban Planning and the British New Right*, London: Routledge.

Larkham, P.J. and Conzen, M.P. (eds) (2014) *Shapers of Urban Form: Explorations in Morphological Agency*, New York: Routledge.

Latham, A. (2004) 'Researching and writing everyday accounts of the city: an introduction to the diary-photo, diary-interview method', in Knowles, C. and Sweetman, P. (eds) *Picturing the Social Landscape: Visual Methods and the Sociological Imagination*, London: Routledge.

Laurie, I.C. (ed) (1979) *Nature in Cities: The Natural Environment in the Design and Development of Urban Green Spaces*, Chichester: John Wiley & Sons.

Lee, D.S., Pitari, G., Grewe, V., Gierens, K., Penner, J.E., Petzold, A., Prather, M.J., Schumann, U., Bais, A., Berntsen, T., Iachetti, D., Lim, L.L., Sausen, R. (2010) 'Transport impacts on atmosphere and climate: aviation', *Atmospheric Environment*, 44(37): 4678–4734.

Lees, L. (2001) 'Towards a critical geography of architecture: the case of an ersatz colosseum', *Ecumene*, 8(1): 51–86.

Lees, L. (2003) 'Policy (re)turns: urban policy and gentrification, gentrification and urban policy', *Environment and Planning A*, 35(4): 571–574.

Lees, L., Shin, H.B., and López-Morales, E. (eds) (2015) *Global Gentrifications: Uneven Development and Displacement*, Bristol: Policy Press.

Lees, L., Shin, H.B. and López-Morales, E. (2016) *Planetary Gentrification*, Cambridge: Polity Press.

Lees, L., Slater, T. and Wyly, E. (2008) *Gentrification*, London: Routledge.

LeGates, R.T. and Stout, F. (2016) *The City Reader*, 6th edn, Abingdon: Routledge.

Leitner, H. and Garner, M. (1993) 'The limits of local initiatives: a reassessment of urban entrepreneurialism for urban development', *Urban Geography*, 14(1): 57–77.

Leonard, M. (1998) *Invisible Work, Invisible Worker: The Informal Economy in Europe and the US*, London: Macmillan.

Levine, M. (2015) *Urban Politics: Cities and Suburbs in a Global Age*, 9th edn, New York: Routledge.

Levy, D., Murphy, L. and Lee, C.K.C. (2008) 'Influences and emotions: exploring family decision-making processes when buying a house', *Housing Studies*, 23(2): 271–289.

Lewis, P.F. (1979) 'Axioms for reading the landscape: some guides to the American scene', in Meinig, D.W. (ed) *The Interpretation of Ordinary Landscapes: Geographical Essays*, New York: Oxford University Press.

Lewis, P.F. (1983) 'The galactic metropolis', in Platt, R.H. and Macinko, G. (eds) *Beyond the Urban Fringe*, Minneapolis, MN: University of Minnesota Press.

Ley, D. (1983) *A Social Geography of the City*, New York: Harper Row.

Ley, D. (1987) 'Styles of the times: Liberal and neo-conservative landscapes in inner Vancouver 1968–1986', *Journal of Historical Geography*, 13(1): 40–56.

Ley, D. (1996) *Gentrification and the Middle Classes*, Oxford: Oxford University Press.

Ley, D. (2017) 'Global China and the making of Vancouver's residential property market', *International Journal of Housing Policy*, 17(1): 15–34.

Ley, D. and Mills, C. (1993) 'Can there be a post-modernism of resistance in the built environment?', in Knox, P. (ed) *The Restless Urban Landscape*, Englewood Cliffs, NJ: Prentice Hall.

Leyshon, A., and Thrift, N. (1995) 'Geographies of financial exclusion: financial abandonment in Britain and the United States', *Transactions of the Institute of British Geographers*, NS, 20: 312–341.

Leyshon, A., French, S. and Signoretta, P. (2008) 'Financial exclusion and the geography of bank and building society branch closure in Britain', *Transactions of the Institute of British Geographers*, NS, 33: 447–465.

Lilley, K.D. (2000) 'Mapping the medieval city: plan analysis and urban history', *Urban History*, 27(1): 5–30.

Lilley, K.D. (2004) 'Cities of God? Medieval urban forms and their Christian symbolism', *Transactions, Institute of British Geographers*, 29: 296–313.

Lilley, K.D. (2009) *City and Cosmos: The Medieval World in Urban Form*, London: Reaktion Books.

Litman, T. (1995) *Transportation Cost Analysis: Techniques, Estimates and Implications*, Victoria: Victoria Transport Policy Institute.

Loftman, P. and Nevin, B. (1998) 'Going for growth: prestige projects in three British cities', *Urban Studies*, 33(6): 991–1019.

Loftman, P. and Nevin, B. (2003) 'Prestige projects, city centre restructuring and social exclusion: taking the long-term view', in Miles, M. and Hall, T. (eds) *Urban Futures: Critical Commentaries on Shaping the City*, London: Routledge.

Logan, J.R. and Molotch, H.L. (1987) *Urban Fortunes: The Political Economy of Place*, Berkeley, CA: University of California Press.

Lojkine, J. (1976) 'Contribution to a Marxist theory of urbanisation', in Pickvance, C. (ed) *Urban Sociology: Critical Essays*, London: Tavistock Press.

Longhurst, R. (2000a) *Bodies: Exploring Fluid Boundaries*, London: Routledge.

Longhurst, R. (2000b) 'Corporeographies of pregnancy: "bikini babes"', *Environment and Planning D: Society and Space*, 18(4): 453–472.

Lucas, K., Grosvenor, T. and Simpson, R. (2001) *Transport, the Environment and Social Exclusion*, York: Joseph Rowntree Foundation.

Luque-Ayala, A. and Marvin, S. (2015) 'Developing a critical understanding of smart urbanism?', *Urban Studies*, 52(12): 2105–2116.

Lynch, K. (1960) *The Image of the City*, Cambridge, MA: MIT Press.

Lynch, K. (2005) *Rural-Urban Interaction in the Developing World*, London: Routledge.

Lyons, G. and Urry, J. (2005) 'Travel time use in the information age', *Transport Research Part A: Policy and Practice*, 39(2–3): 257–276.

Ma, L.J.C. (2002) 'Urban transformation in China, 1949–2000: A review and research agenda', *Environment and Planning A*, 34: 1545–1569.

Maat, K. and Louw, E. (1999) 'Mind the gap: pitfalls of travel reduction measures', *Built Environment*, 25(2): 151–161.

Mackenzie, S. (1988) 'Building women, building cities: toward gender sensitive theory in the environmental disciplines', in Andrew, C. and Milroy, B. (eds) *Life Spaces*, Vancouver: University of British Columbia Press.

MacLeod, G. (2002) 'From urban entrepreneurialism to a "revanchist city"? On the spatial injustices of Glasgow's renaissance', *Antipode*, 34(3): 602–624.

Madden, D. and Marcuse, P. (2016) *In Defense of Housing*, New York: Verso.

Malecki, E. J. (2014) 'Connecting the fragments: looking at the connected city in 2050', *Applied Geography*, 49: 12–17.

Maloutas, T. and Fujita, K. (eds) (2016) *Residential Segregation in Comparative Perspective: Making Sense of Contextual Diversity*, Abingdon: Routledge.

Manley, J.F. (1983) 'Neo-pluralism: a class analysis of pluralism I and pluralism II', *American Political Science Review*, 77(2): 368–383.

Mann, P. (1965) *An Approach to Urban Sociology*, London: Routledge.

Marshall, J.N. (2004) 'Financial institutions in disadvantaged areas: a comparative analysis of policies encouraging financial inclusion in Britain and the United States', *Environment and Planning A*, 36: 241–261.

Marshall, S. (1999a) 'Introduction: travel reduction – means and ends', *Built Environment*, 25(2): 89–93.

Marshall, S. (1999b) 'Restraining mobility while maintaining accessibility: an impression of the "city of sustainable growth"', *Built Environment*, 25(2): 168–179.

Marshall, S., Bannister, D. and McLellan, A. (1997) 'A strategic assessment of travel trends and travel reduction strategies', *Innovation*, 10(3): 289–304.

Massey, D. (1984) *Spatial Division of Labour*, London: Macmillan.

Massey, D. (1994) *Space, Place and Gender*, Minneapolis, MN: University of Minnesota Press.

Massey, D. (1999) 'Cities in the world', in Massey, D., Allen, J. and Pile, S. (eds) *City Worlds*, London: Routledge/Open University.

Massey, D. and Meegan, R. (1982) *The Anatomy of Job Loss: The How, Why and Where of Employment Decline*, London: Macmillan.

May, A.D. (2004) 'Singapore: the development of a world class transport system', *Transport Reviews*, 24(1): 79–101.

May, A.D. (2013) 'Urban transport and sustainability: the key challenges', *International Journal of Sustainable Transportation*, 7(3):170–185.

May, J. (2015) 'Racial vibrations, masculine performances: experiences of homelessness among young men of colour in the Greater Toronto Area', *Gender, Place & Culture*, 22(3): 405–421.

May, J. and Cloke, P. (2014) 'Modes of attentiveness: reading for difference in geographies of homelessness', *Antipode*, 46(4): 894–920.

McCleery, A. (2004) 'So many Glasgows: from "personality of place" to "positionality in space and time"', *Scottish Geographical Journal*, 120(1/2): 3–18.

McEwan, C. (2008) 'Geography, culture and global change', in Daniels, P., Bradshaw, M., Shaw, D. and Sidaway, J. (eds) *An Introduction to Human Geography: Issues for the 21st Century*, Harlow: Pearson.

McFarlane, C. (2008) 'Postcolonial Bombay: decline of a cosmopolitan city?', *Environment and Planning D: Society and Space*, 26(3): 480–499.

McLellan, A. and Collins, D. (2014) '"If you're just a bus community … you're second tier": Motivations for rapid mass transit (RMT) development into mid-sized cities', *Urban Policy and Research* 32(2): 203–217.

McNeill, D. (2001a) 'Embodying a Europe of the cities: the geographies of mayoral leadership', *Area*, 33(4): 353–359.

McNeill, D. (2001b) 'Barcelona as imagined community: Pasqual Maragall's spaces of engagement', *Transactions of the Institute of British Geographers*, 26(3): 340–353.

McNeill, D. (2009) 'Urban politics', in Latham, A., McCormack, D., McNamara, K. and McNeill, D. (eds) *Key Concepts in Urban Geography*, London: Sage.

McNeill, D. (2014) 'Airports and territorial restructuring: the case of Hong Kong', *Urban Studies*, 51(14): 2996–3010.

Meinig, D.W. (1979) *The Interpretation of Ordinary Landscapes: Geographical Essays*, New York: Oxford University Press.

Merriman, P. (2015) 'Mobilities 1: Departures', *Progress in Human Geography*, 39(1):87–95.

Metraux, S. (1999) 'Waiting for the wrecking ball: skid row in postindustrial Philadelphia', *Journal of Urban History*, 25(5): 690–715.

Meyer, W.B. and Esposito, C.R. (2015) 'Burgess and Hoyt in Los Angeles: testing the Chicago models on an automobile age American City', *Urban Geography*, 36(2): 314–325.

Middleton, J. (2016) 'The socialities of everyday urban walking and the "right to the city"', *Urban Studies*, 1–20, 10.1177/0042098016649325.

Miles, M. (1997) *Art, Space and the City*, London: Routledge.

Miles, M. (1998) 'A game of appearance: public art and urban development – complicity or sustainability?', in Hall, T. and Hubbard, P. (eds) *The Entrepreneurial City: Geographies of Politics, Regime and Representation*, Chichester: John Wiley & Sons.

Miles, M. (2002) 'Wish you were here', in Spier, S. (ed) *Urban Visions: Experiencing and Envisioning the City*, Liverpool: Liverpool University Press/Tate Liverpool.

Miles, M. (2004) 'Drawn and quartered: El Raval and the Hausmannization of Barcelona', in Bell, D. and Jayne, M. (eds) *City of Quarters: Urban Villages in the Contemporary City*, Aldershot: Ashgate.

Miles, M. (2007) *Cities and Cultures*, Abingdon: Routledge.

Miles, M. (2008) 'Planning and conflict', in Hall, T., Hubbard, P. and Short, J.R. (eds) *The Sage Companion to the City*, London: Sage, 318–333.

Miller, D., Jackson, P., Thrift, N., Holbrook, B. and Rowlands, M. (1998) *Shopping, Place and Identity*, London: Routledge.

Minton, A. (2001) 'Rising sun', *The Guardian*, 19 September: 5.

Minton, A. (2006) *The Privatisation of Public Space*, London: Royal Institute of Chartered Surveyors.

Minton, A. (2009) *Ground Control: Fear and Happiness in the Twenty-first Century City*, London: Penguin.

Mohan, J. (1999) *A United Kingdom? Economic, Social and Political Geographies*, London: Arnold.

Mokhtarian, P.L. (1990) 'A typology of relationships between telecommunications and transportation', *Transportation Research A*, 24A(3): 231–242.

Mollenkopf, J. (2003) 'How to study urban political power', in LeGates, R.T. and Stout, F. (eds) *The City Reader*, 3rd edn, London: Routledge.

Molotch, H. (1976) 'The city as a growth machine', *American Journal of Sociology*, 82(2): 309–355.

Mom, G. and Kim, N. (2013) 'Editorial', *Transfers: Interdisciplinary Journal of Mobility Studies,* 3(3): 1–5.

Monbiot, G. (2000) 'Complainers who are facing ruin', *The Guardian*, 10 February: 22.

Moran, D., Turner, J. and Jewkes, Y. (2016) 'Becoming big things: building events and the architectural geographies of incarceration in England and Wales', *Transactions of the Institute of British Geographers*, 41(4): 416–428.

Mossberger, K., Clarke, S.E. and John, P. (eds) (2012) *The Oxford Handbook of Urban Politics*, Oxford: Oxford University Press.

Moulaert, F., Rodríguez, A. and Swyngedouw, E. (2003) *The Globalized City: Economic Restructuring and Social Polarization in European Cities*, Oxford: Oxford University Press.

Mukhija, V. and Loukaitou-Sideris, A. (eds) (2014) *The Informal American City: Beyond Taco Trucks and Day Labor*, Cambridge, MA: MIT Press.

Muller, P.O. (1995) 'Transportation and urban form: stages in the spatial evolution of the American metropolis', in Hanson, S. (ed) *The Geography of Urban Transportation*, New York, NY: Guilford Press.

Mumford, L. (1938) *The Culture of Cities*, New York: Harcourt Brace.

Murdie, R.A. (1969) *Factorial Ecology of Metropolitan Toronto, 1951–1961*, Research Paper no. 116, Department of Geography, University of Chicago.

Murdoch, J. (1997) 'Towards a geography of heterogeneous associations', *Progress in Human Geography*, 21(3): 321–337.

Murdoch, J. (1998) 'The spaces of actor-network theory', *Geoforum*, 29(4): 357–374.

Murphy, S. (2009) '"Compassionate" strategies of managing homelessness: post-revanchist geographies in San Francisco', *Antipode*, 41(2): 305–325.

Murray, C. (2001) *Making Sense of Place: New Approaches to Place Marketing*, Stroud: Comedia.

Murray, W. (2006) *Geographies of Globalisation*, London: Routledge.

Musterd, S., van Gent, W.P., Das, M. and Latten, J. (2016) 'Adaptive behaviour in urban space: residential mobility in response to social distance', *Urban Studies*, 53(2): 227–246.

Nairn, K. (1999) 'Embodied Fieldwork', *Journal of Geography*, 98: 272–282.

Nast, H. and Pile, S. (eds) (1998) *Places Through the Body*, London: Routledge.

Németh, J. (2010) 'Security in public space: an empirical assessment of three US cities', *Environment and Planning A*, 42(10): 2487–2507.

Newman, O. (1972) *Defensible Space: Crime Prevention Through Urban Design*, New York: Macmillan.

Newman, P. (1999) 'Transport: reducing automobile dependence', in Satterthwaite, D. (ed) *The Earthscan Reader in Sustainable Cities*, London: Earthscan.

Nijman, J. (1999) 'Cultural globalization and the identity of place: the reconstruction of Amsterdam', *Ecumene*, 6: 146–164.

Nijman, J. (2007) 'Comparative urbanism', *Urban Geography*, 28(1): 1–6.

Normark, D. (2006) 'Tending to mobility: intensities of staying at the petrol station', *Environment and Planning A*, 38(2): 241–252.

Oatley, N. (ed) (1998) *Cities, Economic Competition and Urban Policy*, London: Paul Chapman.

O'Connor, K. and Fuellhart, K. (2016) 'Airports and regional air transport markets: a new perspective', *Journal of Transport Geography*, 53: 78–82.

Oguztimur, S. and Akturan, U. (2016) 'Synthesis of city branding literature (1988–2014) as a research domain', *International Journal of Tourism Research*, 18(4): 357–372.

Oliveira, V. (2016) *Urban Morphology: An Introduction to the Study of the Physical Form of Cities*, Cham, Switzerland: Springer.

Oliver, P. (2006) *Built to Meet Needs: Cultural Issues in Vernacular Architecture*, London: Architectural Press.

Oswalt, P. (ed) (2004) *Schrumpfende Städte. Band 1 Internationale Untersuchung*, Ostfildern-Ruit: Hantje Cantz.

Ottaviano, G.I.P. and Peri, G. (2006) 'The economic value of cultural diversity: evidence from US cities', *Journal of Economic Geography*, 6(1): 9–44.

Pacione, M. (1998) 'Towards a community economy: an examination of local exchange trading systems in west Glasgow', *Urban Geography*, 19(3): 211–231.

Pacione, M. (2001a) 'Models of land use structure in cities of the developed world', *Geography*, 86(2): 97–120.

Pacione, M. (2001b) 'The future of the city – cities of the future', *Geography*, 86(4): 275–286.

Pacione, M. (2009) *Urban Geography: A Global Perspective*, 3rd edn, London: Routledge.

Paddison, R. (ed) (2001) *Handbook of Urban Studies*, London: Sage.

Page, S. (1995) *Urban Tourism*, London: Routledge.

Pahl, R.E. (1969) 'Urban social theory and research', *Environment and Planning A*, 1: 143–153.

Pahl, R.E. (1970) *Whose City? And Other Essays on Sociology and Planning*, Harlow: Longman.

Painter, J. (1995) *Politics, Geography and Political Geography: A Critical Perspective*, London: Arnold.

Paoli, L. (2005) 'Italian organised crime: mafia associations and criminal enterprises', in Galeotti, M. (ed) *Global Crime Today: The Changing Face of Organised Crime*, London: Routledge.

Park, R. and Burgess, E. (1925) *The City*, Chicago: Chicago University Press.

Pasotti, E. (2010) *Political Branding in Cities: The Decline of Machine Politics in Bogota, Naples and Chicago*, Cambridge: Cambridge University Press.

Peck, J. (1995) 'Moving and shaking: business elites, state localism and urban privatism', *Progress in Human Geography*, 19(1): 16–46.

Peck, J. (2005) 'Struggling with the creative class', *International Journal of Urban and Regional Research*, 29(4): 740–770.

Peck, J. and Tickell, A. (2002) 'Neoliberalizing space', *Antipode*, 34(3): 380–404.

Peterson, P.E. (1981) *City Limits*, Chicago, IL: University of Chicago Press.

Pevsner, N. (1943) *An Outline of European Architecture*, Harmondsworth: Penguin Books.

Pickard, R.D. (2002) 'A comparative review of policy for the protection of the architectural heritage of Europe', *International Journal of Heritage Studies*, 8(4): 349–363.

Pike, A. (ed) (2011) *Brands and Branding Geographies*, Cheltenham: Edward Elgar.

Pile, S. (1996) *The Body and the City: Psychoanalysis, Space and Subjectivity*, London: Routledge.

Pile, S., Brook, C. and Mooney, G. (eds) (1999) *Unruly Cities? Order/Disorder*, London: Routledge/Open University.

Pinch, S. (1985) *Cities and Services: The Geography of Collective Consumption*, London: Routledge.

Pinder, D. (2005) 'Arts of urban exploration', *Cultural Geographies*, 12(4): 383–411.

Pinder, D. (2008) 'Urban interventions, art, politics and pedagogy', *International Journal of Urban and Regional Research*, 32(3): 70–76.

Pink, S. (2001) *Doing Visual Ethnography*, London: Sage.

Piven, F.F. and Cloward, R.A. (1977) *Poor People's Movements: Why They Succeed, How They Fail*, New York: Pantheon.

Plaza, B., Tironi, M. and Haarich, S.N. (2009) 'Bilbao's art scene and the "Guggenheim effect" revisited', *European Planning Studies*, 17(11): 1711–1729.

Pocock, D. (1989) 'Sound and the geographer', *Geography*, 74(3): 193–200.

Pocock, D. (1993) 'The senses in focus', *Area*, 25(1): 11–16.

Polsby, N.W. (1963) *Community Power and Political Theory*, New Haven, CT: Yale University Press.

Ponzini, D. and Rossi, U. (2010) 'Becoming a creative city: the entrepreneurial mayor, network politics and the promise of an urban renaissance', *Urban Studies*, 47(5): 1037–1057.

Porteous, J.D. (1985) 'Smellscape', *Progress in Human Geography*, 9(3): 356–378.

Portney, K.E. (2002) 'Taking sustainable cities seriously: a comparative analysis of twenty-four U.S. cities', *Local Environment*, 7(4): 363–380.

Potter, R.B. and Lloyd-Evans, S. (1998) *The City in the Developing World*, London: Prentice Hall.

Pred, A. (1990) *Lost Words and Lost Worlds: Modernity and the Language of Everyday Life in Late Nineteenth-century Stockholm*, Cambridge: Cambridge University Press.

Raban, J. (1975) *The Soft City*, London: Fontana.

Rabinovitch, J. (1992) 'Curitiba: towards sustainable urban development', *Environment and Urbanisation*, 4(2): 62–73.

Rantisi, N.M. and Leslie, D. (2006) 'Branding the design metropole: the case of Montréal, Canada', *Area*, 38(4): 364–376.

Rao, V. (2006) 'Slum as theory: the South/Asian city and globalization', *International Journal of Urban and Regional Research*, 30(1): 225–232.

Rapoport, E. (2014) 'Utopian visions and real estate dreams: the eco-city past, present and future', *Geography Compass*, 8(2): 137–149.

Rees, W. (1997) 'Is "sustainable city" an oxymoron?', *Local Environment*, 2(3): 303–310.

Rees, W. and Wackernagel, M. (2013) 'Urban ecological footprints: why cities cannot be sustainable – and why they are a key to sustainability', in Lin, J. and Mele, C. (eds) *The Urban Sociology Reader*, 2nd edn, Abingdon: Routledge, 157–166.

Relph, E. (1976) *Place and Placelessness*, London: Pion.

Relph, E. (1981) *Rational Landscapes and Humanistic Geography*, London: Croom Helm.

Relph, E. (1987) *The Modern Urban Landscape*, Baltimore, MD: Johns Hopkins University Press.

Rex, J.A. and Moore, R. (1974) *Race, Conflict and Community: A Case Study of Sparkbrook, Birmingham*, London: Oxford University Press.

Rex, J.A. and Tomlinson, S. (1979) *Colonial Immigrants in a British City*, London: Routledge & Kegan Paul.

Riley, E., Fiori, J. and Ramirez, R. (2001) 'Favela Bairro and a new generation of housing programmes for the urban poor', *Geoforum*, 32(4): 521–531.

Ripp, M. and Rodwell, D. (2015) 'The geography of urban heritage', *The Historic Environment: Policy & Practice*, 6(3): 240–276.

Roberts, M. (1991) *Living in a Man-Made World: Gender Assumptions in Modern Housing Design*, London: Routledge.

Roberts, P., Ravetz, J. and George, C. (2009) *Environment and the City*, Abingdon: Routledge.

Robertson, I. and Richards, P. (eds) (2003) *Studying Cultural Landscapes*, London: Arnold.

Robinson, J. (2002) 'Global and world cities: a view from off the map', *International Journal of Urban and Regional Research*, 26(3): 531–554.

Robinson, J. (2005a) *Ordinary Cities: Between Modernity and Development*, Abingdon: Routledge.

Robinson, J. (2005b) 'Urban geography: world cities, or a world of cities', *Progress in Human Geography*, 29(6): 757–765.

Robinson, J. (2011) 'Cities in a world of cities: the comparative gesture', *International Journal of Urban and Regional Research*, 35(1): 1–23.

Rodaway, P. (1994) *Sensuous Geographies: Body, Sense and Place*, London: Routledge.

Rode, P., Floater, G., Thomopoulos, N., Docherty, J., Schwinger, P., Mahendra, A., and Fang, W. (2014) *Accessibility in Cities: Transport and Urban Form*, NCE Cities Paper 03, LSE Cities: London School of Economics and Political Science.

Røe, P.G. (2000) 'Qualitative research on intra-urban travel: an alternative approach', *Journal of Transport Geography*, 8(2): 99–106.

Rogers, R. (1999) *Towards an Urban Renaissance*, London: Department for Environment, Transport and the Regions.

Rogers, R. and Gumuchdjian, P. (1997) *Cities for a Small Planet*, London: Faber & Faber.

Rose, D. (1984) 'Rethinking gentrification: beyond the uneven development of Marxist theory', *Environment and Planning D: Society and Space*, 2(1): 47–74.

Rose, G. (2006) *Visual Methodologies: An Introduction to the Interpretation of Visual Materials*, 2nd edn, London: Sage.

Rossi, P.H. (1955) *Why Families Move*, New York: Macmillan.

Round, J., Williams, G.C. and Rogers, P. (2008) 'Everyday tactics and spaces of power: the role of informal economies in post-Soviet Ukraine', *Social and Cultural Geography*, 9(2): 171–185.

Roy, A. (2009) 'The 21st-century metropolis: new geographies of theory', *Regional Studies*, 43(6): 819–830.

Rudolph, R. and Brade, I. (2005) 'Moscow: processes of restructuring in the post-Soviet metropolitan periphery', *Cities*, 22(2): 135–150.

Samanta, G. and Roy, S. (2013) 'Mobility in the margins: hand-pulled rickshaws in Kolkata', *Transfers*, 3(3): 62–78.

Samuel, R. (1994) *Theatres of Memory: Past and Present in Contemporary Culture*, London: Verso.

Samuels, I. (2010) 'Palladio's children ... and Vitruvius's grandchildren?', *Urban Morphology*, 14(1): 65–69.

Sandercock, L. (1998) *Towards Cosmopolis: Planning for Multicultural Cities*, Chichester: John Wiley & Sons.

Sandercock, L. (2000) 'When strangers become neighbours: managing cities of difference', *Planning Theory and Practice*, 1(1): 13–30.

Sandercock, L. (2003) *Cosmopolis 2: Mongrel Cities of the 21st Century*, London: Continuum.

Sandercock, L. (2006) 'Cosmopolitan urbanism: a love song to our mongrel cities', in Binnie, J., Holloway, J., Millington, S. and Young, C. (eds) *Cosmopolitan Urbanism*, Abingdon: Routledge.

Sarre, P. (1986) 'Choice and constraint in ethnic minority housing: a structurationist view', *Housing Studies*, 1(1): 71–86.

Sassen, S. (1994) *Cities in a World Economy*, Thousand Oaks, CA: Pine Forge Press.

Sassen, S. (2000) *Cities in a World Economy*, 2nd edn, Thousand Oaks, CA: Sage.

Sassen, S. (2002) *The Global City: New York, London, Tokyo*, 2nd edn, Princeton, NJ: Princeton University Press.

Sassen, S. (2008) 'Re-assembling the urban', *Urban Geography*, 29(2): 113–126.

Satterthwaite, D. (2008) 'Cities' contribution to global warming: notes on the allocation of greenhouse gas emissions', *Environment and Urbanization*, 20(2): 539–549.

Satterthwaite, D. (ed) (1999) *The Earthscan Reader in Sustainable Cities*, London: Earthscan.

Saunders, P. (1979) *Urban Politics: A Sociological Interpretation*, Harmondsworth: Penguin.

Saunders, P. (1981) *Social Theory and the Urban Question*, London: Hutchinson.

Saunders, P. (1984) 'Beyond housing classes: the sociological significance of private property rights in the means of consumption', *International Journal of Urban and Regional Research*, 8(2): 202–227.

Savage, M., Warde, A. and Ward, K. (2003) *Urban Sociology, Capitalism and Modernity*, Basingstoke: Palgrave.

Saviano, R. (2008) *Gomorrah: Italy's Other Mafia*, London: Pan Books.

Sayre, W.S. and Kaufman, H. (1960) *Governing New York: Politics in the Metropolis*, New York: Sage Foundation.

Scarre, C. (2005) 'The world transformed: from foragers and farmers to states and empires', in Scarre, C. (ed) *The Human Past: World Prehistory and the Development of Human Societies*, London: Thames & Hudson.

Scott, A. (1988) *Metropolis: From the Division of Labour to Urban Form*, Berkeley, CA: University of California Press.

Scott, A. (1999) 'The US recorded music industry: on the relations between organization, location and creativity in the cultural economy', *Environment and Planning A*, 31(11): 1965–1984.

Scott, A. (2000) *The Cultural Economy of Cities*, London: Sage.

Scott, A. (2002) 'A new map of Hollywood: the production and distribution of American motion pictures', *Regional Studies*, 36(9): 957–975.

Scott, A.J. and Storper, A. (2003) 'Regions, globalization, development', *Regional Studies*, 37(6/7): 579–593.

Sennett, R. (1977) *The Fall of Public Man*, New York: Alfred A. Knopf.

Sennett, R. (1994) *Flesh and Stone: The Body and the City in Western Civilisation*, London: Faber & Faber.

Server, O. (1996) 'Corruption: a major problem for urban management', *Habitat International*, 21(4): 361–375.

Setchell, C. (1995) 'The growing environmental crisis in the world's mega cities: the case of Bangkok', *Third World Planning Review*, 17(1): 1–18.

Sharpe, L. (1995) *The Government of World Cities: The Future of the Metro Model*, Chichester: John Wiley & Sons.

Shaw, A. (2009) 'Town planning in postcolonial India, 1947–1965: Chandigarh re-examined', *Urban Geography*, 30(8): 857–878.

Shearmur, R. (2010) 'Editorial – A world without data? The unintended consequences of fashion in geography', *Urban Geography*, 31(8): 1009–1017.

Sheller, M. and Urry, J. (2000) 'The city and the car', *International Journal of Urban and Regional Research*, 24(4): 737–757.

Sheller, M. and Urry, J. (2006) 'The new mobilities paradigm', *Environment and Planning A*, 38(2): 207–226.

Shen, J., Wong, K-Y. and Feng, Z. (2002) 'State-sponsored and spontaneous urbanization in the Pearl River Delta of South China, 1980–1998', *Urban Geography*, 23(7): 674–694.

Sheppard, E. (2002) 'The spaces and times of globalization: place, scale, networks and positionality', *Economic Geography*, 78(3): 307–330.

Sheringham, M. and Wentworth, R. (2016) 'City as archive: a dialogue between theory and practice', *Cultural Geographies*, 23(3): 517–523.

Shevky, E. and Bell, W. (1955) *Social Area Analysis*, Stanford, CA: Stanford University Press.

Shirgaokar, M. (2014) 'Employment centres and travel behavior: exploring the work commute of Mumbai's rapidly motorizing middle class', *Journal of Transport Geography*, 41: 249–258.

Short, J.R. (1989) 'Yuppies, yuffies and the new urban order', *Transactions of the Institute of British Geographers N.S.*, 14: 173–188.

Short, J.R. (1996) *The Urban Order*, Oxford: Blackwell.

Short, J.R. (2005) *Imagined Country*, Syracuse, NY: Syracuse University Press.

Short, J.R. and Kim, Y.H. (1999) *Globalization and the City*, Harlow: Longman.

Short, J.R. and Pinet-Peralta, L.M. (2010) 'No accident: traffic and pedestrians in the modern city', *Mobilities*, 5(1): 41–59.

Short, J.R., Kim, Y.H., Kuus, M. and Wells, H. (1996) 'The dirty little secret of world cities research: data problems in comparative analysis', *International Journal of Urban and Regional Research*, 20(4): 697–719.

Sibley, D. (1995) *Geographies of Exclusion: Society and Difference in the Urban West*, London: Routledge.

Sibley, D. (1999) 'Outsiders in society and space', in Anderson, K. and Gale, F. (eds) *Cultural Geographies*, Australia: Addison-Wesley Longman.

Silverstone, R. (ed) (1997) *Visions of Suburbia*, London: Routledge.

Simon, D. (ed) (2016) *Rethinking Sustainable Cities: Accessible, Green and Fair*, Bristol: Policy Press.

Sjoberg, G. (1960) *The Pre-industrial City: Past, Present and Future*, New York: Free Press.

Skelton, T. (2013) 'Young people's urban im/mobilities: relationality and identity formation', *Urban Studies*, 50(3): 467–483.

Skinner, C. (2006) 'Falling through the policy gaps? Evidence from the informal economy in Durban, South Africa', *Urban Forum*, 17(2): 125–148.

Sklair, L. (2005) 'The transnational capitalist class and contemporary architecture in globalizing cities', *International Journal of Urban and Regional Research*, 29(3): 485–500.

Smedley, T. (2013) 'Sustainable urban design: lessons to be taken from slums', *The Guardian*, 5 June: www.theguardian.com/sustainable-business/sustainable-design-lessons-from-slums (Accessed 24 March 2017).

Smigiel, C. (2013) 'The production of segregated urban landscapes: a critical analysis of gated communities in Sofia', *Cities*, 35: 125–135.

Smith, D.A. and Timberlake, M. (1995) 'Conceptualising and mapping the structure of the world's city system', *Urban Studies*, 32: 287–302.

Smith, D.P. and Holt, L. (2007) 'Studentification and "apprentice" gentrifiers within Britain's provincial urban locations: extending the meaning of gentrification?', *Environment and Planning A*, 39(1): 142–161.

Smith, N. (1996) *The New Urban Frontier: Gentrification and the Revanchist City*, London: Routledge.

Smith, N. (2002) 'New globalism, new urbanism: gentrification as global urban strategy', *Antipode*, 34(3): 427–450.

Smith, S. (2000) 'Performing the (sound) world', *Environment and Planning D: Society and Space*, 18(5): 615–637.

Smith, S.J., Munro, M. and Christie, H. (2006) 'Performing (housing) markets', *Urban Studies*, 43(1): 81–98.

Soja, E. (1989) *Postmodern Geographies: The Reassertion of Space in Social Theory*, Oxford: Blackwell.

Soja, E. (1995) 'Postmodern urbanism: the six restructurings of Los Angeles', in Watson, S. and Gibson, K. (eds) *Postmodern Cities and Spaces*, Oxford: Blackwell.

Soja, E. (1996) *Thirdspace: Journeys to Los Angeles and Other Real and Imagined Places*, Oxford: Blackwell.

Steger, M.B. (2003) *Globalization: A Very Short Introduction*, Oxford: Oxford University Press.

Steinberg, F. (1996) 'Conservation and rehabilitation of urban heritage in developing countries', *Habitat International*, 20(3): 463–475.

Stevenson, D. (2014) *Cities of Culture: A Global Perspective*, Abingdon: Routledge.

Stoker, G. (1995) 'Regime theory and urban politics', in Judge, J., Stoker, G. and Wolman, H. (eds) *Theories of Urban Politics*, London: Sage.

Stoker, G. and Mossberger, K. (1994) 'Urban regime theory in comparative perspective', *Environment and Planning C: Government and Policy*, 12(2): 195–212.

Stokes, C.J. (1962) 'A theory of slums', *Land Economy*, 38: 127–137.

Stone, C. (1980) 'Systemic power in community decision making: a restatement of stratification theory', *American Political Science Review*, 74(4): 978–990.

Stone, C. (1989) *Regime Politics: Governing Atlanta, 1946–1988*, Lawrence, KS: University of Kansas Press.

Stone, C. (1993) 'Urban regimes and the capacity to govern: a political economy approach', *Journal of Urban Affairs*, 15(1): 1–28.

Stone, C., Orr, M. and Imbroscio, D. (1994) 'The reshaping of urban leadership in U.S. cities: a regime analysis', in Gottdiener, M. and Pickvance, C. (eds) *Urban Life in Transition*, Beverly Hills, CA: Sage.

Strange, I. (1997) 'Planning for change, conserving the past: towards sustainable development policy in historic cities?', *Cities*, 14(4): 227–233.

Strom, E.A. and Mollenkopf, J. (eds) (2003) *The Urban Politics Reader*, London: Routledge.

Stutz, F.P. (1995) 'Environmental impacts', in Hanson, S. (ed) *The Geography of Urban Transportation*, New York: Guilford Press.

Sudjic, D. (2005) *The Edifice Complex: How the Rich and Powerful Shape the World*, London: Allen Lane.

Tagg, J. (1993) *The Burden of Representation: Essays on Photographies and Histories*, Minneapolis, MN: University of Minnesota Press.

Taylor, P. (2004) *World City Network: A Global Urban Analysis*, London: Routledge.

Taylor, P., Hoyler, M. and Verbruggen, R. (2010) 'External urban relational process: introducing central flow theory to complement central place theory', *Urban Studies*, 47(13): 2803–2818.

Taylor, P.J. and Derudder, B. (2015) *World City Network: A Global Urban Analysis*, 2nd edn, New York: Routledge.

The City of Calgary (2007) *Towards a Preferred Future: Understanding Calgary's Ecological Footprint*, Calgary: The City of Calgary.

Theodore, N. and Peck, J. (2012) 'Framing neoliberal urbanism: translating 'common sense' urban policy across the OECD zone', *European Urban and Regional Studies*, 19(1): 20–41.

Thomas, H. (1994) 'The local press and urban renewal: A South Wales case study', *International Journal of Urban and Regional Research*, 18(2): 315–333.

Thrift, N. (2004) 'Driving in the city', *Theory, Culture and Society*, 21(4):41–59.

Till, J. (2009) *Architecture Depends*, Cambridge, MA: MIT Press.

Toffler, A. (1970) *Future Shock*, London: Random House.

Tremlett, G. (2006) *Ghosts of Spain: Travels Through a Country's Hidden Past*, London: Faber & Faber.

Tunbridge, J.E. and Ashworth, G.J. (1996) *Dissonant Heritage: The Management of the Past as a Resource in Conflict*, Chichester: John Wiley & Sons.

UNCHS (Habitat) (1994) *Urban Public Transport in Developing Countries*, Nairobi: UNCHS (Habitat).

UNESCO World Heritage Centre *The States Parties*, http://whc.unesco.org/en/statesparties/ (Accessed 30 March 2017).

UN-HABITAT (2003) *The Challenge of Slums*, London: Earthscan.

UN-HABITAT (2008) *State of the World's Cities 2008/9: Harmonious Cities*, London: Earthscan.

UN-HABITAT (2009) *Planning Sustainable Cities*, London: Earthscan.

UN-HABITAT (2012) *State of the World's Cities 2012/13: Prosperity of Cities*, Nairobi: UN-HABITAT.UN-HABITAT (2013) *Planning and Design for Sustainable Urban Mobility: Global Report on Human Settlements*, Abingdon: Routledge.

UN-HABITAT (2016) *Urbanization and Development: Emerging Futures. World Cities Report 2016*, Nairobi: UNHSP (UN-HABITAT).

Urban, F. (2002) 'Small town, big website: cities and their representation on the internet', *Cities*, 19(1): 49–59.

Urry, J. (2007) *Mobilities*, Cambridge: Polity Press.

Valentine, G. (2001) *Social Geographies: Space and Society*, Harlow: Pearson.

Valins, O. (2003) 'Stubborn identities and the construction of socio-spatial boundaries: ultra-orthodox Jews living in contemporary Britain', *Transactions of the Institute of British Geographers*, 28(2): 158–175.

Van Meeteren, M., Derudder, B. and Bassens, D. (2016) 'Can the straw man speak? An engagement with post-colonial critiques of "global cities research"', *Dialogues in Human Geography*, 6(3): 247–267.

Vance, J.E. (1977) *This Scene of Man: The Role and Structure of the City in the Geography of Western Civilization*, New York: Harper's College Press.

Vandenbulcke, G., Thomas, I., de Geus, B., Degraeuwe, B., Torfs, R., Meeusen, R. and Int Panis, L. (2009) 'Mapping bicycle use and the risk of accidents for commuters who cycle to work in Belgium', *Transport Policy*, 1(2): 77–87.

Vanolo, A. (2013) 'Alternative capitalism and creative economy: the case of Christiania', *International Journal of Urban and Regional Research*, 37(5): 1785–1798.

Vasudevan, A. (2014) 'The makeshift city: towards a global geography of squatting', *Progress in Human Geography*, 39(3): 338–359.

Vaughan, L. (ed) (2015) *Suburban Urbanities: Suburbs and the Life of the High Street*, London: UCL Press.

Vaughan, L., Griffiths, S., Haklay, M. and Jones, C.E. (2009) 'Do the suburbs exist? Discovering complexity and specificity in suburban built form', *Transactions of the Institute of British Geographers*, 34(4): 475–488.

Venturi, R., Scott Brown, D. and Izenour, S. (1972) *Learning from Las Vegas*, Cambridge, MA: MIT Press.

Virilio, P. (1986) *Speed and Politics*, trans. Polizotti, M., New York: Semiotext(e).

Von Mahs, J. (2015) *Down and Out in Los Angeles and Berlin: The Sociospatial Exclusion of Homeless People*, Philadelphia: Temple University Press.

Wackernagel, M. (1998) 'The ecological footprint of Santiago de Chile', *Local Environment*, 3(1): 7–25.

Waldron, R. (2016) 'The "unrevealed casualties" of the Irish mortgage crisis: analysing the broader impacts of mortgage market financialisation', *Geoforum*, 69: 53–66.

Walmsley, D. (1988) *Urban Living: The Individual in the City*, Harlow: Longman.

Walmsley, D.J. and Lewis, G.J. (1993) *People and Place: Behavioural Approaches in Human Geography*, 2nd edn, Harlow: Longman.

Walters, J. (2001) 'War on the car sparks driver rage', the *Observer*, 26 August: 1.

Ward, K. and McCann, E.J. (2006) '"The new path to a new city"? Introduction to a debate on urban politics, social movements and the legacies of Manuel Castells' *The City and the Grassroots*', *International Journal of Urban and Regional Research*, 30(1): 189–193.

Ward, S.V. (1998) *Selling Places: The Marketing and Promotion of Towns and Cities 1850–2000*, London: E. & F.N. Spon.

Ward, S.V. (2002) *Planning the Twentieth-Century City: The Advanced Capitalist World*, Chichester: John Wiley & Sons.

Ward, S.V. (2004) *Planning and Urban Change*, 2nd edn, London: Sage.

Watson, S. (1991) 'Gilding the smokestacks: the new symbolic representations of deindustrialized regions', *Environment and Planning D: Society and Space*, 9(1): 59–71.

Watson, S. (1999) 'City politics', in Pile, S., Brook, C. and Mooney, G. (eds) *Unruly Cities*, London: Routledge.

Watson, S. and Gibson, K. (1995) *Postmodern Cities and Spaces*, Oxford: Blackwell.

Wekerle, G.R. and Whitzman, C. (1995) *Safe Cities: Guidelines for Planning, Design and Management*, London: Van Nostrand Reinhold.

Weltevreden, J.W.J. and Van Rietbergen, T. (2007) 'E-shopping versus city centre shopping: the role of perceived city centre attractiveness', *Tijdschrift voor Economische en Sociale Geografie*, 98(1): 68–85.

Wensing, E. and Porter, L. (2016) 'Unsettling planning's paradigms: towards a just accommodation of Indigenous rights and interests in Australian urban planning?', *Australian Planner*, 53(2): 91–102.

Wheeler, S.M. and Beatley, T. (2014) (eds) *The Sustainable Urban Development Reader*, 3rd edn, Abingdon: Routledge.

Whitehand, J.W.R. (1972) 'Building cycles and the spatial pattern of urban growth', *Transactions of the Institute of British Geographers*, 56: 39–55.

Whitehand, J.W.R. (1987) 'Urban Morphology', in Pacione, M. (ed) *Historical Geography: Progress and Prospect*, London: Croom Helm.

Whitehand, J.W.R. (1992a) 'Recent developments in urban morphology', *Urban Studies*, 29(3/4): 617–634.

Whitehand, J.W.R. (1992b) *The Making of the Urban Landscape*, The Institute of British Geographers Special Publications Series, Oxford: Blackwell.

Whitehand, J.W.R. (1994) 'Development cycles and the urban landscape', *Geography*, 79(1): 3–17.

Whitehand, J.W.R. (2001) 'British urban morphology: the Conzenian tradition', *Urban Morphology*, 5: 103–109.

Whitehand, J.W.R. and Carr, C.M.H. (1999) 'The changing fabrics of ordinary residential areas', *Urban Studies*, 36(10): 1661–1677.

Whitehand, J.W.R. and Carr, C.M.H. (2001) *Twentieth-Century Suburbs: A Morphological Approach*, London: Routledge.

Whitehand, J.W.R. and Gu, K. (2007) 'Extending the compass of plan analysis: a Chinese exploration', *Urban Morphology*, 11: 91–109.

Whitehand, J.W.R. and Morton, N.J. (2006) 'The fringe-belt phenomenon and socio-economic change', *Urban Studies*, 43: 2047–2066.

Whitelegg, J. (1993) *Transport for a Sustainable Future: The Case of Europe*, London: John Wiley & Sons.

Williams, R. (1953) 'The Idea of Culture', *Essays in Criticism*, III(3): 239–266.

Williams, R. (1961) *The Long Revolution*, Harmondsworth: Penguin.

Williams, R. (1973) *The City and the Country*, London: Chatto & Windus.

Wilson, D. (1996) 'Metaphors, growth coalitions and discourses of black poverty neighbourhoods in a U.S. city', *Antipode*, 28(1): 72–86.

Wilson, D. (2007) *Cities and Race: America's New Black Ghetto*, London: Routledge.

Wilson, D.C., Velis, C. and Cheeseman, C. (2006) 'Role of informal sector recycling in waste management in developing countries', *Habitat International*, 30(4): 797–808.

Wilson, E. (1991) *The Sphinx and the City*, London: Verso.

Wirth, L. (1938) 'Urbanism as a way of life', *American Journal of Sociology*, 44(1) 1–24.

Woldoff, R.A. (2011) *White Flight / Black Flight: The Dynamics of Racial Change in an American Neighborhood*, Ithaca, NY: Cornell University Press.

Wolfe, T. (1987) *The Bonfire of the Vanities*, New York: Bantam Books.

Wolman, H. and Goldsmith, M. (1992) *Urban Politics and Policy: A Comparative Approach*, Oxford: Blackwell.

Wolman, H., Ford, C.C. III, and Hill, E. (1994) 'Evaluating the success of urban success stories', *Urban Studies*, 31(6): 835–850.

Women and Geography Study Group (WGSG) (1997) *Feminist Geographies*, Harlow: Addison-Wesley Longman.

Wong, T.C. (2011) 'Eco-cities in China: pearls in the sea of degrading urban environments', in Wong. T-C. and Yuen, B. (eds) *Eco-City Planning: Policies, Practice, and Design*, New York: Springer, 131–150.

Wong, Y.S. (2006) 'Where there are no pagodas on Pagoda Street: language, mapping and navigating ambiguities in colonial Singapore', *Environment and Planning A*, 38(2): 325–340.

Woolley, H. and Johns, R. (2001) 'Skateboarding: city as playground', *Journal of Urban Design*, 6(2): 211–230.

World Commission on Environment and Development (1987) *Our Common Future*, Oxford: Oxford University Press.

Wright, P. (2009) *On Living in an Old Country: The National Past in Contemporary Britain*, Oxford: Oxford University Press.

Wrigley, N. (2002) 'Food deserts in British cities: policy context and research priorities', *Urban Studies*, 39(11): 2029–2040.

Wrigley, N. and Lowe, M. (2002) *Reading Retail: A Geographical Perspective on Retailing and Consumption Spaces*, London: Arnold.

Wu, F. (2002) 'China's changing urban governance in the transition towards a more market-oriented economy', *Urban Studies*, 39(7): 1071–1093.

Wylie, J. (2007) *Landscape*, Abingdon: Routledge.

Wyly, E.K. and Hammel, D.J. (2005) 'Mapping neoliberal American urbanism', in Atkinson, R. and Bridge, G. (eds) *Gentrification in a Global Context: The New Urban Colonialism*, London: Routledge.

Xie, P.F. and Gu, K. (2015) 'The changing urban morphology: waterfront redevelopment and event tourism in New Zealand', *Tourism Management Perspectives*, 15: 105–114.

Young, L. (2003) 'The place of street children in Kampala's urban environment: marginalisation, resistance and acceptance in the urban environment', *Environment and Planning D: Society and Space*, 21(5): 607–621.

Zukin, S. (1989) *Loft Living: Culture and Capital in Urban Change*, 2nd edn, New Brunswick, NJ: Rutgers University Press.

Zukin, S. (1995) *The Cultures of Cities*, Oxford: Blackwell.

Zukin, S. (1998) 'Urban lifestyles: diversity and standardization in spaces of consumption', *Urban Studies*, 35(5–6): 825–839.

Index

Taylor & Francis eBooks

Helping you to choose the right eBooks for your Library

Add Routledge titles to your library's digital collection today. Taylor and Francis ebooks contains over 50,000 titles in the Humanities, Social Sciences, Behavioural Sciences, Built Environment and Law.

Choose from a range of subject packages or create your own!

Benefits for you

» Free MARC records
» COUNTER-compliant usage statistics
» Flexible purchase and pricing options
» All titles DRM-free.

Benefits for your user

» Off-site, anytime access via Athens or referring URL
» Print or copy pages or chapters
» Full content search
» Bookmark, highlight and annotate text
» Access to thousands of pages of quality research at the click of a button.

REQUEST YOUR **FREE** INSTITUTIONAL TRIAL TODAY	**Free Trials Available** We offer free trials to qualifying academic, corporate and government customers.

eCollections – Choose from over 30 subject eCollections, including:

Archaeology	Language Learning
Architecture	Law
Asian Studies	Literature
Business & Management	Media & Communication
Classical Studies	Middle East Studies
Construction	Music
Creative & Media Arts	Philosophy
Criminology & Criminal Justice	Planning
Economics	Politics
Education	Psychology & Mental Health
Energy	Religion
Engineering	Security
English Language & Linguistics	Social Work
Environment & Sustainability	Sociology
Geography	Sport
Health Studies	Theatre & Performance
History	Tourism, Hospitality & Events

For more information, pricing enquiries or to order a free trial, please contact your local sales team: **www.tandfebooks.com/page/sales**

 Routledge
Taylor & Francis Group

The home of Routledge books

www.tandfebooks.com